Non-Clinical Vascular Infusion Technology

Volume II

The Techniques

Non-Clinical Vascular Infusion Technology

Volume II

The Techniques

Edited by
Owen P. Green and Guy Healing

CRC Press
Taylor & Francis Group
Boca Raton London New York

CRC Press is an imprint of the
Taylor & Francis Group, an **informa** business

CRC Press
Taylor & Francis Group
6000 Broken Sound Parkway NW, Suite 300
Boca Raton, FL 33487-2742

Printed on acid-free paper
Version Date: 20131014

International Standard Book Number-13: 978-1-4398-7445-5 (Hardback)

Visit the Taylor & Francis Web site at
http://www.taylorandfrancis.com

and the CRC Press Web site at
http://www.crcpress.com

Contents

Foreword

Good science and good welfare go hand in hand. Innovative science and technology can be used to improve animal welfare. Equally, the 3Rs (replacement, refinement and reduction of animals in research) can provide fresh insight and novel approaches to advance science. By sharing data, knowledge and experience on the science behind infusion models and the refinement of the techniques used there is the potential to have a significant impact on the 3Rs.

Owen Green and Guy Healing have shown the importance of the 3Rs in infusion technology at international meetings and in producing this book. The book will enhance uptake of the latest science behind vascular delivery systems to get better data and help influence decisions around the most appropriate model. It will also contribute to preventing repetition of method development and optimising experiments to answer specific scientific questions with the least impact on animals.

As in other areas of science, the field of infusion technology is constantly evolving. This volume of *Non-Clinical Vascular Infusion Technology* reviews current developments in the field that will support scientists in putting the 3Rs into practice.

Kathryn Chapman

Head of Innovation and Translation, the National Centre for the Replacement, Refinement and Reduction of Animals in Research

Preface

In 2000 the first book covering pre-clinical infusion techniques was published by Taylor & Francis (*Handbook of Pre-Clinical Continuous Intravenous Infusion*, Healing and Smith, editors) and this has become the singular reference source for this technology up to the present time. However, it is now recognised that a number of the techniques have been refined since that time, and also that new equipment and improved equipment were on the market. We therefore accepted the request from the publishers to provide a more current techniques manual, and have taken the opportunity not only to reflect current expertise by including authors who are the current leaders in the field of commercial infusion technology and, therefore, have the most practical experience (hence also providing the most robust background data sets), but have also approached the topic from a fresh angle and structured the chapters differently so that this new volume truly represents a novel approach rather than attempting to simply update the original reference book. To complete the reference material on this technology, there is also a companion volume (*Non-Clinical Vascular Infusion Technology: The Science*) that covers the scientific principles behind the delivery systems, from both physical and physiological standpoints, and also covers formulation-specific considerations.

There are numerous pharmaceuticals on the market or undergoing clinical trials that require intravenous infusion, for either short or longer periods, intermittently or continuously, and so this book should be of interest to many in the pharmaceutical as well as research areas. These applications include chemotherapy (Skubitz 1997; Vallejos et al. 1997; Ikeda et al. 1998; Patel et al. 1998; Stevenson et al. 1999; Valero et al. 1999), and the treatment of various diseases such as HIV (Levy et al. 1999), hepatitis C (Schenker et al. 1997) and cardiovascular disease (Phillips et al. 1997), as well as during and following problematical surgical procedures (Bacher et al. 1998; Llamas et al. 1998; Menart et al. 1998). It is a regulatory and ethical requirement that these pharmaceuticals first be tested on both rodents and non-rodents by the clinical route, and so the range of pre-clinical experimental models is covered. The technique of prolonged intravenous delivery in conscious, free-moving animal models has also broadened the opportunities to study and evaluate the safety and efficacy of those products that have limiting biological or chemical properties such as half-life and formulation issues.

A specific driver for this book was to identify and share best practices across the industry. It is intended that this book will prevent unnecessary method development work and hence potentially decrease animal usage, and also provide guidance on choices for the most acceptable methodologies from an animal welfare perspective, which is particularly pertinent when using higher sentient animals such as dogs and primates.

As stated before, the authors in this book have been selected for their established expertise in the field. It should be noted that many authors' laboratories conduct procedures in more species than those covered in their respective chapter(s), but to gain the widest possible selection of opinions, techniques, and background data, it was necessary to limit each to a specific aspect. There are also variations in the techniques used in different countries, and this has been reflected in the truly international selection of authors. The book is organised by species (those commonly used in pre-clinical studies), namely, rat, mouse, dog, minipig, large primate, and marmoset, and there are also chapters covering juvenile studies and reproductive toxicity studies. Each section is organised in a consistent manner to help find the relevant information quickly, and covers information on the selection of the best model, best practice both surgically and non-surgically, practical techniques, equipment selection, and commonly encountered background pathologies.

References

Bacher H, Mischinger HJ, Supancic A, Leitner G and Porta S. 1998. Dopamine infusion following liver surgery prevents hypomagnesemia. *Magnesium Bulletin* 20: 49–50.

Ikeda K, Terashima M, Kawamura H, Takiyama I, Koeda K, Takagane A, Sato N et al. 1998. Pharmacokinetics of cisplatin in combined cisplatin and 5-fluorouracil therapy: A comparative study of three different schedules of cisplatin administration. *Japanese Journal of Clinical Oncology* 28: 168–175.

Levy Y, Capitant C, Houhou S, Carriere I, Viard JP, Goujard C, Gastaut JA et al. 1999. Comparison of subcutaneous and intravenous interleukin-2 in asymptomatic HIV-1 infection: A randomised controlled trial. *Lancet* 353: 1923–1929.

Llamas P, Cabrera R, Gomezarnau J and Fernadez MN. 1998. Hemostasis and blood requirements in orthotopic liver transplantation with and without high-dose aprotonin. *Haematologica* 83: 338–346.

Menart C, Petit PY, Attali O, Massignon D, Dechavanne M and Negrier C. 1998. Efficacy and safety of continuous infusion of Mononine® during five surgical procedures in three hemophilic patients. *American Journal of Hematology* 58: 110–116.

Patel SR, Vadhanraj S, Burgess MA, Plager C, Papadopoulos N, Jenkins J and Benjamin RS. 1998. Results of two consecutive trials of dose-intensive chemotherapy with doxorubicin and ifosfamide in patients with sarcomas. *American Journal of Clinical Oncology* 21: 317–321.

Phillips BG, Gandhi AJ, Sanoski CA, Just VL and Bauman JL. 1997. Comparison of intravenous diltiazem and verapamil for the acute treatment of atrial fibrillation and atrial flutter. *Pharmacotherapy* 17: 1238–1245.

Schenker S, Cutler D, Finch J, Tamburro CH, Affrime M, Sabo R and Bay M. 1997. Activity and tolerance of a continuous subcutaneous infusion of interferon-alpha-2b in patients with chronic hepatitis C. *Journal of Interferon and Cytokine Research* 17: 665–670.

Skubitz KM. 1997. A phase I study of ambulatory continuous infusion of paclitaxel. *Anti-Cancer Drugs* 8: 823–828.

Stevenson JP, DeMaria D, Sludden J, Kaye SB, Paz-Ares L, Grochow LB, McDonald A et al. 1999. Phase I pharmacokinetic study of the topoisomerase I inhibitor GG211 administered as a 21-day continuous infusion. *Annals of Oncology* 10: 339–344.

Valero V, Buzdar AU, Theriault RL, Azarnia N, Fonseca GA, Willey J, Ewer M et al. 1999. Phase II trial of liposome encapsulated doxorubicin, cyclophosphamide, and fluorouracil as first-line therapy in patients with metastatic breast cancer. *Journal of Clinical Oncology* 17: 1425–1434.

Vallejos C, Solidoro A, Gomez H, Castellano C, Barriga O, Galdos R, Casanova L, Otero J and Rodriguez W. 1997. Ifosfamide plus cisplatin as primary chemotherapy of advanced ovarian cancer. *Gynecologic Oncology* 67: 168–171.

chapter one

Rat

Jennifer Sheehan, BS, SRS, LATg
Huntingdon Life Sciences, USA

Duncan Patten, FIAT, RAnTech
Huntingdon Life Sciences, UK

Vasanthi Mowat, BVSc, MVSc, MRCVS, FRCPath
Huntingdon Life Sciences, UK

Contents

1.1 Introduction

1.1.1 Choice and relevance of the species

The rat is by far and large the most accepted rodent model for pre-clinical testing. With regard to predicting human PK, numerous prediction methodologies, incorporating both *in vitro* and pre-clinical *in vivo* data, have been developed in recent years, each with advantages and disadvantages. Even though the rat is not necessarily a worse predictor for human PK than for example the dog, it is recognised that this species has a higher oxidative metabolism than the human (Martignoni et al. 2006), which may lead to differences in metabolic profile that need to be adequately examined. From a safety testing perspective, retrospective data analysis has shown that when using a rodent model (rat) only, it will generally be predictive for 43% of human toxicities (Olson et al. 2000). Even though the mouse can be regarded as an alternative rodent species, its diminutive size can limit its use as a practical model for prolonged infusion studies, and it is considered a worse predictor of human toxicities (Olson et al. 2000).

1.1.2 Regulatory guidelines

The need for use of the intravenous route is driven by regulatory guidance. In general, the guidelines (EMEA 2000; ICH 2009) specify the need to test the compound via a route that is relevant for human administration. For example, when human subjects are to be dosed via intravenous infusion, the rat should be dosed via the same route as well. Not only does this allow the most adequate assessment of systemic toxicity, but it will also give an insight into potential local effects at the administration site. Other routes of administration may be selected but only when justified on the basis of pharmacological, pharmacokinetic, or toxicological information.

There can be instances where the intended human route is oral exposure, yet an infusion route in the rat is chosen. A good example is compounds that have a fairly good oral bioavailability in humans but not in animals. Various guidelines indeed require a multiple of the intended human exposure to be reached at the high dose level in the test species, and this can potentially be achieved via intravenous dosing. In such cases potential local tolerance or pre-systemic (e.g. gastro-intestinal) effects must be assessed by other means.

1.1.3 Choice of infusion model

The principles and techniques of rat infusion dosing have been refined over the years, and it is an area of constant improvement and technological advancements. Infusion dosing, however, can cause changes in the animals' physiological state that are multi-factorial. When dosing via a centrally implanted catheter, the surgery itself will disturb the normal physiology of the animal. The constant wearing of infusion jackets in ambulatory models or the continuous tethering in non-ambulatory models may cause a more prolonged disturbance. Whereas dosing via a peripheral vein seems less impacting at first glance, it necessitates some form of restraint, which results in stress for the animal. As will be demonstrated in Sections 5.1 and 5.2, both methods of infusion dosing are indeed associated with stress and resultant changes such as a decreased food consumption and body weight, but in general stress is most notable in peripherally dosed, restrained animals. The latter can for instance show increases in liver transaminases as a result of restraint stress, which can be an important confounding factor when interpreting study results.

Another important consideration when deciding between a central and a peripheral route is the fact that evaluation of local tolerance of a test article can be more challenging in rats dosed via the tail vein. This is further discussed in Section 5.2.

To conclude, infusion via a centrally implanted catheter has little impact on animal physiology and well-being when there is an adequate surgical recovery period (about 7 days) and when the procedures and techniques applied are of high standard. The minor surgical procedure itself is associated with stress, but animals show a fast recovery from resultant changes in homeostasis. Peripheral infusion via the tail vein in rodents is associated with pronounced stress during the dosing sessions themselves, resulting in a number of changes. These can all be confounding factors when interpreting effects caused by the studied test article. Given the length of recovery of 7 days after central catheter surgery, the case can be made that peripheral tail vein dosing is not the optimal technique when the dosing period spans more than 7 days. In addition, there are generally limitations to how long a rat can be restrained, and extended dosing cycles would benefit from a central catheter (Sheehan et al. 2011).

1.1.4 *Limitations of available models*

Many different models are available, and sometimes options exist within each particular model. For example, a surgically cannulated rat model can be catheterised via the jugular vein or the femoral vein. The catheter could be exteriorised using a vascular access harness, straight through a tether, or via a tail-cuff. The choice of model often depends on the level of experience and success rates. Each model has pros and cons, and the choice must depend on the users' comfort level, the study design, and/or customer preference.

Limitations often depend on study design, test article properties, and physical limitations of the equipment. Study design will often dictate how a model must be developed. A continuous 24-hour infusion protocol with a relatively high dose volume would not be an ideal candidate for osmotic pumps. A test article known to be a vascular irritant may not be suitable for infusion into a peripheral vessel. Remote dosing, or dosing without the presence of technical staff in the room, should be performed to the greatest extent possible when infusion is the required dose route for a cardiovascular or respiratory safety pharmacology study. Infusion pumps that are programmable, implantable, and refillable are available, but they cannot be recharged and some models have to be completely programmed prior to implantation, which may limit the duration of infusion or impact the accuracy of the dose administration. Osmotic mini-pumps can be used only for small volumes and limited infusion durations.

Limitations will also sometimes depend on animal welfare considerations. Most Animal Care and Use review committees will limit the duration of restraint in all species, which means some intermittent infusion protocols no longer than 2 to 4 hours in duration could be allowed in a restrained model. However, this would also depend on the frequency and duration of dosing: a 6-month, daily, 1-hour infusion in rats may be difficult when trying to attempt to access a tail vein for each dose administration. Though it might be possible to achieve such a study design in a restrained, peripheral model, consideration must be given to study endpoints, technical skill level of the staff, and test article properties.

There are also limitations with surgically cannulated models, especially with long-term studies. Infusion studies require careful monitoring of the infusion equipment, adjustment of the harnesses or jackets to allow for growth, strict aseptic technique, and generally a highly skilled infusion staff. Surgically cannulated models also require a method of maintaining catheter patency, which can be accomplished by continuous infusion of test article or a saline infusion at a keep-vein-open (KVO) maintenance rate between dosing intervals, or by using an anticoagulant locking solution.

1.2 *Best practice*

1.2.1 *Surgical models*

1.2.1.1 *Training*

Performing small animal surgery, particularly with vascular catheterisations, can be challenging. Developing the skills and behaviours necessary to successfully perform the procedures can take time and may even be beyond some people. Manual dexterity, a steady hand, and good eyesight (with glasses or magnification if required) are physical attributes that a surgeon will require. Ideally, aseptic techniques should be incorporated into each stage of the training rather than developed as a separate concept. It may take longer to develop the technical skills, but by the time full competency is achieved asepsis should be second nature.

Like many techniques that incorporate several different elements, it is recommended that learning and the development of competency be broken down into small sections, which are later combined into the complete procedure. For example, it is important that any person trained in aseptic recovery surgical procedures should first be competent in all basic concepts of surgical procedures, such as anaesthetic induction, maintenance and monitoring, surgical site preparation techniques, and analgesia.

There are several elements that can be trained to full competency without using live animals, such as techniques in closing and suturing wounds, securing and exteriorising catheters, and fitting equipment. However, isolating and catheterising the blood vessel is best achieved by working with live animals. The legislation in some countries may preclude the use of live animals for developing technical skills, and this can make training slow and very challenging.

1.2.1.2 *Applicability of the methodology*

The choice of the site of vascular access depends on the preference of the surgeon or investigator. The most common vessels used in rats are the external jugular vein and the femoral vein. For many infused compounds there is little difference in the dilution of test article between the sites, but positioning of the catheter tip is important. For lipid compounds, evidence suggests that the catheter tip should be sited in the widest portion of the venous system, in the inferior vena cava adjacent to renal veins. Superior vena cava infusions or inferior vena cava infusions where the catheter tip is close to the bifurcation are likely to result in significant complications when lipids are infused (Asanuma et al. 2006).

When using the femoral vein, it must first be isolated and cannulated. The catheter is fed through the femoral vein and femoral bifurcation into the inferior vena cava, with the tip of the catheter resting just caudal to the renal veins. Catheters can be purchased with movable retention beads to assist in marking the appropriate insertion length for each individual animal.

The external jugular vein is somewhat easier to isolate since it is not in as close a proximity to an artery or nerve. The catheter is fed through the jugular vein and into the superior vena cava. The tip should rest in the vena cava and ideally not up against or inside the right atrium in order to avoid any cardiac lesions or functional changes.

Either vascular access site is acceptable and feasible, but success depends on the skill of the surgeon and technical staff. The femoral vein is the smaller vein to isolate and cannulate, and there can be some differences depending on the strain of rat being used (for example, in our experience Wistar rats are noted to have smaller, more fragile vessels than Sprague Dawley rats). However, the positioning is more critical when using the jugular vein for insertion: a couple of millimetres difference in tip positioning in the superior vena cava may have a larger impact than a couple of millimetres difference in the tip positioning in the inferior vena cava. On the other hand, cannulation of the jugular vein requires less tunnelling of the catheter between the insertion and exteriorisation sites than does the femoral approach. Yet again, depending on the method of securing the catheter, the jacket or harness may interfere with the surgical incision site when using the jugular approach, whereas the incision required for the femoral approach is a sufficient distance from the exteriorisation site.

The surgical procedure for the femoral approach will be discussed in greater detail in this chapter; however, most information is applicable to either vascular access site.

As with most surgical procedures for implantation of a device, antibiotics and analgesics are often used. These are discussed in detail in Section 3.1.1.1.

Following the surgical procedure, rats must recover from anaesthesia and be equipped with the materials necessary to access or protect the catheter. This may be a vascular access harness, in which case the catheter must be filled with a locking solution and the animal fitted with a harness or jacket. Or this may be a simple exteriorisation directly into a tether, in which case the animal is fitted with a harness or jacket and the catheter is fed through the tether and then connected to a swivel and infusion pump programmed to deliver the KVO saline rate throughout the recovery period. The KVO maintenance rate in rats is approximately 0.5 mL/hr, which is sufficient to maintain positive flow through the catheter and catheter patency. This rate can be maintained throughout the post-surgical recovery period or in-between dosing sessions without impacting the haemodynamics.

If using a vascular access button or vascular access port, the catheters will usually require a locking solution, but external devices (such as a jacket or harness) are not required. Osmotic and other mini-pumps are completely implantable and, when primed, tend to begin pumping the test article upon implantation.

External pumps used for rodents are usually syringe infusion pumps and can handle a wide range of infusion rates accurately. This is an important feature when selecting syringe infusion pumps since they should be able to handle very low rates for slow intermittent or continuous infusion durations as well as larger volumes. Syringe infusion pumps have been the most common type of pump for rat infusion studies; however, some ambulatory pumps are now able to handle lower volumes and infusion rates as accurately as syringe infusion pumps. A detailed overview of different infusion pumps is presented in Section 4.1.4. The vendor should be able to assist with the selection of the infusion pump that is most appropriate for your needs.

Monitoring the accuracy of dose administration is important because unlike other dose routes, infusion dosing is usually performed by a programmed infusion pump, and mechanical devices can sometimes fail. In a GLP environment it is necessary to demonstrate that the correct amount of test article has been delivered to the animal.

On study, infusion accountability must be performed after every infusion interval or syringe/reservoir change. Since this practice also monitors the performance of the infusion pumps and at the same time confirms dose-delivery accuracy, one should not wait to the end of the study to perform all the accountability calculations. Otherwise it is possible, for instance, to get to the end of a 28-day study only to realise that a pump has been malfunctioning since day 2. The accountability data can be reported on an interval-by-interval basis or on a per-study basis. Assessing pump accuracy is discussed in Section 4.1.4.

Once the animals are surgically prepared and have recovered and pump accuracy is established, the test article administration can begin. The duration of infusion can vary greatly depending on the study design and requirements. Rats can be maintained on continuous infusion for up to 6 months and sometimes longer. Intermittent infusion designs also vary greatly and can range from a single dose to twice daily dosing for 6 months. The duration of the infusion cycle can also vary, and as a general rule anything infused for longer than 2 or 3 minutes would benefit from the use of an infusion pump.

A number of publications deal with the rates and volumes that should be considered good practice when dosing rats (or other species) via continuous intravenous infusion (Diehl et al. 2001; Smith 1999; Morton 2001). As a general guide, the maximum volume administered intravenously should be around 5% of the circulating blood volume over an hour. The mean circulating blood volume in a rat is 64 mL/kg, but there is considerable variation in this value, probably an artefact of the different measuring techniques used. The total duration of infusion is also a determining factor, and the volumes and rates that are considered good practice will be dependent upon the infusion cycle time. Typical ranges as described in literature are presented in Table 1.1.

In general practice, infusion rates in rats are typically in the range of 1 to 4 mL/kg/hr. Even though these values give an indication of standard good practice, there may be large differences in tolerated volume by intravenous infusion, depending on the vehicle used. Vehicles should be biologically inert, have no effects on the biophysical properties of the compound, and have no toxic effect on the rat. If a vehicle displays some of these effects, the dose should be adjusted, and the infusion rates as described may not be valid. Furthermore, vehicles for injectables should be compatible with blood (not causing haemolysis or coagulation) and not cause degenerative or inflammatory reactions in blood vessel walls or the surrounding perivascular tissue.

1.2.1.3 Risks/advantages/disadvantages

The highest risk associated with surgical models is lack of experience or poor training. The surgical procedure is not overly complex or difficult to learn; however, having a successful model is multi-factorial. It is important to understand all aspects of anaesthesia, surgery, asepsis, post-operative care, maintenance of infusion systems, associated equipment, troubleshooting, and dosing techniques. Training is important to help technical staff understand how each aspect of an infusion study can potentially affect the outcome and results. Additionally the scientists, toxicologists, and pathologists involved in infusion studies must be familiar with and experienced enough to distinguish background findings typical for infusion studies from drug-related changes.

Table 1.1 Typical Infusion Rates and Volumes

Daily infusion period	Total daily volume (mL/kg)	Infusion rate (mL/kg/hour)
4 hours	20	5
24 hours	48–60	2–2.5

Advantages of surgical models are that there is little risk of not being able to access a vessel for dosing, that long-term or continuous infusion dosing can be achieved, and that the infusion site is in a central vessel rather than a peripheral vessel. A surgically implanted catheter will guarantee access to a central vessel provided that the surgical procedure was properly performed and there are no patency or animal interference issues. If the test article is a vascular irritant, it may be less optimal to use a catheter in a peripheral vessel or to access it repeatedly. Another advantage is that the animal is free to move about its cage, and there is very little stress involved with dosing procedures since the animal is not restrained. And finally, there are no reliable non-surgical alternatives for achieving long-term continuous infusion.

A disadvantage of surgical models is the fact that animals have to undergo a surgical procedure, but with proper technique, analgesics, and post-operative recovery care the cost to the animal is minimal. Other disadvantages include the constant wearing of a jacket or harness, the single-housing requirement for tethered models, and risk of contamination or infection of the infusion system. The risk of contamination or infection can be mitigated by well-defined catheter maintenance procedures, standard operating procedures designed to minimise the risk of cross contamination, and well-trained, dedicated staff that understand the model and associated risks.

Another consideration is the cost of the equipment necessary to perform large-scale infusion studies. Each animal requires an infusion pump, a swivel, a swivel mount, and special caging. Another cost consideration is the time invested into surgically preparing the animals, availability of a surgical suite, and the consumables that are involved with general surgical procedures (drape material, anaesthetics, analgesics, catheters, etc.).

One final disadvantage is the possibility of completely invalidating a study if the animals are improperly prepared or maintained. Poor materials, bad technique, or inexperience can lead to pathology findings that may mask or be mistaken for test-article effects. Common findings in infusion study control animals and measures that can be taken to minimise the incidence of such changes will be discussed later in this chapter.

1.2.2 *Non-surgical models*

1.2.2.1 *Methods of restraint*

Non-surgical infusion models generally require a method of restraint to achieve the desired dose administration. Since non-surgical models are used for intermittent infusion study designs, the duration of restraint can last for minutes up to hours. Depending on the duration of infusion, this can be accomplished using manual restraint (simply holding the animal for a short infusion period) or by placing the animal in a restraint device.

Many types of rat restraint devices are available and will be discussed in this chapter. Animals should be habituated to the restraint device.

Animal Care and Use review committees must approve the duration of restraint for infusion purposes. If the infusion interval is extended (>2–4 hours), additional justification is usually necessary. In the case of prolonged restraint, consideration must be given to the provision of food and water at regular intervals, and animals should be released from restraint devices at least daily and allowed unrestrained activity to prevent muscle atrophy and skin necrosis, unless this interferes with achieving the experimental goals and is documented and approved by the Animal Care and Use review committee.

1.2.2.2 *Applicability of the methodology*

Peripheral vein infusion is a very common and acceptable method of infusion in rats. It is most often used for intermittent infusion study designs where the rat is restrained for the duration of dose and a temporary indwelling catheter is placed in the lateral tail vein. When the study design is such that there is an option to use either a surgical model or a peripheral model, the decision must be made based on critical study endpoints (see Section 1.3).

The frequency and duration of access to lateral tail veins depend largely on the skill of the staff and the recommendation of the Animal Care and Use review committee. If the study design requires repeated (daily) access to the lateral tail vein, the staff must be experienced in the procedure and able to proficiently place intravenous needles and catheters. The lateral vein can be accessed starting at the distal portion of the tail and moving proximally as necessary. However, there are often limits to how many times either vessel can be accessed, and this is usually determined by local legislation and/or Animal Care and Use review committees.

1.2.2.3 *Risks/advantages/disadvantages*

A risk when choosing a peripheral dosing method is the possibility of not being able to access the vessel repeatedly. If the peripheral vessels become excessively damaged by haemorrhage, local tolerance issues, or scarring, it may not be possible to establish venous access. This would generally precipitate a dose holiday or lead to removal of the animal from study.

When choosing a restrained model it is important to keep in mind the physiological effects that may be brought about by the act of restraint itself, and, as stated above, consideration should be given to the overall goals of the study. Physiological changes inherent to both techniques are discussed in Sections 5.1 and 5.2.

There are also some advantages to peripheral dosing. Peripherally dosed models do not generally undergo a surgical procedure, which is considered by some to be an advantage in itself. Not having to undergo surgery means that there is no requirement to utilise technical staff with surgical expertise or experience in long-term maintenance of infusion systems. Additionally, non-surgical models do not require any catheter maintenance between dosing sessions.

Peripherally dosed studies can usually be started up fairly quickly without the lead time necessary for surgery and recovery periods. Peripheral dosing does not require the extensive equipment (such as harnesses, tethers, and swivels) necessary for surgical models, and space is maximised in the animal room since the infusion pumps do not have to be housed in the room or on the caging systems. Additionally, animals can remain pair- or group-housed.

With the advances in equipment and techniques to accommodate surgically prepared animals, however, the benefits of a clean, reliable surgical model will many times outweigh a peripherally dosed model for critical infusion toxicity study endpoints.

1.3 Practical techniques

1.3.1 Surgical models

In this section, we shall describe the surgical catheterisation of a rat via the femoral vein with the catheter exteriorised at the nape of the neck and protected with a harness. This appears to be the most common surgical model used for infusion at the time of writing; however, many of the principles are equally applicable to catheterisation via the jugular vein and with other methods of protecting the catheter.

1.3.1.1 Preparation

The ideal surgical suite will have dedicated animal holding, preparation, and recovery areas associated with the surgery area. Pre-operative analgesia and antibiosis can be given by injection in the animal holding area. When it is time for the animal to be prepared for surgery it can be brought into the prep room and anaesthetised (usually in an induction chamber if inhaled anaesthetics are being used).

1.3.1.1.1 Premedication. Antibiotics are generally given as a prophylactic measure since strict aseptic technique should always be applied to surgical procedures for implantation of vascular catheters. The choice of antibiotic is not extremely critical; however, it should be a bacteriocidal agent that is effective against Gram-negative and Gram-positive bacteria. A broad-spectrum antibiotic (for example, enrofloxacin 5 mg/kg) is usually adequate.

Since most infusion catheter surgical procedures are considered minor surgery, a non-steroidal anti-inflammatory drug (NSAID) such as carprofen or flunixin meglumine (5 mg/kg and 2 mg/kg, respectively) administered subcutaneously (flunixin may also be administered intramuscularly) are adequate and should be given at least 30 minutes prior to surgery to ensure adequate intra-operative pain relief (Flecknell and Waterman-Pearson 2000). The onset time of analgesics indeed needs to be considered to ensure adequate intra-operative and post-operative analgesia (highly skilled surgeons may complete the surgical procedure in under 10 minutes, which can be less than the onset time of analgesia).

Whichever antibiotic or analgesic is selected, its half-life should be such that it has been completely excreted by the system by the time dosing starts. It is a good practice to avoid any treatment that could be an inducer or inhibitor of drug metabolising enzymes so as to avoid potential drug–drug interactions.

Although some form of pain relief is universally accepted, not all researchers use antibiotics. Excellent asepsis alone should be sufficient to minimise post-operative infection, and researchers routinely perform successful surgery without providing antibiotics. However, many feel that it is better to provide antibiosis as routine in case of an accidental breakdown in asepsis rather than trying to treat animals post-infection. Most will continue to provide daily antibiotics for 3 days following surgery. Irrespective of which view is supported, it is generally accepted that antibiotic treatment should not be provided just to compensate for poor asepsis.

With competent surgeons, good quality materials, and strict aseptic technique, rats will recover from the surgical implantation of an infusion catheter approximately 7 days post-surgery.

1.3.1.1.2 Anaesthesia. Both injectable anaesthetics and gaseous anaesthetics are commonly used. Injectable anaesthetics are relatively cheap and can be used by researchers without having to invest in expensive gaseous anaesthetic equipment such as vaporisers, flow regulators, and scavenger units. Injectable anaesthetics give greater flexibility for manipulating animals during surgery and may make asepsis easier to maintain through the absence of tubing and associated equipment. They may provide post-operative analgesia and sedative effects (unless antagonists are used). On the downside, they are less controllable than inhaled anaesthetics, with a greater individual variability in effect. Some frequently used combinations of drugs can keep animals sedated for some time following surgery. Thermoregulation can be an issue, and injectable anaesthetics may lead to longer post-operative recovery.

Many researchers who perform surgery on a regular basis prefer to use inhaled anaesthetics. These are very easily controlled, even at different stages of procedures. An experienced anaesthetist will have animals regain full consciousness and display normal behaviour within minutes of the completion of surgery. This high level of control and rapid recovery of conscious state when combined with appropriate analgesia is often considered to outweigh the expense and technical challenges. Effective anaesthesia is provided by 2–3% isoflurane carried in oxygen at 1 L/min.

1.3.1.1.3 Presurgical preparation. Ideally the preparation of the animals is performed by a dedicated team who can monitor the animals until the surgeon is ready. The sites of venous access and exteriorisation are clipped free from hair using electric clippers (Figure 1.1). Although not common, there are occasions where catheterisation in one leg is unsuccessful and the femoral vein in the other leg needs to be accessed; to allow for this eventuality the inguinal region of both hind limbs is clipped prior to surgery. Wet shaving the skin is not considered necessary or good practice and may increase the incidence of infection compared to clipping. The use of vacuum clippers will keep the preparation area free from loose hair. In females, at least two nipples on each side of the animal will be in the area being clipped, and great care is necessary to avoid damaging the nipples.

Once free from hair, the surgical sites are sanitised using a suitable antimicrobial skin sanitiser such as povidone-iodine or chlorhexidine gluconate. For maximum effect these agents should be left in contact with the skin during surgery and rinsed off following wound closure. In fact, some skin sanitising agents will not be effective unless they have had adequate contact or drying time. With both a dorsal and a ventral area being clipped and sanitised it can be easy to contaminate one of the areas already prepared, and therefore sterile drapes should be used. There are many types of sterile surgical drapes available for rat surgery, and the use of whole-body sterile wraps (such as those used in food preparation) is becoming more widespread amongst those performing this procedure.

Figure 1.1 Areas clipped free of hair and sanitised ready for surgery.

Ocular lubricants are an important aspect of the peri-operative care. An animal lying without a blinking reflex on a heated operating table may develop eye lesions unless the eyes are protected from drying out. New surgeons developing competence (and taking longer to perform the procedure) or those using long-lasting injectable anaesthetics may increase the likelihood of eye damage unless adequate protection is provided. Some researchers, in addition to using ocular lubricants, protect the eyes with moistened sterile swabs.

1.3.1.2 *Surgical procedure*

The term femoral catheterisation (or cannulation) in this context describes the surgical isolation of the femoral vein through which a catheter is introduced and passed through the iliac vein, past the bifurcation, with the catheter tip resting in the inferior vena cava (IVC). There has been a lot of discussion about the exact positioning of the catheter tip. Some researchers feel that there is little difference in results so long as the tip is past the bifurcation and in the IVC. Others advocate a tip position caudal to the renal veins. Some will introduce the catheter a set distance through the femoral vein (ranging from 40 mm to 60 mm irrespective of animal size), whereas others will approximate the distance for each animal according to its size usually based on proximity to the kidneys/renal veins (Figure 1.2). However, growth of the animal following surgery often leads to a catheter tip position more caudal in the vena cava at termination than when first implanted. With most classes of infused compounds

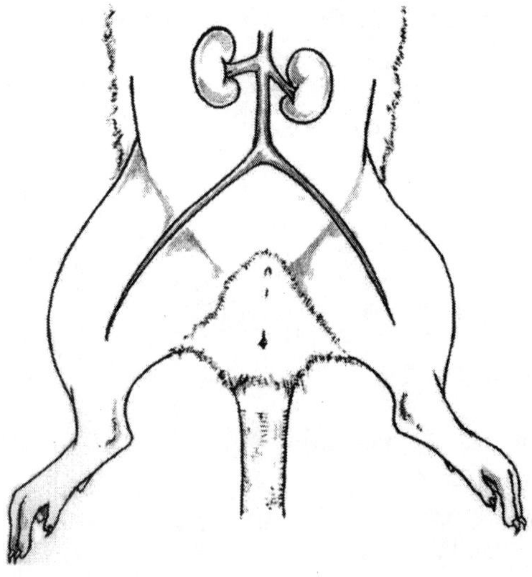

Figure 1.2 Target location for catheter tip.

these variations are not significant, especially when compared to the more precise positioning required for a successful jugular catheterisation. However, initially placing the catheter tip adjacent to the renal veins is generally considered best practice (Asanuma et al. 2006). In order to achieve this, catheter retention beads are best moved into the correct position based on the size of each individual animal. This will set the maximum distance of insertion and will usually be performed by the surgeon as each animal is presented for surgery. If researchers are conducting a programme of work with animals of a consistent size, then they may be able to establish a uniform distance of insertion and possibly use catheters with retention beads fixed at set distances. This may help to reduce the need for moving the catheter post-surgery.

Rats are considered more resistant to wound infection than some other laboratory species and very rarely show clinical signs of infection, yet very few still believe that because of this, the basic principles of asepsis do not apply. Good aseptic technique has been shown to reduce the time needed for an animal to regain its pre-operative weight, improve wound healing, extend catheter patency (Popp and Brennan 1981), and reduce adverse pathology (discussed later in this chapter). Many researchers now adopt as stringent aseptic procedures for their rat surgery as would be adopted for large-animal or human surgery. It is not the intention of this chapter to describe aseptic techniques in great detail, but measures for enhancing researchers' asepsis protocols will be discussed.

Once in surgery, the animal is laid upon a draped, heated operating table (and connected to a face-mask if gaseous anaesthetics are used), and the depth of anaesthesia is assessed both visually (skin colour, breathing) and physically (pedal response) and additional anaesthesia provided if required.

The depth of anaesthesia and intra-operative animal health are usually assessed visually during the surgery. Heated operating tables with thermocoupled rectal probes will provide the animal with additional warmth should its temperature start to fall, and some researchers use veterinary or paediatric pulse oximeters to enhance the monitoring of the animal during surgery.

An assistant secures the hind limbs of the animal to the operating surface (Figure 1.3). Surgical tape is often used. It is important to accurately position the animal to minimise the risk of damage to the blood vessel. The target limb should be stretched down the table to create as close to a straight line as possible between where the initial incision will be made and the intended site of the catheter tip. If possible, the leg should be turned outwards slightly to create a flat surface for the incision to be made.

If resistance is felt when passing the catheter into the blood vessel, the catheter should not be forced, as this will only cause damage to the blood vessel at some level.

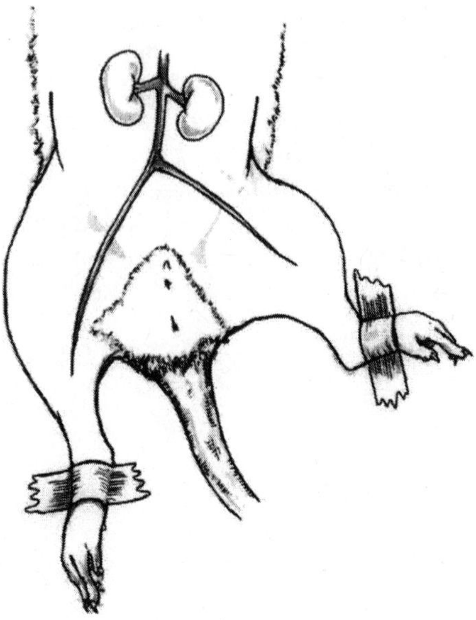

Figure 1.3 Positioning of leg.

1.3.1.2.1 Initial incision. If sterile wrap is not being used, then the animal should be covered with a sterile drape. An aperture is cut through the drape to expose the target site. With sticky drapes or cling wrap it is possible to make the initial incision straight through the covering material. The initial incision, made with a scalpel or surgical scissors, should be directly over the target vessel and close to the abdominal wall (Figure 1.4).

Surgeons often aim to keep the incision small (sometimes as small as 5 mm) to reduce closing and healing time, often equating this with good surgical performance. Although catheter introduction is perfectly possible through an incision of this size, it makes the subsequent securing of the catheter in place very challenging: the skin must be stretched dramatically, and visualising the muscle bed to which the catheter is sutured is very difficult. It is recommended that a larger incision, of 10–15 mm, be made to help ensure accurate catheter placement, reduce overall tissue trauma, and eliminate potential kinking of the catheter due to improper technique when securing the retention loop to the muscle.

1.3.1.2.2 Subcutaneous pocket. Using haemostats or blunt-nosed scissors, a pocket is made on the inside of the thigh; this will be used later to allow room for a retention loop in the catheter (Figure 1.5). During this

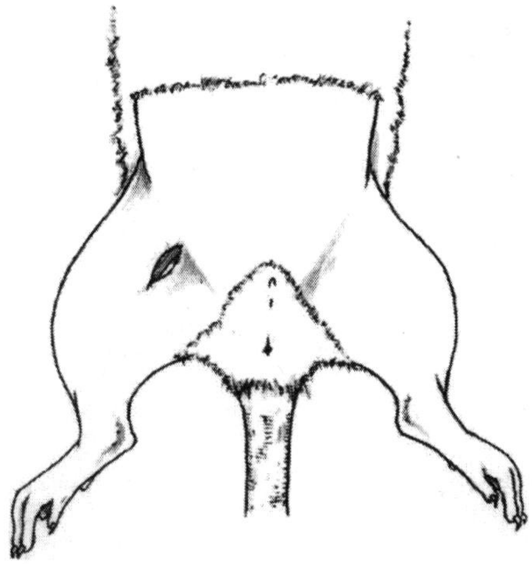

Figure 1.4 Initial incision (wider, direction of incision).

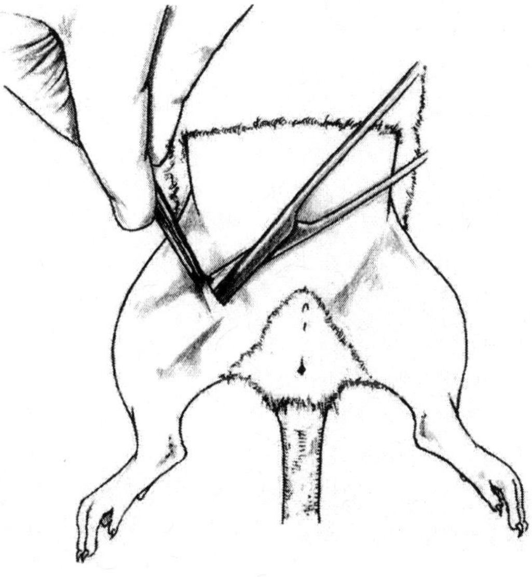

Figure 1.5 Pocket being made under the initial inguinal skin incision.

process much of the overlying adipose tissue comes away from the blood vessel, and any that remains can be blunt dissected away.

Some researchers apply a topical anaesthetic (such as 0.05 mL of bupivicaine) directly onto the exposed blood vessels to reduce vessel constriction caused by trauma during separation and provide post-operative pain relief.

The target for isolation is the portion of the femoral vein lying between the abdominal wall and the bifurcation with the superficial saphenous vein (Figure 1.6). In large male animals the abdominal wall can extend farther over the target area, reducing the space available for catheterisation.

1.3.1.2.3 Separating the blood vessel from the artery and nerve. There are several techniques for isolating the femoral vein from the associated artery and nerve (the jugular vein has no associated artery/nerve and obviates this step of the procedure). The tips of forceps or fine-tipped haemostats can be run between the artery and vein and then separated by placing fine forceps under the blood vessel (a variation of this is to run them just along the side of the vein, and then use the tips of the forceps to come up between the vein and artery). Alternatively, forceps can be placed under the whole bundle, and then the nerve and artery are separated out individually until the forceps are just under the vein (Figure 1.7). Any remaining connective tissue can constrict the vein, as will excess manipulation (such as tugging and pulling) of the vessel. Ideally a dilated vessel full of blood will be exposed, but those new to the procedure might be left with a constricted vessel that is very difficult to catheterise. Should this occur, with the distal portion of the vessel occluded, blood can be gently massaged from the inside of the thigh to refill the vessel.

Figure 1.6 Target vessel.

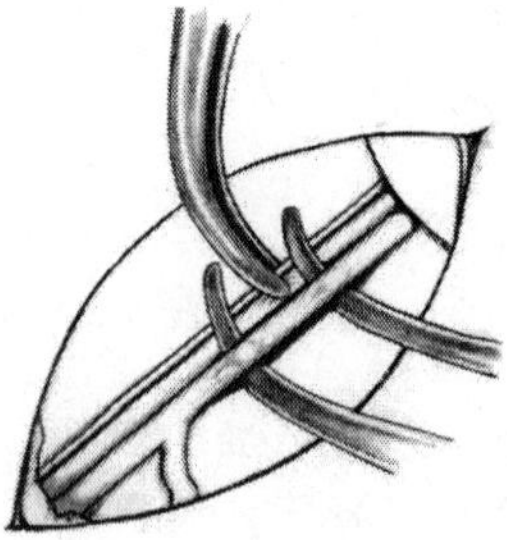

Figure 1.7 Vessel separation.

A doubled strand of suture silk (4.0) is passed under the isolated vein. It is important to lift the vessel when passing the suture silk to prevent twisting or rolling of the vessel, accidental trauma, or inadvertent occlusion of the femoral artery. Silk has been known to cause a localised inflammatory reaction, but very rarely does this have any impact on the evaluation of the dose site or the model in general.

Cutting the end of the silk suture allows two ligatures to be formed (Figure 1.8). These ligatures are usually tied loosely at this stage, ready to be tied off completely following catheter insertion.

Haemostats are clamped to the ends of the ligatures to provide tension that controls blood flow and raises the vessel, allowing easier access. Some surgeons use surgical tape to secure in place the lower haemostat controlling blood flow into the vessel; this helps prevent accidental movement of this haemostat and bleeding. Other surgeons rely on the weight of the haemostat alone to control blood flow and prefer the ability to move the haemostat if required.

With the blood flow into the vein controlled, a venotomy through the wall of the isolated portion of the blood vessel needs to be made (Figure 1.9). Two common methods used employ the use of micro-scissors or a 23G hypodermic needle bent at a 45° angle.

Skilled surgeons advocate the method of venotomy with which they are most comfortable. The use of micro-scissors appears to make location of the venotomy site and subsequent catheter insertion easier, but the size of the incision can vary; a too large incision will weaken the vessel, which

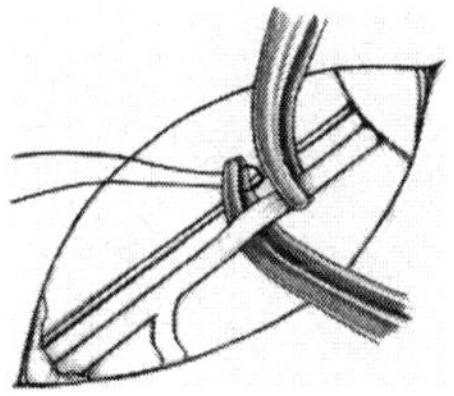

Figure 1.8 Ligated vessel.

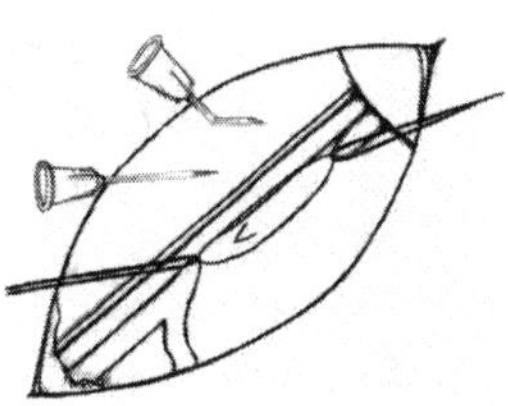

Figure 1.9 Venotomy.

may break during catheter introduction. In extreme cases the surgeon may cut completely through the vessel when attempting to make the initial incision. The use of a hypodermic needle reduces this variability; the size of the incision remains relatively constant and can result in a more robust vessel for catheterisation. However, this smaller incision can be more difficult to locate with the naked eye, and trying to locate this incision with catheter introducers can cause more trauma to the vessel. Additionally, if the angle of penetration is not precise, there is the risk of puncturing the opposite side of the vein, which can make catheter introduction very difficult.

Whichever method is employed, the initial incision should be made close to the proximal ligature. This allows a greater length of catheter to be inserted into the blood vessel before the tension on the distal ligature needs to be released to allow further progress of the catheter. Initial incisions made close to the distal ligature often result in the catheter slipping out of the vessel when the surgeon attempts to release the tension of the ligature.

1.3.1.2.4 Introducing the catheter. The catheter will have been prepared by attaching to a 1 or 2 mL syringe and blunted needle, and filled (primed) with saline to remove air. Moveable suture-retaining beads will have been correctly positioned on the catheter.

A variety of catheter introducers are available to help insert the catheter into the vessel. Fine 'picks' (often skin-grafting hooks) inserted into the vessel incision can be used to lift the wall of the vessel. This method is often used where the incision has been made using micro-scissors. Purpose-made vessel dilators or angled fine-pointed forceps can have the tips inserted into the incision and used to open up the vessel. The catheter is then inserted between the points and into the vein (Figure 1.10).

The elasticity of the vein in comparison to an artery really assists this process; however, surgeons have suggested that there are slight vascular differences between strains of rats. Fisher rats, though of a similar weight as the Sprague Dawley, have slightly smaller vessels, making standard

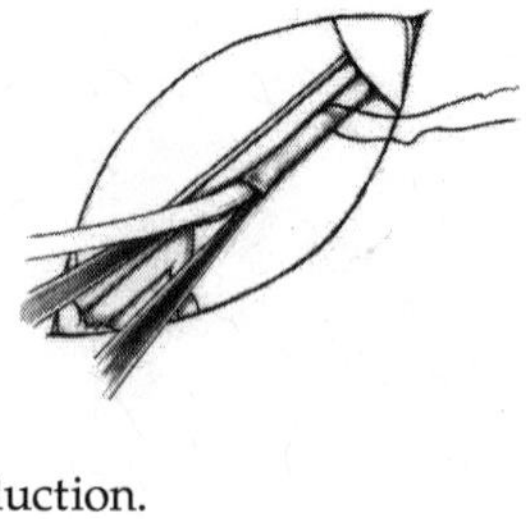

Figure 1.10 Catheter introduction.

catheter progression more challenging, and the femoral vein in Wistar rats appears slightly more fragile.

Once the incision has been made and the tip of the catheter introduced, the catheter should be progressed into the vessel until the first suture bead is reached. When passing the catheter along the blood vessel it is advisable to try to position the catheter so that the curvature takes it up the body rather than towards the opposite leg. Progress through the blood vessel should be smooth and without resistance. Any resistance encountered will most often be as the catheter tip reaches the bifurcation into the inferior vena cava. Applying light pressure to the abdomen directly above the bifurcation is usually sufficient to channel the catheter in the right direction. Readjusting the curvature of the external portion of the catheter or straightening the leg may also help overcome any resistance; the catheter should never be forced. Continually encountered resistance is usually associated with the initial positioning of the animal's hind limbs.

Catheter patency is usually checked once the catheter has been fully inserted. It should be possible to freely aspirate blood into the catheter.

1.3.1.2.5 Securing the catheter. The suture bead should be up against the incision in the blood vessel. The proximal ligature controlling blood flow into the vein should be tied off around the blood vessel, completely sealing it off. It should then be tied behind the suture bead to keep the catheter in place. It is advisable to secure this suture in place first as this will prevent any movement of the catheter out of the vessel as well as prevent any accidental bleeding. The distal ligature should then be tied around both the blood vessel and the now internal catheter. This will permanently ligate the vessel. Other vessels will take up the blood flow, so this permanent occlusion of the vein will not adversely affect the animal. Up to two further sutures may be added around the vein distal to the first suture-retaining bead to hold the catheter securely in place (Figure 1.11). This is recommended when using catheters with moveable rather than fixed suture beads. Until fully proficient it is good practice to check patency after each suture is tied to ensure that the catheter has not been occluded by too tight sutures.

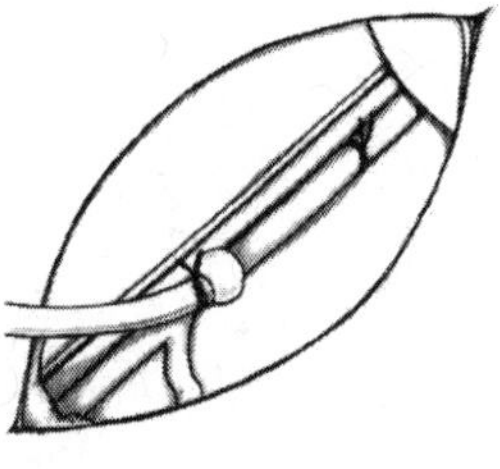

Figure 1.11 Catheter sutured in place.

1.3.1.2.6 Exteriorising the catheter. Most femoral catheterisation models will have generally followed the procedure described up to this point. However, the different models based on the method chosen to protect the externalised portion of the catheter introduce some variations to the surgical procedure. The procedure outlined hereafter will continue to focus on the model employing a harness, and significant differences between this method and those for other models will be highlighted.

Sterile gauze is placed over the femoral incision site, and the animal is turned over. A 20-G needle or scalpel is used to make an incision through the skin directly between the raised scapulas of the rat. This is where the catheter will exit from the animal, and it is important that this incision is precise so that the catheter can go straight up the tether or be attached in a straight line to the pin in the dome of the harness saddle. If this incision is not correctly sited, the catheter could become attached to the harness or travel up the tether at a slight angle and may be subject to increased stress and possible leakage. In extreme cases the point of exteriorisation might not be protected by the harness and could be subjected to animal interference. The exit point on the skin can be indicated with a marker pen by the animal preparation team.

There are a great variety of tunnelling tools that can be used to exteriorise catheters, ranging from blunt-handled or blunt-tipped tools (shown in Figure 1.12), and long-handled crocodile haemostats, to sharp tunnelling needles (which have been used for both tail and scapula exit points). It is difficult to promote one design over another; however, greater care is required with very sharp tunnelling tools to prevent penetrating the skin during tunnelling or even piercing the abdominal wall. Less-sharp tunnelling tools may be more difficult to progress subcutaneously but reduce the risk of accidental damage.

Figure 1.12 Subcutaneous tunnelling.

Depending on the circumference of the tunnelling tool being used, many surgeons will insert the tips of fine-nosed haemostats or scissors into this incision and gently open them to increase the size. Some will go further and make a small subcutaneous pocket below the incision to place a loop of catheter. This loop gives a greater length of catheter that will aid in securing the catheter to the harness saddle.

The tunnelling tool is passed subcutaneously along the animal's back and side, avoiding fat pads, to exit at the femoral incision site.

To exteriorise through the tail, a sharp fine tunnelling needle is used. It is passed from the dorsal surface of the tail along the side of the tail to exit at the femoral incision. The distance travelled is far less, but great care must be taken to avoid causing trauma to the skin and underlying tissues during this transit.

The catheter is clamped shut part way along its length to prevent blood travelling up the catheter (or the accidental introduction of air) once the syringe is removed for exteriorisation. It is advisable to use coated vessel clamps or a sterile gauze swab to protect the catheter from being damaged. The catheter is then fed along the tunnelling tool from the femoral incision site to the exteriorisation site, the tunnelling tool is removed, and then the catheter is reconnected to the syringe before releasing the clamp (Figure 1.13).

A loop of catheter is secured to the muscle on the inside of the leg with two sutures. Usually one suture is positioned at the base of the loop and one behind a second suture-retaining bead. As well as securing the catheter in place, this loop is essential to maintain catheter patency while

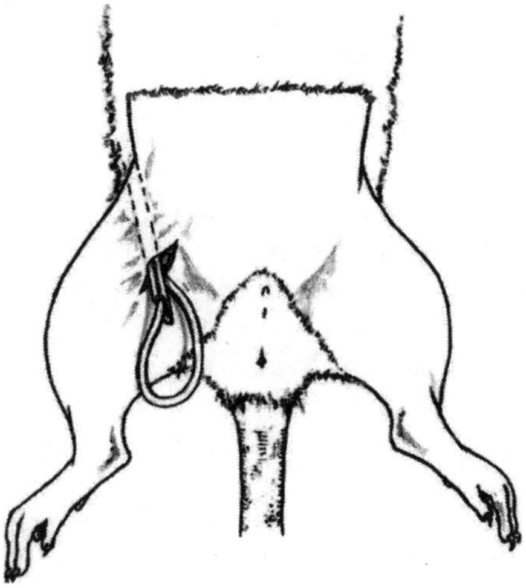

Figure 1.13 Exteriorisation.

the animal moves around. The size of this 'loop' doesn't need to be too generous, and a large loop may in fact increase the potential for catheter occlusion. It is recommended that the surgeon manipulate the catheterised leg of the animal into various positions while checking patency to identify any potential positional occlusions prior to closing the femoral incision.

The scapular incision is usually closed with a single absorbable suture. If a particularly thin tunnelling tool is used and a subcutaneous scapular pocket is not formed, then the need for this suture may be obviated. It is worth noting that the point of exteriorisation, like any permanent penetration through the skin, is probably the weakest point of this model for the introduction of infective organism. Getting a good seal around catheter and skin will reduce this risk, and some researchers apply an antibiotic gel to form a physical barrier to prevent infection.

The femoral incision may be closed using a variety of methods. Wound clips are a fast and expedient method of closing an incision, requiring a lower degree of skill compared to conventional suturing; however, some institutions do not endorse their use because of the potential increase in tissue trauma. Simple interrupted sutures are often used; absorbable sutures will not need removing. Buried subcutaneous sutures may reduce the risk of animal interference but require a higher level of skill to be performed competently. Some researchers also pull together the overlying fat tissue and suture this back together prior to closing the skin; however, the need for this step is questionable.

1.3.1.3 Recovery procedures

The rapid recovery of animals from gaseous anaesthesia makes this the method of choice for most researchers, especially those performing large numbers of surgeries. Isoflurane anaesthesia, appropriate NSAID analgesia, and short-duration surgery performed by skilled surgeries will often result in animals resuming normal behaviour within 10–15 minutes of surgery. For such animals, minimal post-operative weight loss is observed.

Appropriate post-operative analgesia and antibiosis is often debated, with changing views of what is necessary and what is 'best practice'.

Many researchers will give an NSAID injection prior to surgery and for up to 3 days post-operatively, and this appears to be effective in controlling pain and promoting normal behaviour. Others will combine this with a pre- (or post-) surgery opioid (such as buprenorphine 0.01–0.05 mg/kg s/c). Although this will provide excellent pain relief when combined with the NSAID, it can lead to a reduced food intake and greater weight loss immediately following surgery. It may also increase the incidence of post-operative wound interference in some animals (Patten 2011). The use of premedication is discussed in more detail in Section 3.1.1.1.

Injectable anaesthetics are often accompanied by protracted recovery times, compared to gaseous anaesthetics, even when antagonists are used. This can lead to an extended working day to monitor animals as they recover from surgery (or, alternatively, reduce the available hours during which surgery is performed). During this protracted recovery time it is important to provide supporting heat to the animals (lamps, pads, whole room) and possibly fluids to promote recovery.

A minimum recovery time of 7 days is recommended following surgery. Careful monitoring of behaviour, wound sites, clinical condition, tethering systems, body weights, and food consumption is routinely performed until normality is ascertained.

1.3.1.4 Housing

There are a variety of housing systems being used successfully across industry, from modified versions of standard caging to purpose-built dedicated infusion caging. Space is often at a premium as the need for a tethering system and space for the infusion pumps can reduce normal room occupancy. Add to this the possible need to collect biological samples (such as urine), and the amount of space required increases even further.

The cages used for most infusion models must be able to accommodate the tether and swivel. Many different designs of tethers, swivels, and swivel arms are available to accommodate most common cage designs. Most vendors will also modify the equipment based on individual specifications. The most important issues are that the tether is long enough to allow the animal full movement within the cage, and the swivel can function unimpeded to avoid twisted or broken tethers. There must also be a way to connect the catheter to the pump, which means there must be provision for the pump to be located close to the cage. This can be accomplished by housing the pumps on the actual rack of caging or by attaching them to a pole or other housing device close to the rack of caging.

It appears uniform to house tethered animals individually. Some researchers will keep animals fitted with the quick disconnect harness in pairs or groups prior to tethering, but many feel that the risk of interference with the harness from cage-mates outweighs the benefits to the animals and will also house such animals individually.

Some systems will employ a counter-balance over each cage that takes away the weight of the tether from the animal by responding to changes in tension from the tether. This is considered a benefit to the animals and is in common use for mice but not uniformly used for rats.

Enrichment varies from institution to intuition. Contact bedding is used by many researchers without major issues; however, the tail-cuff model appears to suffer most from the presence of contact bedding, which can work its way into the cuff and cause irritation or, when combined with urine or faeces, lead to ulceration and infection. Where tail-cuffs and contact bedding are combined, technicians take great care in removing bedding or other debris from the cuff on a daily basis.

Some contact bedding will create a higher dust environment. This can make the maintenance of strict asepsis during repairs or dose administration more challenging. Dust-free bedding may reduce this challenge.

Shelters used for conventional rats are rarely suitable for tethered animals; however, other enrichment products are readily provided. Nylon chew toys, wooden chew blocks, and nestlets are provided for tethered rats without any reported issues.

1.3.1.5 Maintenance and troubleshooting

The care and attention afforded to the maintenance of these surgically prepared animals is as important to the success of the model as the surgery itself.

The major challenge for researchers is maintaining catheter patency. In most cases patency to infuse can be extended way beyond the patency to withdraw blood. Some catheter occlusions are temporary and are cleared when the reason for the blockage is removed or when the piece of faulty equipment is repaired or replaced. Some occlusions are permanent and, in such cases, will lead to the removal of the animal from study if the catheter cannot be repaired or replaced.

Occlusions are often associated with (and indicated by) various other system issues, including leaks in the infusion system. For example, an animal escaping from its harness may have been caused by a blocked catheter creating a leak in the system that has subsequently loosened the belly bands.

In rare cases, stress points in the catheter (around sutures or kinks) may result in internal leaks manifesting themselves as fluid-filled subcutaneous swellings in the inguinal region.

The most common patency issues involve the build-up and accumulation of thrombi and fibrin inside the catheter or around its tip. Often this results in the inability to withdraw blood, even though it is still possible to infuse fluids. Stringent asepsis, complete flushing after checking patency, and care during catheter introduction will all help minimise thrombus formation and fibrin accumulation and increase the duration of patency.

It is essential that animals and systems be checked carefully through the day and any external system leaks identified. Even when all fittings are secure, leaks can occur at any of the delivery system joints, especially at the swivel, which has tubing going up to the syringe and down to the animal. The seal on the swivel itself may also break either through wear and tear or, with plastic disposable swivels, through over-tightening the holders used to secure the swivel. Leaking swivels will often deposit the infusate down the tether and onto the animal. In very slow delivery, these leaks can be difficult to detect. Wetting of the animal's fur or crystallisation on the tether during saline KVO is usually the first signs of a leak within the tether or from the swivel.

Tethering systems are designed to allow the animal full movement around its cage while protecting the catheter from animal interference. Whichever system is employed to secure the tether to the animal, there will be occasions when the animal removes itself from the tethering system. Sometimes the catheter may remain intact or have sufficient viable catheter still exposed to effect a repair using an extension, so that the infusion system can be re-established.

However, animals may pull on the catheter once exposed, dislodging the catheter from its initial position or, if sufficient force is used, pulling the catheter completely out of the vessel (this is often suggested when swellings appear along some portion of the catheter tract). Other animals may chew through an exposed catheter, which may regress under the skin.

Where catheter patency cannot be re-established, many institutions will remove animals from testing. However, some researchers have the necessary skill and institutional authorities to perform reparative surgery. This may involve repairing a catheter or refitting a tail-cuff or skin button under general anaesthesia or may extend to the complete removal and replacement of a damaged or blocked catheter.

1.3.1.6 Record keeping

Surgical records are kept to indicate the administration of pre-, peri-, and post-operative treatments. In GLP environments the surgical records should minimally include the surgeon, surgical assistant(s), time and date of surgery, post-operative care, and any other relevant information. Equipment is available to monitor vital signs of rats during anaesthesia (respiration rate, body temperature, heart rate, response to stimuli, anaesthetic level, etc.), and the surgical record should include periodic documentation of these values. For relatively short or basic procedures, rats may be monitored by visual assessment. Post-surgical pain/distress assessments should also be documented, as well as any associated supplementary care.

The dosing accuracies and dose accountability calculations discussed in Section 2.1.2 apply to both surgical and non-surgical (or peripherally dosed) infusion models as well and should be performed in the same manner. This is generally done for other dose routes by weighing the dose container before and after dosing each group of animals and comparing the actual weight with the theoretical weight. Infusion accountability is done in much the same way; however, it can be done on an individual animal basis. The syringes or test article reservoirs are filled, labelled, and weighed prior to infusion dosing. Following the dose administration period, the syringe or reservoir is removed and weighed again. The pre- and post-weights are documented, and a simple calculation is performed to demonstrate accuracy of delivery. The weights and calculations are based on the assumption that the solution has a specific gravity close to 1, and therefore it is accepted that 1 mL = 1 g. The data can be presented as the absolute difference (mL) between theoretical and actual volumes, or as a percentage difference between the theoretical and actual volumes:

Pre-weight (g) – post-weight (g) = actual weight (g)

Theoretical weight = infusion rate (mL/hr) * [time infusion (min)/60]

Percent difference = [(actual weight (g)/theoretical weight (g)) * 100] – 100

1.00 mL = 1.00 g

The test article administration is generally considered within acceptable limits when the actual volume delivered is less than or equal to +/– 10% of the theoretical volume. This is subject to individual institution requirements and may be adjusted as needed, especially when infusing very low volumes.

The quality and relevance of all records maintained may well help in identifying causes of problems that may manifest later.

1.3.2 Non-surgical models

1.3.2.1 Methods of restraint and acclimatisation

There are many types of restraint devices for infusion in rats, but the ideal design will be made of transparent material to allow for visual assessment of the rat, have enough ventilation so the rat can breathe easily and not become hyperthermic, and allow easy access to the tail while keeping the rat from turning or escaping. Animals in restraint devices should be monitored continuously.

Regardless of what design of restraint device is used, rats should be habituated to the device prior to the commencement of dosing. Habituation aids in lowering the amount of stress caused by restraint, as restraint stress can introduce an experimental variable, and so this should be minimised whenever possible. The habituation should be performed in increasing time intervals up to at least half of the expected total time of restraint, and in some cases up to the full duration of the intended infusion period.

1.3.2.2 Methods of vascular access

The peripheral veins most commonly accessed for non-surgical intravenous infusion in the rat are the lateral veins on the left and right sides of the tail. These vessels are located subcutaneously and are relatively easy to access in an unanaesthetised rat. The distal end of the rat tail is best suited for access, since the structures of the musculoskeletal system gradually diminish towards the end of the tail, while the sizes of the blood vessels do not decrease in proportion; the blood vessels become most prominent near the tail's tip (Staszyk et al. 2003).

It is recommended to warm the tail prior to access so the blood vessels become dilated to facilitate insertion of the needle or catheter. This can be accomplished by warming the rat using a warming chamber or lamp or by immersing the tail in warm water (not to exceed 45°C) for 5 to 10 seconds. The temperature of the animal must be closely monitored during these procedures since a relatively small increase in core temperature can have significant effects. A 1.5°C elevation of body temperature above normal core temperature can result in increased rates of embryonic death, malformation, or resorption (Graham et al. 1998).

Once the tail is warmed and the vessels are sufficiently dilated, the rat can be placed in the restraint device for access. The lateral vessels can be visualised by lifting the tail and slightly rotating it in either direction. The catheter or injection set should be primed with saline. The intended insertion site may be wiped with alcohol, and while holding the tail with one hand the other hand will guide the needle through the skin almost parallel to the tail and into the vessel. Depending on the type of needle or catheter used, usually there will be a small flash of blood in the hub to verify proper needle placement. If not, an attempt to aspirate a small amount of blood can be made. Alternatively, a small amount of saline can be injected to confirm placement. If accurate placement cannot be confirmed, then the needle can be removed and a site on the same vessel in a more proximal location can be attempted.

Once confirmed access to the vessel is established, the needle or catheter can be connected to the infusion pump using extension tubing, and the infusion can commence.

When the infusion interval is complete, the needle or catheter is flushed with saline and removed. Slight pressure is applied to the injection site until the bleeding has stopped. The animal can then be removed from the restraint device and returned to its home cage.

1.3.2.3 Housing

There are generally no housing specifications or restrictions for non-surgical restrained models other than those defined by local regulatory agencies. There are no permanent or external devices that would preclude them from the standard housing and enrichment detailed in the facility's animal care programme.

1.3.2.4 Checks and troubleshooting

The most difficult part of non-surgical, peripheral dosing is establishing venous access, especially for repeat dosing. Once venous access is confirmed, it is important to ensure the catheter or needle remains in the vessel for the duration of the infusion interval. This is done by utilising proper restraint techniques and appropriate equipment. With an indwelling catheter, the rat will have more freedom to move its tail during the infusion process. Using a metal needle will usually require immobilisation of the tail.

Signs that the needle may have become dislodged from the vessel are swelling at the injection site, occlusion of the infusion system, and a reaction from the animal indicating discomfort. If this occurs, the dosing needle or catheter must be removed and replaced. The injection site must be closely monitored throughout the dosing period.

1.4 Equipment

1.4.1 Surgical models

1.4.1.1 Surgical facilities

The surgical suite should comprise the following:

Preparation room. This area should have an anaesthetic induction chamber if gaseous anaesthesia is used, a method for maintaining anaesthesia during preparation (this may require several outlets for busy surgeries with multiple animals being prepared), and clipping facilities for hair removal. Often preparation areas are used to store surgical instruments, equipment, and consumables, as well as an autoclave to sterilise instruments for surgery. Ideally the preparation room will be adjacent or in close proximity to an animal holding room. If this is not the case, the preparation area should include a private holding area with uninterrupted access to water for animals awaiting a surgical procedure.

Depending on the system employed, animals are often returned to the prep-area for fitting of harnesses or jackets, whereas tail-cuffs and skin-buttons are secured as part of the surgical procedure.

Surgeon preparation (scrub) room. An area for surgeons to scrub and become gown ready for surgery.

Surgery. An area with dust-free surfaces free from cupboards. Set up with operating tables or stations with piped gases and vaporisers for gaseous anaesthesia and focussed cold lighting. Adequate waste anaesthetic gas scavenging systems and excess gas evacuation systems. Positive pressure airflow with HEPA filtration is highly desirable.

Recovery room. A quiet area with dimmer switches and additional heating facilities for animal recovery and monitoring. It should be large enough to house recovery caging as necessary.

Animal room. Once recovered, animals are transferred to the animal room. Most standard animal rooms can be used, although additional power sockets will need to be provided and the room will need to be set up to cope with the additional electrical load of all the infusion pumps that will be used. Additionally, although most infusion pumps have battery backup, it is recommended that the pumps be connected to an uninterrupted power supply (UPS) in the event of a power outage. These can be commercially purchased, and depending on the specifications of the individual unit most can support multiple infusion pumps for a prolonged period of time. Large studies with multiple infusion pumps may require more than one UPS unit in the animal room.

1.4.1.2 Catheters

There are several manufacturers offering catheters of varying designs, dimensions, and materials (polyurethane, silicone and, less commonly, polyvinylchloride or polyethylene).

It is difficult to advocate a particular design of catheter over another, as researchers are performing successful studies with a variety of designs and are sometimes limited in choice of material by test-article compatibility issues. However, the design that has been used successfully for studies up to 6 months in duration in our laboratory has the following:

- 3-Fr medical-grade polyurethane catheter
- OD 1.00 mm
- ID 0.39 mm
- 5-mm soft polyurethane rounded tip, heat welded
- Two moveable suture beads

Polyurethane is a durable material with a good tensile strength but can cause more vascular damage than softer material. Whereas silicone catheters may be less thrombogenic, they have a very thin, soft wall and can more easily kink or occlude or even leak more readily than medical grade polyurethane; their very softness can also make them more difficult to pass into the vessel. Looking to combine the positive elements of both materials, the ideal catheter is polyurethane along its length but has a 5-mm softer polyurethane tip heat-welded to the portion of the catheter that will lie inside the blood vessel.

This design of polyurethane catheter will perform well in most situations; however, polyurethane (and silicone) has a relatively high gas permeability. If used with prolonged locking intervals or very slow infusions, there is the potential for the locking solution or infused material to evaporate through the catheter wall, allowing blood to enter the catheter and leading to potential patency complications. Catheters channelled directly from the animal through the tether to the swivel increase the likelihood of this occurring because of the greater surface area of catheter wall in contact with the locking solution or infusate. Reducing catheter length to a minimum by attaching it directly to a vascular access harness or port is recommended in such situations (Nolan et al. 2008). Catheter tips can be rounded, square cut, or cut at an angle (bevelled). Rounded tipped catheters are considered the least damaging to the intimal lining of the blood vessel and likely to result in longer patency than square cut or bevelled catheters of the same material (Vemulapalli and Fredenburg 2007; Nolan et al. 2008). Some 'rounded'–tip catheters are ground to produce a smooth surface (which can be angled rather than truly round); however, this process can leave irregularities in the surface and has the potential to introduce shards of catheter material into the animal.

Bevelled catheters may be easier to implant into the vessel and are usually used with silicone catheters to overcome the difficulty of inserting a soft material. However, bevelled (and square-cut) polyurethane catheters can cause significant damage to the blood vessel both during and after insertion. This may lead to undesirable histopathological results and decrease the patency of the catheter.

To extend the period of patency, some polyurethane catheters can be supplied coated (covalently bonded) with certain materials to improve haemocompatibility. These coatings prevent blood components from adhering to the catheter's surface that could potentially occlude the catheter (Foley et al. 2002). To increase the durability of the exteriorised portion of the catheter, some designs are tapered, with the external portion having a significantly thicker wall than the part of the catheter in the blood vessel. Also available are 'double skinned' exteriorised catheters. The retention beads used to secure the catheter in place are typically (although not always) of the same material as the catheter and may be heat-welded in place or moveable. Moveable retention beads are tight fitting and rely on the traction of the materials to keep them in place after being situated in the appropriate position.

1.4.1.3 External equipment

1.4.1.3.1 Tethers. Tethers are in place to prevent animals from interfering with the infusion line or catheter. They have taken on a fairly standard design, typically a tightly coiled stainless steel spring through which the exteriorised catheter or infusion line is passed (Figure 1.14). It is connected to the side or above the cage and will be long enough for the animal to move freely around and reach each part of its cage. Lengths can vary according to the cage design/dimensions and mounting system in place, but 30 cm appears quite typical. Some tethers (like those used with the VAH Quick Disconnect system) are supplied with integral tubing for infusion. Tethers can be single, dual, or (rarely) multi-channelled. All will have some method of attaching them to the animal's harness, jacket, tail-cuff, or skin-button.

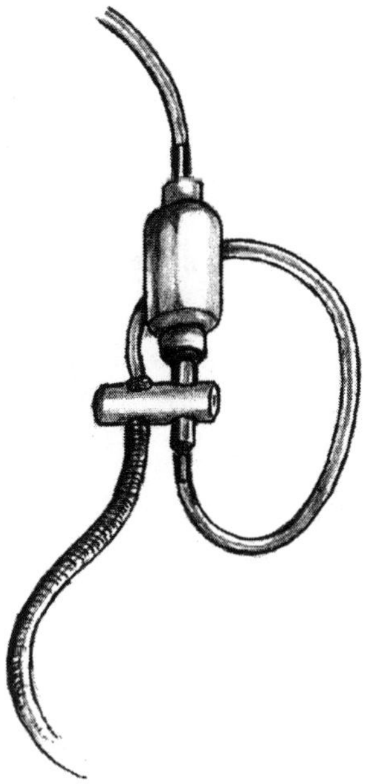

Figure 1.14 Tether.

1.4.1.3.2 Swivels. Tethers are routinely attached to a swivel that rotates with thc movements of the animal in its cage, preventing the tether from tangling and possibly occluding the infusion line or catheter that it is protecting (Figure 1.15). The swivel is mounted above (or to the side of) the cage in a swivel mount. Several types are available. Brass and stainless steel swivels are excellent and are often cleaned, sterilised (they may be autoclaved), and reused. Disposable polysulfone swivels that have stainless steel pins and a Teflon seal are widely used and are discarded following use to prevent cross contamination. Most disposable swivels can be autoclaved once only.

The catheter or connective tubing passing up the tether is attached directly to the swivel's stainless steel pin; connective tubing also connects the swivel to the syringe or infusion pump.

Swivels are available with single, dual, or multiple channels, and specialised swivels with a quartz-lined fluid path are available for use with infusates that may react with stainless steel.

1.4.1.3.3 Vascular access harness (VAH). The vascular access 'quick disconnect' harness is relatively simple to fit and can be easily adjusted as the animal grows (or loses weight). It comprises a domed 'saddle' through which a pair of belly bands are looped (Figure 1.16). The 'quick

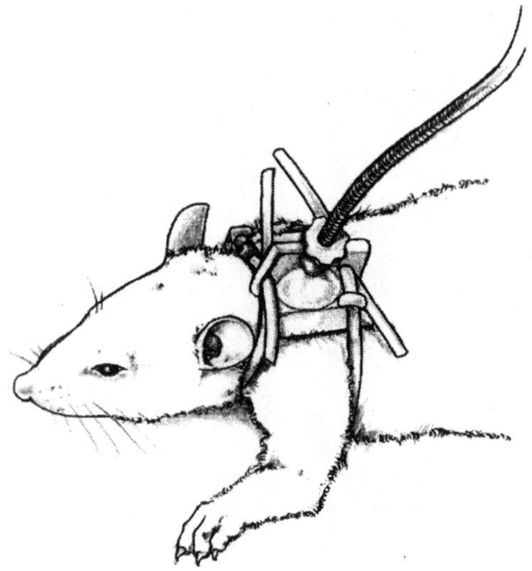

Figure 1.15 Swivel.

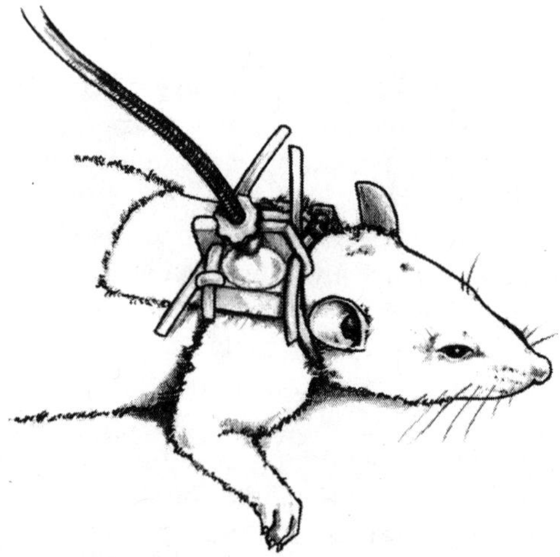

Figure 1.16 Rat in a VAH.

disconnect' version allows a tether with integral connective tubing to be plugged easily into the saddle to allow infusion. This system allows the animal to be disconnected from the tether and infusion system during non-infusion periods while still maintaining a closed system, and is easily connected to the infusion system as needed.

The natural traction of the hollow silicone bands creates a secure system, and it is usually necessary to moisten the bands to first fit the harness. This is also one of the potential drawbacks of the system, as the traction of the silicone bands can remove the hair and abrade the skin around the forelimbs. Too much movement of the harness can result in an animal chewing the belly bands or escaping from the harness. Conversely, care has to be taken not to over-tighten the harness.

Fitting the catheter to the pin under the dome of the saddle can be a challenge unless excess exteriorised catheter is available that will be pushed back into the sub-scapula pocket after fitting. The catheter is then locked with an appropriate locking solution and the animal allowed to recover untethered. This system lends itself well to intermittent infusions where the animal can be disconnected from the tethering system when not being dosed and the catheter kept patent with a locking solution.

Another version of the harness allows the catheter to travel from the animal straight up through the tether and is well suited to KVO or continuous infusion regimens.

1.4.1.3.4 Jackets and tethers. Jackets and tethers were amongst methods developed to protect the exterior portion of the catheter and, with little change in design, still continue to be successfully employed by many institutions. With this method the catheter travels straight up the tether from the exteriorisation site while the tether is held in place with a Velcro pad to a mesh-cloth jacket (Figure 1.17).

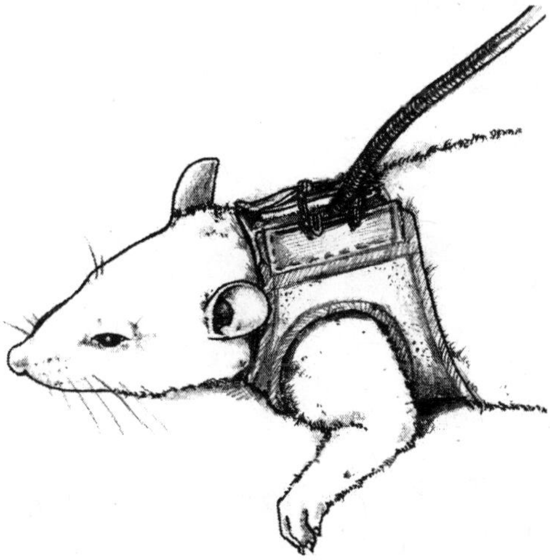

Figure 1.17 Jacket and tether.

Increasing sizes of jackets can be fitted to accommodate animal growth, and minor adjustments to individual jacket sizes can also be made. Cost may be less than the harness, and many researchers using this method consider it a more secure method with less opportunity for escape. Potential drawbacks include general wear and tear of the jackets, with the risk of animals trapping their teeth in the jacket fibres. Like the harness, lesions can develop around the forelimbs. However, experienced technicians can minimise such complications by carefully fitting the appropriate size jackets and by close monitoring.

Some researchers will habituate or acclimate animals to the jackets or harnesses and tethers for several days prior to surgery, considering there to be less stress to the animals if introduced gradually; others will fit the tethering system immediately following surgery without any habituation. Researchers assessing behavioural changes or measuring cardiovascular parameters appear to place a greater emphasis on habituating rats to the tethering system. Accepting that such parameters may be adversely affected by a lack of acclimation strongly suggests that habituation to the tethering system prior to surgery should be adopted as routine best practice.

1.4.1.3.5 Tail-cuffs. Aluminium or stainless steel tail-cuffs are screwed to a tether via an angled connection tube on the cuff (Figure 1.18). The tail-cuff method is potentially the most secure model but probably the one requiring the greatest amount of skill to fit properly. The tension (and thickness) of the wire usually used to secure the cuff to the tail is critical to the longevity of the infusion system. A tail-cuff secured by sutures rather than steel wire has been successfully used. Aside from the high skill level required to fit the tail-cuff, the resultant pathology from the tail-cuff and the securing wire is considered by some to be a significant drawback with this model.

Figure 1.18 Tail-cuff.

1.4.1.3.6 Skin Buttons. Obviating the need for a harness/jacket or tail-cuff, the skin button (Figure 1.19) is secured directly to the animal. Stainless steel buttons have been overtaken in design by solid or mesh plastic buttons. The method and site of securing to the animal varies, but for researchers looking to exteriorise the catheter in the scapular region, the skin button is usually held in place by suturing it directly to the fascia under the skin or using a subcutaneous in-growing mesh collar. Developments in the materials used for buttons have led to far less irritation and inflammation from their use. The tether is screwed to the button, or a 'quick disconnect' system can be used similar to that available with the harness.

This system is not widely used, although it has some benefits for applications where the use of a jacket or harness is undesirable (e.g., respiratory evaluations). Complications with this system include swelling and serous discharge at the button site due to movement caused by the connection of the tether to the button, often leading to infection. These systems can also lead to animal discomfort if not properly fitted or monitored.

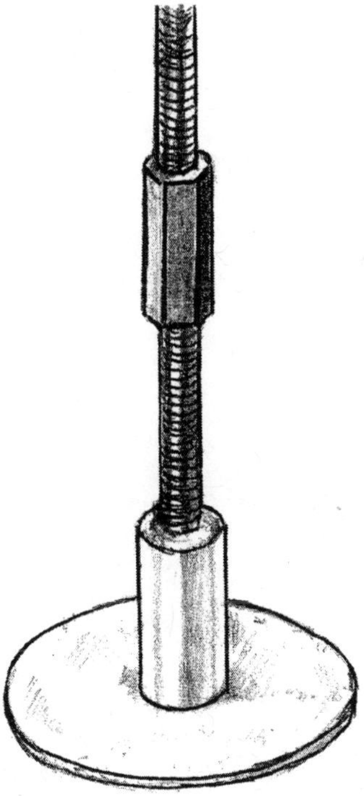

Figure 1.19 Skin button.

1.4.1.3.7 Vascular access ports (VAPs). Many different designs of vascular access ports are commercially available for rats. VAPs comprise a dome with a septum connected to a catheter (Figure 1.20). The dome is implanted subcutaneously, and the catheter is implanted as detailed above. The catheter is accessed by penetrating the skin with a non-coring needle that remains in the port for infusion. Although these systems are completely implantable, they are not commonly used in rodent infusion since the above listed alternatives are cheaper, easier, and more reliable systems.

1.4.1.3.8 Infusion lines. The infusion device needs to be connected to the swivel by an infusion line (connecting tubing). Ideally this tubing will be of the same material as the catheter to minimise the potential issues with infusate incompatibility. Such lines will vary in length depending on cage design and pump location, but it is imperative that their internal diameter allows them to be fitted securely to the pins of the swivels being used. Some facilities use very long infusion lines (sometimes coiled) to allow animals to be removed fully from their cages for procedures (such as blood sampling or ophthalmic examinations) while still connected to the infusion pump. This is sometimes an important consideration when a limited amount of test article is available. The infusion line will also need to be connected to the syringe on the pump. This can be through an integral female luer adapter fitted to the tubing or through a blunted needle. Both the luer adapter and blunted needle are usually attached to an injection cap that is then pierced by a needle attached to the syringe. This keeps the system closed when syringes are changed or if the animal is removed from the infusion system. It is important to use non-coring needles to pierce the injection cap; hypodermic needles will cut small slivers of rubber from the injection cap septum, which may enter the animal or cause the injection cap to leak.

Black or coloured infusion lines are available for the administration of light-sensitive compounds.

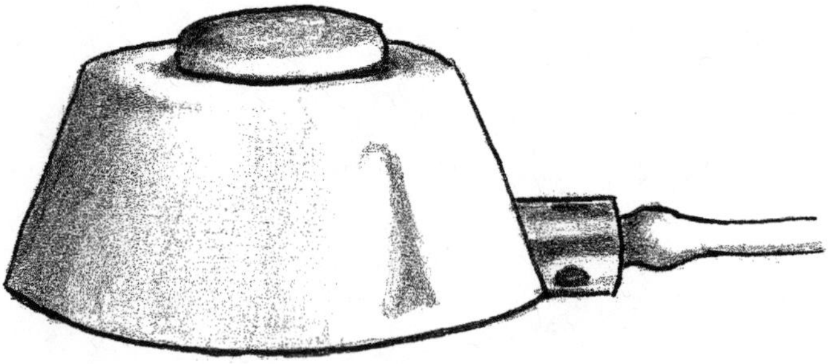

Figure 1.20 A vascular access port (VAP).

1.4.1.3.9 Filters. Some researchers incorporate filters into their infusion lines (0.2-micron disk filters, for example). These will help to maintain asepsis but can adversely affect some compounds. Many consider that with good asepsis during compound preparation and syringe filling filters should not be required; however, in-line filters will prevent the accidental introduction of contaminants should there be a breakdown in aseptic technique.

1.4.1.4 Infusion pumps

The type of infusion pump selected depends on many variables, including type of test article to be infused, volume, rate, duration, and required accuracy. The study design will usually determine what the optimal setup will be, but listed below are some recommended options for several types of protocols.

1.4.1.4.1 Tethered infusion pumps. Despite the fact there is a myriad of pumps commercially available for tethered infusion in rats, these pumps are almost all one of three types: syringe pumps, peristaltic pumps, and piston pumps. Syringe pumps are the most commonly used pump for rat infusion because of their accuracy, ability to deliver small volumes at very low rates, small size, and ease of use.

Pump accuracy. Most infusion pumps are supplied with certification and do not require routine preventative maintenance for approximately 5 years. After 5 years of regular use, best practice is to have a yearly preventative maintenance programme for all infusion pumps. Additionally, the performance of all pumps must be monitored, and any questionable performance must be documented and the pump removed from service and repaired before returning to service. Even though the accuracy of the above-described infusion pumps has significantly increased in recent years, it is good practice to check pump accuracy before the start of each experiment. An accuracy test is a pre-study run where the syringes are filled with saline or water and the pumps are programmed to deliver at an infusion rate and duration similar to that which will be used on study.

Automation. Especially when dealing with rat infusion studies, the number of animals can be in the hundreds. Individual programming of infusion pumps, calculating flow rates, and monitoring alarms can be a logistical nightmare. And all of this needs to be combined with procedures such as blood sampling for PK, clinical observations, and sampling for biomarkers. This leaves the door wide open for errors, and the constant presence of technicians in the animal room can be a stressor for the animals.

Recently, several software and hardware systems have become available that network pumps, allowing centralised control and monitoring.

The network can be hard-wired or wireless, and some can handle up to 300 pumps. These systems allow for a significant amount of automation, as the programme for an entire study can be fed to the network of pumps with a single action. The flow rates for each individual pump can be directly calculated by the software from imported body weights. It is even possible to send pump information, such as alarms, via email or text message to researchers via the central monitoring computer. Especially in a GLP setting these systems have the additional advantage to reduce the massive amounts of paperwork generated by the classical infusion pump setups. Part 11-compliant systems can even eliminate all the paperwork by storing all events electronically. Obviously, this type of automation necessitates a type of pump that can support the necessary network interface, and it may therefore be necessary to replace an existing pump inventory with new pumps when a networked system is considered. Nonetheless, these systems have an enormous potential to reduce workload and errors and increase animal welfare when dealing with large-scale rat infusion studies.

1.4.1.4.2 Tetherless infusion pumps. Until recently, tetherless infusion in rats meant the pump in question had to be small and light enough to be implantable; the ambulatory setup for large animals where the pump is worn in a jacket was not available for rats. But technology has evolved, and a similar system is now commercialised and will be discussed further on.

Historically, tetherless infusion in rats was mainly achieved via the use of passive osmotic minipumps. They can simply be implanted subcutaneously, but they can also be subsequently attached to a catheter to provide targeted delivery for intravenous infusion. In such case they are typically implanted in a pocket on the back of the animal in the mid-scapular region, and the catheter is subcutaneously tunnelled to the venous access site. Osmotic minipumps operate through an osmotic difference between a compartment within the pump and the tissue environment in which the pump is implanted. Even though osmotic minipumps represent an elegant solution for tetherless infusion, they have one large disadvantage, namely the volume of drug their reservoir will hold combined with a low, fixed flow rate. The largest reservoir pump commercially available is 2 mL with a length of 5.1 cm and a diameter of 1.4 cm. This pump is implantable in the rat but not in the mouse. At a flow rate of 2.5 µL/hr, a 2 mL pump will infuse for 4 weeks. The low flow rate and limited capacity makes these pumps unsuitable for the more classical kind of small molecule toxicity studies as well as for use in larger animal species. Replacement pumps can be implanted, but this requires a new surgical procedure.

Implantable, refillable, and programmable infusion pumps have been used in humans for applications such as pain control and tumour treatment. Until recently, only passive, non-refillable constant-rate implantable osmotic infusion pumps were available for use in animals. Now, however, an implantable infusion device using a microprocessor-controlled peristalsis mechanism has become available for use in rats. This device is 'intelligent'. It can be programmed with simple or more complex infusion protocols, and the accuracy is better than 5%. The programming is managed through designated computer hard- and software and is uploaded to the pump via infrared communication, which also activates the pump. This can be done while the pump is still in the packaging blister to maintain sterility. Flow rates can be from 1.0 µL/hour up to 30.0 µL/hour. Next to programmability, a huge advantage of this type of pump is that it is refillable via a percutaneously accessible septum port. However, the battery power requirements are dependent on the total power consumption of the pumping mechanism. Sufficient power is available for six months at the 1 µL/hour infusion rate but as short as one week for the 30 µL/hour infusion rate. Lifetime of the battery allows a total volume of solution(s) from 4.5 mL to 6.5 mL to be infused depending on the infusion rate and the protocol. This type of pump is single-use as well. Despite the limitation of a rather short battery life at a high infusion volume, there is no doubt that devices of this kind represent a new milestone in tetherless infusion in rats. For example, the programmability allows intermittent infusion protocols for which the restricted battery life may be much less of an issue. Another potential application may be to use them in juvenile animal studies and to programme the flow rate to be in sync with the growth rate of the animal.

1.4.2 Non-surgical models

1.4.2.1 Restraining devices

The restraining device must ensure the safety of the animal, allow access to the desired vein, and prevent the rat from moving or dislodging the catheter during the infusion period.

Restraint tubes should be made of a transparent, durable plastic material to allow visualisation of the rat, be easy to disinfect, have adequate ventilation, and provide for easy access to the lateral tail veins (Figure 1.21). Usually the tubes have an adjustable plug on the loading side to allow for different-sized animals, with an opening for the tail. The tubes themselves are supplied in various sizes, and the appropriate size should always be used for maximum comfort and safety.

The tube must be securely fastened to the dosing surface to prevent slippage or movement. Some tubes come with suction cones on the bottom, and some have to be secured to the dosing surface using a clamp.

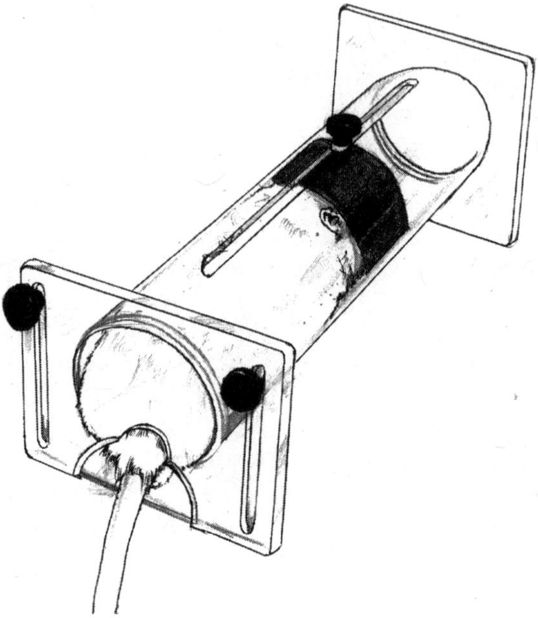

Figure 1.21 Restraint tube.

Restraint tubes are most commonly used for longer-term restraint and infusion. The various types of restraint devices are more than adequate for short-term or specialised procedures other than infusion.

Restraint bags are usually made of plastic and designed to be disposable. These are best suited for indications where it is not feasible to disinfect and reuse the tubes (as with certain test articles, radiolabeled materials, etc.). They are cheap and easy to use and can be purchased in many sizes and styles. The disadvantage is that they are not well ventilated and cannot be used with certain study designs, such as collecting respiratory data.

Snuggles are made of a durable, nylon-coated canvas material with Velcro fasteners. These also are available in various sizes and are easy to fit to individual animals. They are washable and reusable. The snuggle can keep the rat comfortable and secure and can be used for infusion; however, it does not allow for full visualisation of the animal, and caution must be taken to ensure the eyes are not damaged during longer durations of restraint.

Restraint boards are commonly used for jugular access for bleeding, but this is not a common vessel for restrained infusion intervals. Although it allows access to the jugular vein, the rat is restrained in an unnatural and very vulnerable position. These are not recommended for long-term restraint and should not be used for infusion.

1.4.2.2 Vascular access

For short-term infusions, a 24G or 25G butterfly catheter can be used. This is a needle with butterfly 'wings' for gripping and an extension line that can be connected to an infusion pump. The extension lines come in various lengths to accommodate the distance between the animal and the infusion pump. The disadvantage of a butterfly catheter is that the needle must remain in the vessel for the duration of the infusion, which means that slight movements of the rat's tail can compromise the integrity of the infusion site. The needle can puncture the vessel wall and lead to perivascular infusion of the test article, which would also negatively affect assessment of local tolerance. For these reasons the tail would have to be almost completely immobilised.

For longer-term infusions a 24G percutaneous indwelling catheter can be used. These catheters usually have a 24G polyurethane catheter over a 27G introducer needle. The needle is used to penetrate the skin and enter the vessel. Once the catheter is in the lumen of the vessel, the introducer needle is backed out and the softer polyurethane catheter is advanced into the vessel. The catheter is then capped using an injection cap. It is important to prime the injection cap with saline, since many have considerable dead space and air could be introduced into the vein upon injection.

Extension lines are available that connect directly to a syringe and can accommodate a needle on the end for insertion into the injection cap.

As with the surgical models, any of the common syringe, peristaltic, or piston pumps can be used to infuse the test article provided that they are accurate enough for the volume and rate of infusion. Once adequate venous access has been established, the primed infusion line (either butterfly line or extension set) is connected from the catheter to the infusion pump. A detailed overview of the various infusion pumps is given in Section 4.1.4.

1.5 Background data

1.5.1 Surgical models

1.5.1.1 In-life changes associated with the model

Following central catheter implantation, it is common to see a body weight loss (as a result of decreased food consumption) the first 2 days following surgery (Figure 1.22). Usually body weight normalises again 6–7 days later (Figure 1.23). Not taking potential test-article-related effects into account, central catheter dosing itself is not associated

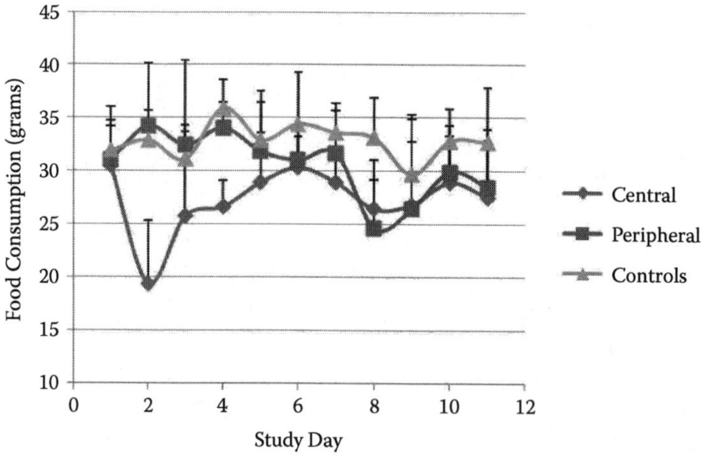

Figure 1.22 Typical food consumption patterns in Sprague Dawley rats dosed intravenously via a central catheter (non-restrained) (n = 8) and restrained animals dosed via the peripheral tail vein (n = 8) compared to untreated controls (n = 4). Surgery in central catheter animals was on Day 1; dosing started on Day 8. Dosing duration was 5 days, and length of infusion was 1 hour/day. A sharp decrease in food consumption can be noted the first few days after surgery, which normalizes by Day 6. Dosing during restraint is associated with a similar, but less pronounced, decrease.

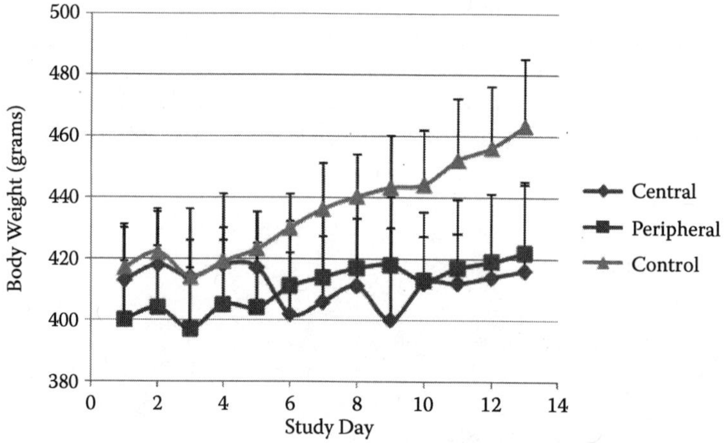

Figure 1.23 Typical body weight patterns in Sprague Dawley rats dosed intravenously via a central catheter (non-restrained) (n = 8) and restrained animals dosed via the peripheral tail vein (n = 8) compared to untreated controls (n = 4). Surgery in central catheter animals was on Day 1; dosing started on Day 8. Dosing duration was 5 days, and length of infusion was 1 hour/day.

with changes in body weight or food consumption, as it is relatively stress-free. Continuous intravenous dosing via a centrally implanted catheter does not affect body temperature or heart rate, either (Sheehan 2010).

1.5.1.2 Pathology

1.5.1.2.1 Clinical pathology. Infusion studies using saline as a control do not readily demonstrate changes in haematology or urinalysis when dosing physiologically acceptable volumes (Morton et al. 1997) and when surgery is performed aseptically. Surgery for catheter implantation, however, is associated with post-surgical stress. The latter was demonstrated by measuring serum corticosterone in animals dosed via a central vein after surgery or via the tail vein during restraint (Sheehan et al. 2011) (Figure 1.24). The surgical procedure required for central catheter implantation causes an increase in serum corticosterone concentration, which reflects stress caused by surgery. As described for food consumption and body weight, corticosterone tends to normalise within 5 to 8 days post-surgery and is not elevated by the dosing procedure.

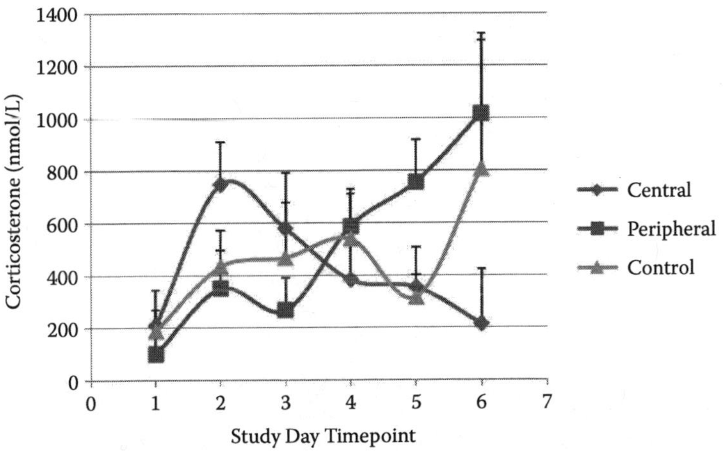

Figure 1.24 Serum corticosterone in Sprague Dawley rats dosed intravenously via a central catheter (non-restrained) (n = 8) and restrained (n = 8) animals dosed via the peripheral tail vein compared to untreated controls (n = 4). Dosing duration was 5 days, and length of infusion was 1 hour/day. Surgery for catheter implantation occurred on Day 1 and was associated with increased corticosterone levels in the centrally dosed rats. Normalization occurs around Day 6 post-surgery, and the actual dosing phase itself (Days 8–12) is not associated with increased corticosterone levels. The opposite is noted in peripherally dosed rats, where an increase is noted once dosing and restraint starts.

Marginal changes in red and white blood cell parameters, not reaching statistical significance, are sometimes associated with intravenous infusion but do not interfere with interpretation of treatment-related changes. A slight rise in white blood cell numbers often manifests shortly after surgery, due primarily to an increase in neutrophils with sometimes a lesser increase in lymphocytes. These increases generally show a return to baseline levels within two to three weeks. They are likely to be a consequence of the tissue damage and inflammation associated with surgery. Slight reductions in red blood cell counts and haemoglobin, associated with an increase in reticulocyte counts, have been seen in intravenous infusion studies conducted at our laboratory. These marginal alterations often persist throughout the duration of the study. However, when compared with pre-infusion values, these reductions are minimal and not considered to be of biological significance. Similar minimal reductions in red blood cell parameters (particularly red blood cell counts and haemoglobin and haematocrit values) have been recorded in the literature (Cave et al. 1995; Wirnitzer and Hartmann 2011). These haematological changes are possibly due to a minimally increased rate of red blood cell destruction, associated with cell membrane damage. A minimal increase in blood volume has also been suggested as a possible cause of these decreases (Cave et al. 1995).

1.5.1.2.2 Macroscopic pathology. In general there are few macroscopic changes associated with good intravenous infusion technique when using either the jugular or the femoral vein. Findings commonly observed at our laboratory include thickening at the interscapular region around the exteriorisation site of the catheter due to granulation tissue formation (observed at an incidence of 5–10%); and slight thickening and white material at the catheter tip in the jugular vein or vena cava, also occasionally observed at incidences of 5–10%, but should be minimal in nature and not cause obstruction of the catheter.

Unacceptable macroscopic findings can result from poor aseptic technique during surgery, unsuitable equipment, or poor infusion technique, or various combinations of these factors. If the skin-catheter seal is breached at the exteriorisation site and contamination occurs, abscesses or localised suppurative inflammatory exudate may develop there, appearing grossly as thick viscous material, whitish or greenish in colour, in the subcutaneous tissue. Multiple pale areas in different parenchymatous organs are likely to indicate multifocal abscessation or suppurative inflammation due to infection and haematogenous spread of bacterial emboli. Grossly the spleen appears enlarged from the markedly increased haemopoiesis. These changes, if seen in more than one or two animals on a study, indicate unacceptable levels of contamination at some stage of the procedure.

Marked dilation and thickening of the blood vessel at the catheter tip indicate an undesirable level of local reaction, generally with thrombus formation. These can result from large intravenous thrombi and/or fibrosis of the vessel wall with granulation tissue formation. Such changes may be associated with ridged or sharp catheter edges, or excessive movement of the catheter tip if the wrong size of catheter is used. These changes should not occur if the optimal catheter design is used.

1.5.1.2.3 Histopathology

1.5.1.2.3.1 Sampling procedures. Areas to be evaluated histologically should as a minimum include the vein exteriorisation site if the catheter is tunnelled under the skin, the catheter entry point into the blood vessel, and the vein at the tip of the catheter. Also, to identify any test-article-related changes that may be masked by procedural lesions, a section of the blood vessel 0.5–1 cm in front of the catheter tip should be sampled and evaluated. This area represents the first point of contact between the test material and the blood vessel endothelium.

1.5.1.2.3.2 Acceptable levels of procedure-related pathology. A low severity and incidence of common procedural lesions is observed in most infusion studies. These changes have no functional impact on the study and are therefore considered acceptable for interpretation. The pathologist should be familiar with these lesions and capable of assessing whether these are within acceptable levels. Consistency of terminology and grading thresholds is essential between pathologists at one laboratory and ideally between all laboratories for standardisation of levels of these pathological findings.

Procedural changes may be divided into localised findings, which are specific to the route of infusion, and systemic changes related to the general infusion procedure. Changes related to the mechanical presence of the catheter are seen where the catheter enters the skin, along its subcutaneous track, at the point of access where it enters the vein and at the cannula tip. The areas and vessels affected depend on whether the jugular or femoral vein is cannulated.

Common procedural lesions at the exteriorisation site (if interscapular) and along the catheter track include slight to moderate levels of granulation tissue with a low severity of inflammation. These changes reflect a reaction to the mechanical presence of the catheter. At the jugular or femoral vein access site, suture granulomata, reflecting a foreign body reaction to the suture material, are seen where the catheter is sutured in position. At the tip of the catheter, the blood vessel may exhibit minimal to mild endothelial proliferation, intimal thickening, and small thrombi. Thin fibrin sleeves are often present around the catheter. These often persist in the lumen after the catheter is removed and are seen in sections of the vena cava. Because all these changes represent physiological responses to a foreign material, they are likely to be seen at high incidences of up to 100%, but are of no pathological significance.

Irrespective of the route of infusion, systemically the parenchymatous organ most commonly affected is the lungs. Perivascular granulocytic cuffing is commonly recorded in the lungs, sometimes appearing more marked towards the periphery of the lung lobes. Low numbers of small granulomata, often containing foreign material, may also be seen in the lungs. These can contain fibres from various components of the infusion apparatus or endogenous materials such as hair shafts or epithelial fragments. Small pulmonary thromboemboli are also occasional findings and, if seen as isolated occurrences, are of no concern. Perivascular granulocytic cuffing can be seen at quite a high incidence (up to 70%), but at minimal to slight severities this finding is of no concern.

Other lesions relating to infusion may be seen in the parenchymatous organs as an increase in the level of common background pathologies. An increase in the level of small inflammatory cell foci in the liver, usually comprising lymphocytes and macrophages, often accompanies the infusion procedure. Cortical scarring in the kidneys, seen macroscopically as small depressions on the renal surface, reflects resolution of small subcapsular inflammatory lesions by fibrosis and subsequent contraction.

At minimal to slight severities all these lesions have no functional implication and are acceptable. In general, the longer the study duration, the higher the incidence of lesions even in a well-run study, and the greater the likelihood of problems (Weber et al. 2011).

1.5.1.2.3.3 Procedural pathology of significance. Whereas very few animals even in a well-run study may show evidence of sepsis, as described below, for several animals to display these lesions is unacceptable. Contamination, or appearance of infection, is always a problem, as the importance of aseptic technique is well recognised (Popp and Brennan 1981), but occasionally changes indicative of infection may be seen that are due to lapses in maintaining sterile precautions, either at surgery or during infusion. Catheter contamination is the commonest cause of infection. Bacterial invasion results in abscess formation at the exteriorisation site or vein entry sites on the catheter or at the catheter tip. The subsequent haematogenous transport and seeding of bacterial and septic emboli may result in abscess formation and septicaemia. The lungs, heart, and kidneys are sites of predilection for the development of suppurative inflammation and abscesses. Inflammation is predominantly neutrophilic, with large numbers of dead and degenerating neutrophils and variable amounts of fibrin. Abscesses can be seen in the lungs and kidneys. In addition the kidneys often develop descending pyelonephritis as a sequel, manifesting morphologically as cortical wedge-shaped areas of suppurative inflammation in the cortex, which then extend down into the medulla and papilla. These lesions are often noted grossly as pale areas, sometimes raised if the lesions are recent. In lesions of some chronicity focal tubular necrosis may follow, with subsequent fibrosis and scarring of the affected areas, causing the kidneys to appear misshapen or shrunken at necropsy. Widespread purulent or inflammatory lesions induce haemopoiesis as a secondary change in the spleen, liver, and bone marrow (Healing and Smith 2000). This finding is often predominantly granulocytic, with the spleen reacting rapidly initially, followed by the bone marrow.

The valves of the heart are a predilection site for development of valvular endocarditis. Seeding of emboli containing bacterial colonies on the valves commonly occurs in animals with abscess formation at the infusion sites. Suppurative inflammation and thrombi on the surfaces of the valve leaflets, often containing large bacterial colonies, are the characteristic histological manifestations of this change. These lesions can become very extensive, resulting in compensatory dilation of the heart chambers and hypertrophy of the ventricular walls. Heart failure follows, eventually resulting in death or necessitating sacrifice of the animal. Accompanying impaired heart function, pigmented macrophages may accumulate in the lungs as a result of passive congestion.

A catheter with a sharp or ridged, rather than a rounded, tip causes injury to the vessel wall. This predisposes to thrombus formation secondary to endothelial cell damage and denudation. As the thrombus becomes organised it becomes more involved with the vessel wall, which displays progressive fibrosis due to chronic trauma. The entire lesion appears grossly as a thickening of the vessel from the extensive granulation tissue formation involving both the thrombus and the vessel wall. This reduces the lumen and may result in blockage of the catheter and impaired blood flow. Such lesions should not occur when the optimal catheter design is used. Thromboemboli from the catheter tip are likely to break off because of altered haemodynamics and turbulence and lodge in the pulmonary capillaries. Whereas lungs infarcts are rare in rodents, they are more commonly seen in the dog.

High or fluctuating levels of infusion pressure, even if intermittent, can result in marked pathological changes in the cardiovascular system. Changes related to a high workload, presumably secondary to high pulmonary arterial pressure, may develop in the heart. Multifocal myofibre degeneration, predominantly involving the right ventricle (Figure 1.25), has been seen in association with other changes indicative of pulmonary hypertension with 5% dextrose perfusion; whether dextrose has a role in the development of these lesions has to be considered. Periarterial fibrosis and inflammation of the pulmonary artery, as mentioned above,

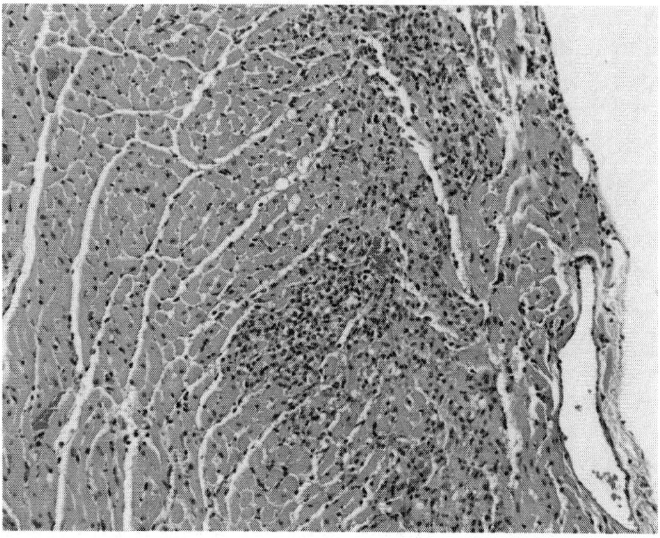

Figure 1.25 (See colour insert.) Right ventricle: Myocardial degeneration.

can also be present. Although the incidence of small inflammatory foci in the heart is often increased in infusion studies and is of no consequence, widespread multifocal myocardial change and pulmonary arterial changes can complicate interpretation and lead to premature death of the animals.

Medial hypertrophy and hyperplasia (Figure 1.26) of the pulmonary blood vessels, with the vessels appearing two to three times their normal diameter, may be a consequence. In addition, mild to marked perivascular inflammatory cell infiltrate, composed predominantly of granulocytes, may be seen, and is sometimes accompanied by perivascular oedema. Increases in perivascular granulocytic infiltration are known to be associated with higher volumes and speed of saline infusion (Morton et al. 1997; Wirnitzer and Hartmann 2011). Perivascular fibrosis of the pulmonary arteries, often associated with adventitial fibrosis of the pulmonary artery as it arises from the heart, is an unusual finding and is also likely to reflect high infusion pressure. This change is sometimes accompanied by mixed cell inflammation of the adventitia, as distinct from the commoner granulocytic infiltration associated with infusion. The pathogenesis of these cardiopulmonary changes is unclear. Pulmonary hypertension is associated with medial thickening of the pulmonary arteries in humans, both spontaneously and in association with treatment using sympathomimetics. These agents are thought to interfere with clearance of serotonin, which has powerful vasoconstrictive effects. Hypoxia in rats is associated with temporarily increased

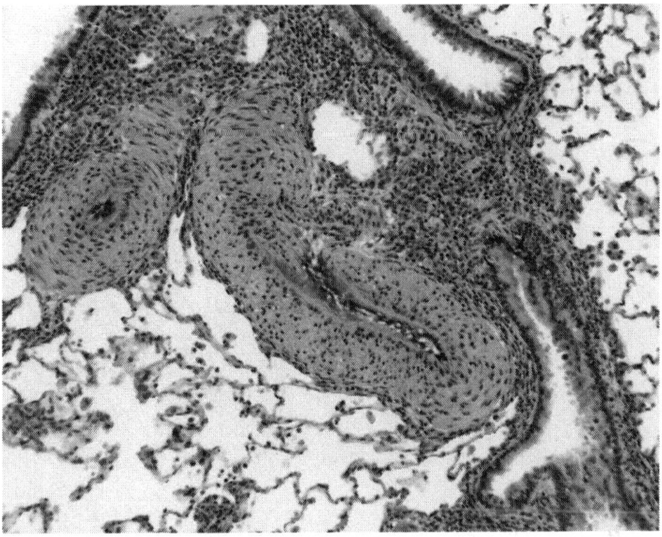

Figure 1.26 (See colour insert.) Medial hypertrophy of pulmonary arteries.

pulmonary arterial pressure, followed in a few days by medial hypertrophy and hyperplasia of the pulmonary arteries. Infusion of large volumes of isotonic saline is also associated with pulmonary arterial medial thickening (Morton et al. 1997). A low level of minimal to slight arterial medial hypertrophy (0–20%) is acceptable, whereas higher incidences and severities of this finding merit investigation. Perivascular granulocytic cuffing, of no concern at low severities, is undesirable at moderate to marked severities, particularly if accompanied by oedema and/or fibrosis.

Large or multiple thromboemboli in the lungs may result from a combination of marked thrombus formation at the catheter tip and frequent flushing or poor infusion technique, resulting in large numbers of circulating thromboemboli, which are most visible in the lungs. If numerous or very large, these are potentially a cause of pulmonary infarction. Because of the lungs' effective collateral circulation, necrosis is rarely a sequel to infarction, though it can occasionally occur (Figure 1.27).

1.5.2 Non-surgical models

1.5.2.1 In-life changes associated with the model

Despite the absence of a surgical procedure, restrained rats dosed via the tail vein show a decrease in food consumption and body weight gain during the dosing phase itself (Figures 1.22 and 1.23). This is typically not

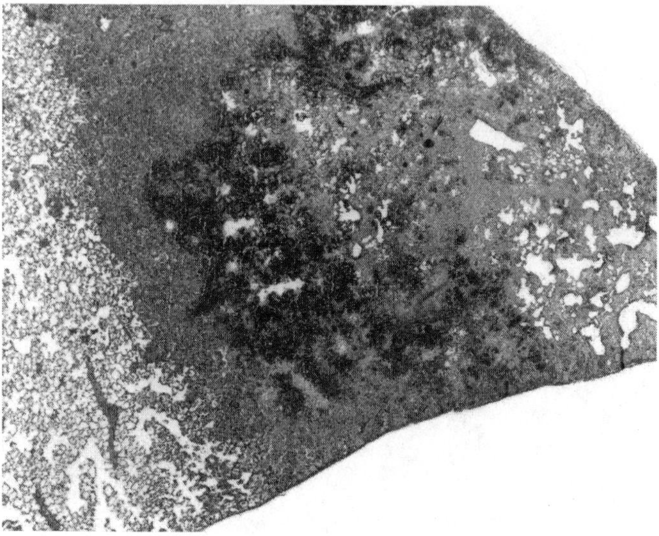

Figure 1.27 (See colour insert.) Infarction of rodent lung.

associated with a body weight loss, as is usually observed the first 2 days following central catheter surgery, but the latter is not associated with reduced body weight and food consumption during the dosing phase. In addition, tail vein dosing increases heart rate and body temperature during restraint. Increased heart rates are likely linked to the restraint stress, but increased body temperatures could also be related to a diminished ability to regulate body temperature while being in the confined space of a restraint chamber (Sheehan 2011).

1.5.2.2 Pathology

1.5.2.2.1 Clinical pathology. Dosing via the tail vein during restraint has recently been shown to cause mild alanine aminotransferase (ALT) elevations (Figure 1.28) (Sheehan 2011). This is probably caused by immobilisation stress, which causes leakage of hepatocytic enzymes following cellular injury due to the release of catecholamines. It has indeed been shown that repeated daily restraint and tail vein catheterisation for intravenous dosing cause an increase in serum corticosterone concentration (Figure 1.24). These changes occur distinctly during the dosing sessions, demonstrating the impact of restraint stress and repeated tail vein catheterisation on this variable.

Fibrinogen can also be elevated in peripherally dosed rats (Figure 1.29). Fibrinogen is a positive acute phase protein and a marker of inflammation, likely a result of inflammation at the level of the infusion site.

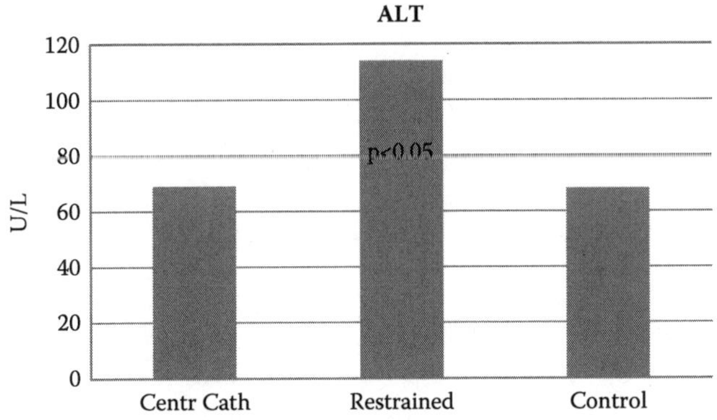

Figure 1.28 Mean alanine aminotransferase in Sprague Dawley rats dosed intravenously via a central catheter (non-restrained) (n = 8) and restrained (n = 8) animals dosed via the peripheral tail vein compared to untreated controls (n = 4). Dosing duration was 5 days, and length of infusion was 1 hour/day. Blood samples were taken at the end of the 5-day dosing period. An increase in ALT is noted in restrained animals, but animals dosed via a central catheter are not affected.

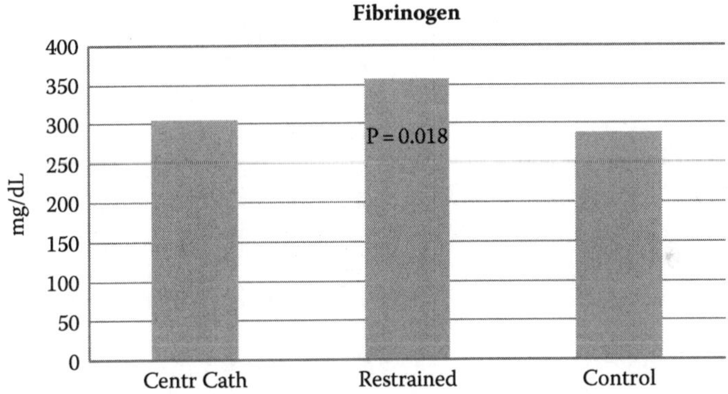

Figure 1.29 Mean fibrinogen values in Sprague Dawley rats dosed intravenously via a central catheter (non-restrained) (n = 8) and restrained (n = 8) animals dosed via the peripheral tail vein compared to untreated controls (n = 4). Dosing duration was 5 days, and length of infusion was 1 hour/day. Blood samples were taken at the end of the 5-day dosing period. An increase in fibrinogen is noted in restrained animals, but animals dosed via a central catheter are not affected.

1.5.2.2.2 Macroscopic pathology. Scabs on the tail and slight swelling at the vein exteriorisation site are often observed when infusion is performed by this route. Any contamination occurring may lead to abscess formation, with localised accumulation of greenish material. More extensive bacterial infection results in diffuse inflammation, and may cause marked swelling of the infusion site, which can involve a large area of the tail. The inflammatory process can involve the groin area if the infusion site is near the tail base. Marked irritation of the tail veins from repeated venipuncture or irritation from the catheter tip predisposes to thrombus formation and vascular thickening. If tail veins are largely or completely occluded, the epidermis covering the distal portion of the tail can become necrotic. The skin appears dark or black and the tail tip is shrunken and may eventually be lost.

1.5.2.2.3 Histopathology. One of the most important aspects of safety evaluation of intravenous drugs is to assess local tolerance at the site of administration. In this respect, a few considerations must be made when deciding to use a peripheral (tail) vein versus an indwelling catheter in the abdominal vena cava. The latter vessel is significantly larger and has a higher blood flow. This represents a situation more closely resembling that in humans. The smaller tail vein, together with a more limited blood flow, is more susceptible to local irritation effects because the compound will be less diluted in the bloodstream and will be in relatively closer contact with the venous intima than when using the vena cava. Furthermore, repetitive puncture of the tail vein will cause mechanical damage or even thrombus formation, which may make it more difficult to assess any compound-related effects. On the other hand, the prosector will need to carefully determine where the tip of the catheter was situated when performing autopsy on a rat with an indwelling catheter to harvest that section of the vena cava that is most relevant for local tolerance evaluation. For this reason, pathologists will often decide to harvest sections that are proximal and distal from this site as well.

If the tail vein infusion procedure is carried out competently, only minimal localised changes are seen histopathologically. These comprise localised minimal to mild granulocytic infiltration and haemorrhage in the connective tissue around the blood vessel. Small areas of mural necrosis in the vessel wall, at the area where the catheter enters the vein, are also part of the common background of lesions seen with this procedure. These lesions are easily related to the procedure and have no functional impact. Contamination around the catheter entry site sometimes results in formation of small localised abscesses in the skin and underlying connective tissue. Inflammation extending into the underlying tissues can involve the muscles and bone, resulting in a variety of changes, including suppurative inflammation, necrosis of muscle and fascia, reactive formation of new bone, necrosis of the coccygeal vertebrae, and osteomyelitis if the bone marrow cavity is involved.

1.6 Conclusion

Infusion via a centrally implanted catheter has little impact on animal physiology and well-being with an adequate surgical recovery period (about 7 days) when the procedures and techniques applied are of high standard. The minor surgical procedure itself is associated with stress, but animals show a fast recovery from resulting changes in homeostasis. Peripheral infusion via the tail vein in rodents is associated with stress and changes in body weight, food consumption, heart rate, body temperature, and even liver transaminases during the dosing sessions themselves. These can all be confounding factors when interpreting effects caused by the studied test article.

Training and practice in accessing the lateral tail vein is essential, as access can be difficult if using improper technique. Improper technique could lead to bruising, vessel damage, or inability to establish repeat access to the same vessel, all of which will cause stress to the animal and possibly impact the overall success of the study. Furthermore, improper technique or multiple, frequent vessel access intervals over a prolonged period may affect the ability to assess local tolerance at the site of administration. It is important to understand the critical endpoints of each particular study when choosing the method of infusion.

Considering the same background changes can occur irrespective of whether a central or peripheral vein is used for infusion, the importance of an adequate number of control animals is equally important for the latter route. Furthermore, when there is suspicion that the test article may cause mild irritant effects at the injection site, it may be wise to opt for a central vein, as this will allow a more careful histological assessment of local tolerance (see Section 1.5.2.2.3).

References

Asanuma K, Komatsu S, Sakurai T, Takai R and Chiba S. 2006. Total parenteral nutrition using continuous intravenous infusion via the posterior vena cava in rats. *Journal of Toxicological Sciences* 131(2): 139–147.

Cave DA, Schoenmakers AC, van Wijk HJ, Enninga IC and van der Hoeven JC. 1995. Continuous intravenous infusion in the unrestrained rat: Procedures and results. *Hum. Exp. Toxicol.* 14(2): 192–200.

Diehl K-H et al. 2001. A good practice guide to the administration of substances and removal of blood, including routes and volumes. *Journal of Applied Toxicology* 21: 15–23.

European Agency for the Evaluation of Medicinal Products (EMEA). 2000. CPMP: Note for guidance on repeated dose toxicity. 3BS2a. London, 27 July 2000.

Flecknell P and Waterman-Pearson A. 2000. Management of postoperative pain and other acute pain, Chapter 5. In: *Pain Management in Animals*. London: WB Saunders.

Graham JM et al. 1998. Teratogen update: Gestational effects of maternal hyperthermia due to febrile illnesses and resultant patterns of defects in humans. *Teratology* 58: 209–221.

Healing G and Smith D. 2000. Common pathological findings in continuous infusion studies, Chapter 20, pp. 253–264. In: *Handbook of Pre-clinical Continuous Intravenous Infusion*. Boca Raton, FL: CRC Press/Taylor & Francis.

ICH. 2009. Guideline M3(R2): *Non-Clinical Safety Studies for the Conduct of Human Clinical Trials and Marketing Authorization for Pharmaceuticals*. June 2009. CPMP/ICH/286/95.

Martignoni M et al. 2006. Species differences between mouse, rat, dog, monkey and human CYP-mediated drug metabolism, inhibition and induction. *Expert Opin. Drug Metab. Toxicol.* 2(6).

Morton D et al. 1997. Effects of infusion rates in rats receiving repeated large volumes of saline solution intravenously. *Laboratory Animal Science* 47(6) December.

Morton D, Safron JA, Glosson J, Rice DW, Wilson DM and White RD. 1997. Histologic lesions associated with intravenous infusions of large volumes of isotonic saline solution in rats for 30 days. *Toxicol. Pathol.* 25: 390–394.

Morton DB et al. 2001. Refining procedures for the administration of substances. *Laboratory Animals* 35: 1–41.

Nolan T, Loughnane M and Jacobson A. 2008. Tethered infusion and withdrawal in laboratory animals. *ALN Magazine* July 2008.

Olson H et al. 2000. Concordance of the toxicity of pharmaceuticals in humans and in animals. *Reg. Tox. Pharm.* 32: 56.

Patten D. 2011. Post-operative wound interference in adult female Sprague Dawley rats. Infusion Technology Organisation Conference, Barcelona.

Popp MB and Brennan MF. 1981. Long-term vascular access in the rat: Importance of asepsis. *Am. J. Physiol.* 241: H606–H612.

Ruiter M. 2011. Infusion Tips: The Three Rs of catheters: Reliability, resistance, and ROI. *ALN Magazine*. September 25.

Ruiter M. 2011. Infusion tips: French, gauge, OD/ID, mm, inches: What does it all mean? *ALN Magazine*. August 26.

Sheehan J et al. 2011. Intravenous infusion dosing in the rat: Considerations when selecting the optimal technique. *The Infusionist* 1: 44–57.

Smith D. 1999. Dosing limit volumes: A European view. Refinement Workshop. New Orleans, March 1999.

Staszyk C et al. 2003. Blood vessels of the rat tail: A histological re-evaluation with respect to blood vessel puncture methods. *Laboratory Animals* 37: 121–125.

Vemulapalli T H and Fredenburg N J. 2007. Evaluation of catheter tip configuration and lock solutions in a rat jugular vein catheterization model. 58th AALAS National Meeting, Charlotte, NC. October, 2007.

Weber K et al. 2011. Pathology in continuous infusion studies in rodents and nonrodents and ITO (Infusion Technology Organisation)-recommended terminology for tissue sampling and method-related lesions. *The Infusionist* 1(1): 86–98.

Wirnitzer U and Hartmann E. 2011. Effects of infusion administration volume on clinicochemical parameters and histopathology in vehicle control rats. *The Infusionist* 1(1).

chapter two

Mouse

Hans van Wijk, FIAT, RAnTech, Biotechnicus
Covance Laboratories Ltd, UK

Alice J M Fraser, BVM&S, MRCVS, FRCPath
Covance Laboratories Ltd, UK

Contents

2.1 Introduction

2.1.1 Choice and relevance of the species

The mouse, *Mus musculus*, including genetically modified models, is a commonly used laboratory animal. In preclinical safety assessment studies (toxicology, metabolism, pharmacology, and pharmacokinetic studies) the mouse is the most common alternative rodent species to the rat. The mouse can offer physiological and metabolic characteristics that are more relevant to the human condition than the rat.

Another advantage of the mouse is its size, which enables test articles in early development (and thus in short supply) to be tested in smaller quantities.

This chapter describes the mouse as an alternative to the rat for vascular infusion.

2.1.2 *Regulatory guidelines*

Regulatory guidelines indicate that where continuous intravenous infusion is the route of administration intended for a therapeutic agent, this should be mimicked in preclinical studies to ensure exposure comparable to the intended use (ICH M3(R2)). The rat is normally the first species of choice, with the mouse the second.

2.1.3 *Choice of infusion model*

A variety of infusion models are available. The choice between intravenous (i.v.) bolus tail vein injection and continuous intravenous infusion models using surgically vascular catheterised mice will be discussed in this chapter.

2.1.3.1 *Restraint: Reason for developing a surgical model for sustained or frequent intravenous dosing*

For many short-term multi-dose intravenous infusion toxicity studies, the test article can be administered as a slow bolus injection via the tail vein with a needle and syringe or infusion set and infusion pump. This technique requires minimal preparation; however, the animals need to be restrained, and the success of the study is dependent on the technical expertise of the study personnel. This approach is recommended only for short-term studies (up to seven consecutive days) and for administration durations of less than five minutes, because of the adverse effects on tail quality with longer dosing (personal observation).

Repeated intravenous bolus administration of a test article that elicits histaminergic-type reactions poses a particular problem in rodents, since infiltration of an irritant solution into perivascular tissues can lead to inflammation, connective tissue necrosis, and scarring and make further injections difficult or impossible. These might not be caused by the irritant solution, but to the actual pharmacology of dosing.

For this reason a mouse catheterised continuous intravenous infusion technique was further developed and validated at our laboratory using a surgically implanted catheter to perform daily repeat-dose infusion studies featuring administration of test article for longer than 5 minutes per dosing session and for more than 7 consecutive days.

2.1.3.2 *Advantages of the surgical model*

The main reasons for continuous infusion are that it

- Is the intended clinical route of test article administration
- Ensures systemic exposure to the compound when this would be low by other routes of administration (absorption issues)
- May have lower irritant effects at point of injection than when the test article is given by other routes
- Allows assessment of pharmacokinetic parameters under steady-state conditions

2.1.4 *Limitation of available surgical models: species perspective*

A possible disadvantage of using the mouse as a surgically vascular catheterisation model is its small size. The mouse has small veins that are technically challenging to access. Fully implanted systems such as vascular access ports (VAPs) used in large laboratory animals (monkeys, dogs) minimise risk of infection whilst not in use, but once a dosing needle is passed through the skin there is a point of access for bacteria to enter subcutaneous tissue. VAPs of sizes suitable for implantation into mice do exist, but there is no tether-connection device to protect the infusion line from being damaged by the mouse so these VAPs cannot be used for successful continuous intravenous infusion of mouse models. Models where the catheter is channelled subcutaneously from the vascular catheterisation site to exit at the tail (tail-cuff model) or via the scapular region (jacket and harness/skin-button model) are currently the most successful model for rodent continuous intravenous infusion, primarily in rats. However, these are constantly at risk of infection by bacteria tracking through the wound or through the catheter moving into or out of the exit site.

The mouse is more active than the rat, and it has a thinner skin and tail sheath than the rat. These features cause increased post-surgical complications at the catheter exit site and at the connection of the catheter-protecting device (jacket, skin-button, harness, or tail-cuff) on the animal. These complications can limit the length of the infusion period, particularly in toxicity studies where the animal needs quite frequent handling for study procedures such as blood sampling, body weight, and clinical examinations. Such handling of mice usually involves scruffing to restrain the mouse, compared with rats, which usually only need lifting and holding for similar study procedures. Thus in the mouse there is more tension and wear and tear on the catheter exteriorisation site and infusion equipment. It is a prudent precaution to minimise handling by obtaining data for several parameters each time the animal is handled.

The low body weight of the mouse increases the chances of blood clotting in the catheter especially during handling procedures because the flow rate of infusate is low in animals with a small body weight dosed in mL/kg/hr. For this reason the standard infusion rate used in rats has been doubled, resulting in a standard rate of 4 mL/kg/hr for mice (equivalent to a flow rate of 0.1 mL/hr, which is the normal 'keep-vein open' infusion rate for a 25-gram mouse). The need for blood sampling can also be a deciding factor for the choice of infusion model (see advantages and disadvantages of available models in Section 2.3.1).

2.1.4.1 *Limitation of available models from a technical point of view*

The mouse has emerged in recent years as a viable option for undertaking continuous intravenous infusion studies owing to the successful miniaturisation of the equipment used to infuse the rat (Jones and Hynd 1981; Cave et al. 1995). An understanding of the expected effects of the technical procedures (both surgical and manipulative) is important when assessing adverse effects of test materials following multi-dose toxicity studies (van Wijk and Robb 2000). A mouse efficacy study was used to validate our infusion model (van Wijk et al. 2006).

There is limited published data available concerning comparison of the reliability of the tail-cuff, jacket, and harness/skin-button infusion models. In this chapter, results from intravenous infusion using the slow bolus tail-vein injection technique (non-surgical) will be compared with the most recently developed continuous intravenous infusion model using surgical vascular catheterised mice (surgical model).

For the tail-cuff continuous intravenous infusion method, data are presented from a small validation study together with control data from three intermittent intravenous infusion studies using surgically vascular catheterised mice with study durations of 7, 14, and 28 days with different infusion rates. For longer-term infusion in the mouse, the harness exteriorisation method, which had proved to be successful for continuous intravenous infusion studies in the rat for up to 91 days, combined with a skin-button at the exteriorisation site, was selected for trial. Review of the data from these studies may enable improvements in animal welfare during infusion and allow more accurate prediction of the number of animals required to be surgically prepared (to allow for any losses between surgery and the start of study).

2.2 *Best practice*

Best practice can be considered to refer to a process or method that delivers a consistently repeatable outcome while minimising complications by application of accumulated experience in specific situations. This might change with time through experience gained by process improvement and refinement (Webb 2011).

The surgical vascular catheterisation models in the mouse described in this chapter are still being developed, and as yet there is no one model that fits all studies. This is partly because of variations in the requirements between the different types of preclinical safety assessment infusion studies (toxicity, metabolism, pharmacology, and pharmacokinetic), such that the number and types of samples to be obtained and/or parameters to be measured differ with respect to timing, frequency, number, volume, etc. Hence, in this section on best practice we present and discuss the advantages and disadvantages of each mainstream infusion method based on experience at our laboratory, with animal welfare being a primary consideration. An appreciation of the expected effects of the technical procedures is critical when assessing adverse effects of test materials following multi-dose toxicity studies (Van Wijk and Robb 2000).

Ethical Review is mandatory under animal protection legislation in most countries and is a central component of best practice since it seeks to ensure that the animal welfare cost to animals used is minimised by application of Replacement, Refinement, and Reduction. In addressing the design of proposed experiments, the ethical review process looks at maximising benefits from experimental results while minimising welfare costs to animals (Webb 2011).

2.2.1 *Surgical models*

2.2.1.1 *Training*

Preparation of the surgically vascular catheterised infusion model requires two teams: the surgical team and the continuous infusion team. Trainees start in the infusion team; suitable individuals are then selected from the infusion team to train for the surgical team. The head laboratory animal surgeon trains the surgical team how to look after surgically vascular catheterised animals and also trains the infusion team how to carry out infusion dosing and checks during the study.

Surgical training is undertaken using cadavers until the trainee is deemed competent in the procedure. The laboratory animal surgeon trainee is then fully supervised by an experienced laboratory animal surgeon whilst experience is gained on actual studies. UK law dictates that training not be carried out on 'spare' animals. Therefore, all training is carried out under strict supervision on actual studies and covers all aspects of the surgery, anaesthesia, handling of tethered animals, husbandry, dosing, and flushing via implanted catheters.

The training of new laboratory animal surgeons takes time in order to build up their experience so that they are consistent in their work and can produce a reliable and consistent surgical vascular infusion preparation. The actual time taken for a trainee to reach this stage depends on the number of studies being done over time. Trainees begin with rats and must become competent with this species before embarking on mouse surgery.

At our laboratory there are five technicians in the surgical team, three of whom are also in the infusion team, and eight in the infusion team. The surgical team is responsible for surgical preparation and the pre- and post-operative care of the mice; working closely with the infusion team, they are responsible for care and welfare of the animal throughout the study.

Careful planning and management of each study is important as the success rate can be affected by the number of animals expected to be surgically prepared each week; the surgical team should not be put under pressure that would compromise the success of surgery.

The infusion team undertake all the in-life phases of the infusion study, including aseptic syringe changes and assessing exteriorisation sites.

Good communication between the infusion and surgical teams throughout the surgery, recovery period, and study is essential. On completion of each study the reasons for surgical failures and infusion failures during a study are recorded and discussed by both teams to facilitate improvement of the model and to improve success rates.

The technical resources available to perform continuous infusion studies vary between laboratories. The number of staff required depends on study design, the size of the study, duration and frequency of syringe changes, and the numbers of studies that are required.

2.2.1.2 Applicability of the method

2.2.1.2.1 Choice of site of vascular access (factors influencing study design). In mice there are two main sites for inserting permanent venous catheters: the jugular vein, from where the catheter tip is advanced through the cranial vena cava so that it lies just inside the right atrium; and the femoral vein, from where the tip is advanced into the abdominal vena cava (*vena cava caudalis*).

2.2.1.2.1.1 Jugular vein. As the jugular vein is a relatively large-diameter blood vessel, commercially available, standard catheters can be used. The catheter should be specially designed to limit irritation or damage to the vein (refer to Section 4, Equipment). However, the proximity of the catheter tip to the right atrium poses the potential risks of physical damage to intracardiac structures and of direct exposures of the heart and lungs to high concentrations of test article. Great care is therefore required to ensure that jugular-inserted catheters are precisely positioned by using point-referenced bifurcations of blood vessels as landmarks (Webb 2011). The number of mice used in toxicity studies can be up to two hundred, which means a wide range in body weights, making it difficult to standardise lengths of catheters. Information on choosing catheter length according to mouse body weight is provided in Table 2.1.

Table 2.1 Measurements of Catheters Inserted into the Blood Vessel in Relation to Body Weight

Strain	Reference	Reason	Sex	Vessel	Catheter information		
					Detail	ID	OD
Swiss Albino	Mokhtarian et al.1993	Blood sampling/ IV infusion		Jugular vein Carotid artery	Silastic 602-105	0.30 mm	0.64 mm
C57BL/6J	Kelley et al. 1997	IV administration	Male	Jugular	Infusion part (replaceable)	0.011 in	0.024 in
FVB	Bardelmeijer et al. 2003	Blood sampling	Female	Jugular	Fixed to skin; bevelled and punctured silicone	0.3	0.6
Cr1:CD-1(1CR) BR	Macleod and Shapiro 1988	Blood sampling	Both	Arterial	Silastic	0.012 in	0.025 in
	Hodge and Shalev 1992	Blood sampling/ IV infusion		Jugular	Silastic	0.3 mm	0.64 mm
C3H	Popovic et al. 1968	Blood sampling and circulatory studies	Female	Jugular/ carotid	Polyethylene (PE10) mechanical stretching decreased	0.28 mm	
C57BL/6J C3H DBA/2	Barr et al. 1979	IV administration	Male	Jugular	Silica No. 602-105	0.30 mm	0.64 mm
CD-1	Van Wijk et al. 2007	Blood sampling	Male	Jugular	2-French silicone	0.30 mm	0.64 mm

Position of catheter tip	Catheter inserted: Body weight	Catheter inserted: Distance	Patency	Sampling details
Junction of anterior jugular, acromiodeltoid and cephalic veins	20 g	11.5 mm	To 6 weeks, average 3 weeks	5 x 0.1 ml in 2½ hours
	30 g	12.7 mm		
	40 g	14 mm		
	21–24 g	7–10 mm	Can be 72 days	
11–12 mm from external jugular vein?	(20–25) (25–30)	11–12 mm	1 day	
A 0.2 cm segment of PE-50 is positioned 1.5 cm from the proximal (cardiac) end	30 g	11–12 mm	21.5 days	100 µl every 20 min for 8 hours. 250 µl once daily for 3 days followed by a day of every 20 min for 8 hours.
Atrium	20–23 g	12.0 mm		100 µl every 10 min for 4 hours. 50% died probably from hypoproteinemia
	23–29 g	12.5 mm		
	29–33 g	13.0 mm		
	33–36 g	13.5 mm		
	36–40 g	14.0 m		
2 mm from proximal end of cannula	25–30 g		Injection unlimited, withdrawal limited to 10–15 days	
3 mm above where jugular vein courses with pectoral muscles. Nb. 1 mm above atrial valve	18–20 g	Y = 12 mm	At least 14 days	Of 244 mice used 11 died during or following the operation and 3 were excluded because of inoperative cannulae
	21–24 g	Y = 13 mm		
	25–30 g	Y = 14 mm		
	18–20 g	Y = 11 mm		
	21–24 g	Y = 12 mm		
	25–30 g	Y = 13 mm		
	18–20 g	Y = 10 mm		
	21–24 g	Y = 11 mm		
	25-30 g	Y = 12 mm		
	35 g	12 mm	NA	50 µl on 7 occasions over a 24-hour period in 4/5 animals

For the catheterisation of the jugular vein, 2-French silicone catheters are frequently used in short-term (24 hours) automated blood sample studies. The advantage of softer catheters in terms of reduced potential tissue damage has to be countered by the increased tendency for them to be damaged and leak when exteriorised. Also when small silicone catheters are used for vascular infusion studies, care has to be taken when they become clotted or blocked. Unblocking silicone catheters can cause the catheter to balloon, creating the risk of a breach of the catheter wall. The smaller catheters pose a greater risk. For this reason polyurethane catheters might be preferred for jugular catheterisation in vascular toxicity infusion studies.

A further disadvantage is that the tip of catheter can be close to the atrium. Thus the recommended sample tissue locations (Weber et al. 2011) such as 3 mm distally from the point of delivery of the catheter, which are required at necropsy in mouse vascular toxicology infusion studies, becomes difficult to achieve. Therefore, it is recommended to take two tissue samples, one at the catheter tip and one 3 mm distal to that point. These techniques have not been used at our laboratory.

2.2.1.2.1.2 Femoral vein. The femoral vein is a narrower vessel than the jugular vein, necessitating use of smaller gauge catheters that may need in-house modifications (see Section 4). The catheter tip lies in the *vena cava caudalis*; its exact location is not critical.

In deciding where to access the venous system for infusion, other study procedures must be considered, in particular blood-sampling sites, including frequency, sample volume, and timing with respect to the duration of infusion. For example, in studies where blood samples must be taken from the jugular vein by needle and syringe, the vein used for sampling should not be the vein containing a catheter for infusion, to obviate the risk of catheter damage by needles.

2.2.1.2.2 Recommended infusion rates and volumes. Immediately after surgery, physiological saline is infused at a rate of 0.1 mL/h, which is equivalent to an infusion rate of 4 mL/kg/hour for a mouse of 25 g. This appears relatively high with respect to a mouse's body weight (in comparison with the ideal infusion rate for rats of 2 mL/kg/hour) but is recommended to avoid blood clotting in the catheter during animal-handling procedures.

At our laboratory, in two commercial studies over 400 animals were successfully infused using the tail-cuff infusion model with the test article in saline (0.9% NaCl) for 1 hour/day or 2 hours/day, for at least 14 days at a rate of 4 mL/kg/hour and 8 mL/kg/hour, respectively. For the remaining 23 or 22 hours of each day the infusion system was maintained with saline alone at the same infusion rate.

In another commercial tail-cuff infusion study at this laboratory, single infusions (one infusion on one day) at an infusion rate of 1.44 mL/hour (equivalent to 50 mL/kg/hour) for 30 minutes to mice of average body weight of 30 grams has been achieved without any adverse effects. In addition, low infusion rates of 0.03 mL/hour, equivalent to 1 mL/kg/hour, for a period of 24 hours to mice with a body weight range of 26–29 grams have also been successfully achieved.

In a 20 animal, 91-day, harness/skin-button infusion validation study, 8 animals were successfully infused for 24 hours a day at a continuous rate of 4mL/Kg/hour. Six out of the 8 animals that successfully completed the study showed adverse effects to the eyes. This was not contributed by the infusion rate, but caused by the constriction effect of the harness. The tail-cuff infusion method has been successfully validated for a maximum period of 28 days.

A surgically implanted catheter is the optimal model for daily repeat dose infusion studies featuring administration of test article for longer than 5 minutes per dosing session and for more than 7 consecutive days (but no longer than 28 days).

2.2.2 *Non-surgical models*

2.2.2.1 *Methods of restraint*

To avoid body weight loss on Day 1 of study, when the animals are first restrained in a tube, they are acclimatised to the restraining tube for a few days prior to dosing.

For example, if the infusion period is 1 hour, then on the 4th day before Study Day 1 the animals are acclimatised to 1 minute in the tube, on the 3rd day for 15 minutes, on the 2nd day for 30 minutes, and for 1 hour the day before the study starts.

2.2.2.2 *Applicability of method*

The recommended administration volume for an i.v. bolus injection is 5 mL/kg with a maximum dose volume of 25 mL/kg (by slow injection) (Diehl et al. 2001).

At this laboratory the standard maximum volume of a manual bolus injection is 10 mL/kg at a rate not exceeding 2 mL/min. An infusion pump may be used for slow bolus infusion at a maximum dose volume and rate of 20 mL/kg and 2 mL/min, respectively.

Mouse infusion for more than 5 minutes has not been performed in this laboratory by this technique.

2.2.2.2.1 Choice of peripheral vein access. The lateral tail vein is the most suitable peripheral vein to access in the mouse.

2.2.2.2.2 Frequency and duration of vascular access/dosing. A historical overview of daily repeat i.v. bolus administration and slow bolus (60 seconds to 5 minutes) in mice shows that 3 studies have been performed over the last 12 years in the toxicology department of this laboratory.

Two studies were performed for up to 14 days and one study for up to 28 days (in which the animals were dosed every other day).

Other infusion studies performed at this laboratory entailed once weekly dosing for up to 13 weeks. This meant that the tail vein was able to recover between i.v. injections from penetrating injury to the vein and any subcutaneous damage.

Neither daily repeat dose infusion nor intermittent infusion studies featuring administration of test article for longer than 5 minutes per dosing session and for more than 7 consecutive days were performed at this laboratory before the development of the surgically vascular catheterised infusion model.

The main limiting factor on repeat intravenous dosing is that the infusion line or needle in the mouse is difficult to secure within the vein without the needle perforating the blood vessel wall, through the spontaneous movement of the tail. The mouse tail is more difficult to secure compared with that of the rat. The technicians have to stay with the mouse and hold the needle in the tail for the duration of infusion.

A catheter can be advanced into the tail vein to avoid perforation of the tail vein by the needle (see Section 4).

It is considered (personal communications) that multiple insertions of this type of catheter into the tail vein can lead to additional adverse pathology findings, as the needle used to insert the catheter is larger than the normal i.v. bolus needle in order to guide the catheter into the blood vessel. In addition, the catheter itself can be a portal of entry of a foreign body or infection into the circulation. The number of times that the catheter can be inserted in the lateral tail vein will be about the same as for bolus injections.

The second limiting factor on repeat intravenous dosing by use of a needle is the cumulative destructive effect of repeated physical penetration of the vein.

Inflammation, local scarring, and haemorrhage from the vein tend to obscure and divert the path of the vein, so there is a progressive increase in difficulty in delivering test formulation into the lumen of the vein. Given the limited length of tail vein available, there is a decreasing possibility of finding areas of tail unaffected by damage associated with earlier needle tracts.

Lesser factors against intravenous dosing by conventional use of hypodermic needle are (1) the possibility of physical injury to the tail, limbs, or snout as the animal is manoeuvred into the confined space of restraint tube and (2) the stress of confinement in the restraint tube.

2.2.3 *Advantages of available models—Surgical vs non-surgical*

2.2.3.1 *Surgical model*

2.2.3.1.1 Issues with blood sampling of mice. The intravenous infusion of a test substance via an implanted catheter can provide some significant advantages in situations where blood sampling is required in mice. Efficacy, safety, and drug metabolism pharmacokinetic (DMPK) studies often require repeated blood sampling for the measurement of drug, drug metabolites, or biomarkers. Frequently this will call for blood samples at, or close to, the Cmax. In the conventional situation, where both drug administration and blood sampling are made from the tail vein, contamination from residue in the vessel can result. This issue can be overcome in continuous intravenous infusion studies by infusing the test article via an indwelling catheter in the femoral vein and sampling from a tail vein or, for larger sample volumes, from the orbital sinus or jugular vein via needle and syringe under anaesthesia. Using these techniques blood samples may be collected both during and after infusion without risk of sample contamination. Jugular catheterisation has also been used with the tail-cuff exteriorisation technique for repeated blood sampling (van Wijk et al. 2006). The disadvantage is that there is a longer length of catheter required, which increases the dead space of the catheter. When combined with the dried blood spot technique (Clark et al. 2010), which only requires small blood volumes (20 µL) (Dainty et al. 2012), multiple samples may be taken, allowing a whole PK/TK profile to be determined from a single mouse. With a suitable pharmacodynamic measurement this can provide an optimal PKPD model (Andes et al. 2003). Further, using dual catheterisation, more complex studies can be performed. For example, Ayala et al. (2011) have nicely demonstrated the application of a hyperinsulinemic-euglycemic clamp in conscious unrestrained mice via catheterisation of both the jugular vein and carotid artery. Hence, depending on the objective of the experiment, indwelling catheterisation in mice can significantly impact the 3Rs: reducing the number of animals required where repeated blood sampling is required, refining the dosing and sampling procedure to allow a more accurate time course, and replacing techniques that require multiple venepuncture.

2.2.3.1.2 Advantages of surgical models.
Surgical models have a number of advantages:

- No warming required for vasodilation of tail vein or for access to lateral tail vein.
- Mice have access to food and water during infusion period.
- Accurate daily repeat infusion is achievable for 24 hours a day for up to 28 consecutive days (providing the infusion system is patent).
- No body weight loss during the dosing phase of the study.
- Tail vein damage from perivascular injection is avoided.
- Test article enters the circulation via a larger blood vessel (*vena cava caudalis* compared with lateral tail vein), decreasing the potential for pathology of the vessel wall (e.g. necrosis).

- Clinical and behavioural signs can be clearly monitored with the surgical models as the animal is freely moving.
- Speed of dosing.
- Restraint of animal not necessary in continuous infusion and daily handling not required for dosing, therefore less handling and less restraint. The disadvantage of the mouse being continuously tethered is outweighed by the decreased handling or 'time-in-tube' needed, as the mouse is more sensitive than the rat to being handled.
- Large (total) volume of test article can be infused over a period of time compared to test article volume administered by i.v. bolus injection.

Tail-cuff exteriorised animals were easier to handle for animal room procedures such as body weight and ophthalmoscopy compared with animals with the harness/skin-button.

In terms of animal welfare, the tail-cuff model provides the mouse with greater freedom of movement and move around the cage, thus allowing the mouse to exhibit its normal behavioural and sleeping patterns. This method therefore has significant welfare advantages over other methods.

The continuous vascular infusion surgical model can be dosed more quickly than by conventional tail vein i.v. bolus injection via the lateral tail vein, as there is no requirement to warm the mice (approximately 15 minutes) or place them into restraint tubes before administration. In large continuous infusion studies, all groups of animals (up to 200) can be dosed at the same time, whereas with conventional i.v. bolus studies, dosing must either be staggered (by group) or a larger number of expert staff is required to dose all groups at the same time. In situations where timing/duration of dosing is critical, such as if the drug formulation could be affected by the delay of animals being in the warming cabinet for tail vein vasodilation, the surgical method is preferred. This is the case for test articles with short half lives (e.g. less than 1 hour before expiry, or stem cells with limited cell viability) as they may expire before the infusion period ends. Vasodilation must be achieved at the time of dosing, otherwise the success rate for accessing the tail vein is reduced; the timing is therefore crucial, as the vasodilatory effect is lost once the animals are removed from the warming chamber.

2.2.3.1.3 Advantages of surgical models performed in-house. Performing surgery in-house rather than importing surgically prepared animals reduces the amount of time required before the start of the study. The transport of animals that have been surgically prepared incurs higher risks (this may lead to the animals expressing the disease state before commencement of the study), and they require more recovery time than non-prepared transported animals. In addition, more unusual strains, one not available surgically prepared, can be used in studies.

2.2.3.1.4 Animal-handling procedures. Ophthalmoscopy was performed on both infusion systems, and the handling was different as a result of the catheter exteriorisation methods.

For the tail-cuff method the animals could be scruffed with one single hand, but for the harness method two hands were needed to restrain the animals. Regular scruffing of the animals could have caused dislodging of the skin button. The handling procedure used for the tail-cuff method was preferable for the mouse and the technicians. This was also observed in a comparison study of exteriorisation methods in rats (McCarthy and McGregor 1999).

Two hands were needed for restraining both the harness and jacket exteriorised infusion mouse for procedures such as ophthalmoscopy.

The tail-cuff exteriorisation method provides greater freedom to move around the cage. Mice were noted to climb up and along the cage lid and onto the food hopper in the same manner as non-tethered mice, demonstrating how little the tail-cuff tether affects their movements (Figure 2.1).

For the performance of oncology studies, the tail-cuff infusion model was preferred to the jacket/harness system, as tumour measurement was simplified when the animals did not wear jackets during the study.

2.2.3.1.5 Risks and disadvantages of surgical models

- Risk of bacterial infection during surgery, during syringe changes, via the catheter, and through wounds or sores caused by the infusion equipment. A high level of asepsis has to be maintained during these procedures.

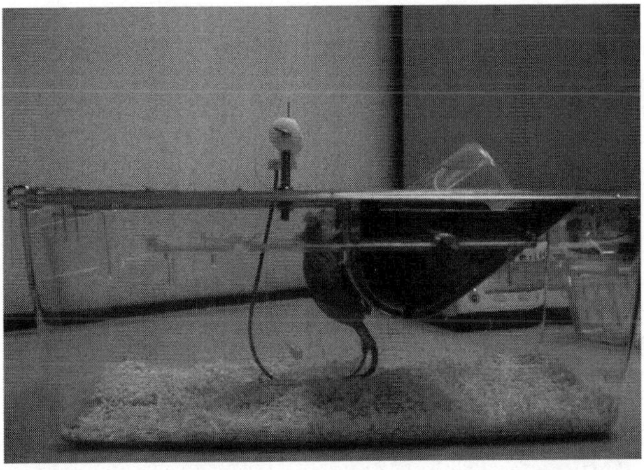

Figure 2.1 Mouse with tail-cuff exteriorisation demonstrating the freedom of movement. (Reproduced by kind permission of Animal Technology & Welfare.)

- There are animal welfare concerns surrounding the continuous infusion methods, as the model necessitates an invasive procedure to implant the catheter under general anaesthesia, a well-known stress factor in rodents and other laboratory animals (Fagin et al. 1983).
- Cost of study.
- Requires experienced laboratory animal surgeons to perform aseptic surgical implantation of venous catheters and aseptic syringe changes.
- Requires trained staff for the infusion team, too! They need a higher level of specialised skills than a tech looking after a conventional i.v. injection study. They must be available day-by-day throughout the study.
- Problems associated with the surgical model equipment: The harness exteriorisation method can lead to sores and lesions in the skin caused by the straps of the harness rubbing the skin. It also can cause circulation problems to the anterior body circulation when the straps are too tight around the body. The tail-cuff can cause swelling of the tail and sores and lesions around the tail-cuff where it is held in place by a surgical wire suture.
- Additional animals may have to be surgically prepared (tail-cuff infusion) to replace those with malfunctioning or non-patent infusion systems for the dosing phase on study Day 1.
- Animals are tethered all the time.
- Need to have sufficient background data to distinguish surgical procedural effects from test article effects.
- The study should be evaluated by a pathologist with experience of the model and the procedural findings.
- Femoral vein surgically prepared mice not always available from animal suppliers.

2.2.3.2 *Non-surgical model*

2.2.3.2.1 *Advantages of the non-surgical model*

- No surgery needed.
- Minimal preparation of animals.
- Little investment necessary to perform studies.
- Performing of once weekly dosing of slow bolus infusions up to 13 and 16 weeks period is possible.
- Low risks of complications.

2.2.3.2.2 *Disadvantages of the non-surgical model*

- Large number of skilled laboratory technicians needed for i.v. dosing in tail vein for full-sized toxicology studies.

- Warming of animals in a chamber (overheating/dehydrating) can lead to extra stress (and occasionally more marked signs).
- Restraining of animals in tube for dosing (stress effect, nose and teeth damage possible in restrainers).
- No access to water/food whilst in the restrainers.
- Needle sticks can be painful and stressful.
- The mouse has a small, fragile tail vein that is sensitive to irritant test articles. Additionally, repeat intravenous (i.v.) bolus administration of drugs that elicit histaminergic-type reactions pose a particular problem in mice, as infiltration of an irritant solution around the tail vein (perivascular) can potentially lead to necrosis and make further injections difficult or impossible.
- Collecting urine samples during infusion can be difficult.
- Gradual loss of vascular access after 7 consecutive days, after which the likelihood increases of having to stop dosing for a few days to let the tail recover from vascular damage.
- Blood sampling is difficult during infusion, particularly if infusion and sample points are not separate.
- Difficult to secure infusion line in the vein because of spontaneous movements of the tail of the mouse.

2.3 Practical techniques

2.3.1 Surgical model

2.3.1.1 Preparation

Pre-surgical planning commences with liaison between all members of the surgical and infusion teams, including the surgeon, anaesthetist, veterinarian, surgical technicians, animal care staff, and investigator. A surgical plan is formulated with a checklist to confirm adequacy facility resources, identification and scheduling of staff and their roles, proper equipment, and availability/preparation of consumables. A company veterinarian is involved in discussions surrounding the selection and dosage of anaesthetic agents and post-operative analgesia.

Presurgical planning also specifies the requirements for post-surgical monitoring, care, and record keeping.

2.3.1.1.1 Premedication. The analgesic carprofen (Rimadyl Large Animal Solution, Pfizer, 50 mg/mL diluted to 5 mg/mL, i.e., 1 mL Rimadyl to 9 mL sterile water for injection) is given before start of surgery and again 24 hours after surgery. The dose given at this laboratory to mice is 1 mL/kg (5 mg/kg), subcutaneously, to provide pain relief.

At the same time, the antibiotic cefuroxime (Zinacef, GSK) is given once prior to surgery. On the day of use one 250 mg vial is dissolved in 2 mL of sterile water for injection to make a solution of 125 mg/mL. The dose volume used at this laboratory is 1.0 mL/kg (125 mg/kg) administered intravenously or intramuscularly.

2.3.1.1.2 Presurgical preparation. All equipment needed for surgery is collected and checked to ensure nothing is damaged or missing. Any equipment used is serviced regularly and calibrated where required.

On the day before surgery, surgical tools and consumables required for the surgical procedure, including spares, are sterilised in an appropriate autoclave. The surgery room and pre- and post-surgical rooms are prepared by cleaning all work surfaces, floor, anaesthetic trolleys, and non-disposable equipment with a suitable disinfectant, such as Virkon solution, by technicians who must wear hats, face masks, and over-shoes. All equipment, sterile surgical instruments, sterile consumables, skin disinfectants, etc. are laid out and checked ready for the surgeon and the pre-/post-operative technician(s). Once the surgical area has been disinfected, this area is a restricted zone accessible only by authorised personnel.

On the day of surgery, all surgery work surfaces, instrument trolleys, anaesthetic trolleys, heated mats, and associated equipment (where safe and practical) are swabbed with surgical spirit. Rechargeable equipment is switched on.

Anaesthetic gas systems, waste gas extract systems, and oxygen supplies are checked, and any electronic recording devices are switched on and checked by the allocated anaesthetist. Any other electric equipment (warming blankets, microscopes, lights, etc.) is switched on and checked by the theatre assistant.

Pre-/post-operative drugs are prepared and labelled. Animal body weights (BWs) are received, and dose volumes calculated and checked.

2.3.1.1.3 Anaesthesia. Isoflurane is the preferred anaesthetic as it provides adequate muscle relaxation, analgesia, and rapid control of the depth of anaesthesia and allows rapid recovery. Isoflurane in oxygen is administered by inhalation within a clear-walled induction chamber. When voluntary movement has ceased, the animal is removed from the chamber, anaesthesia is continued through a nose cone, and the animal is placed on a heated pad at a temperature of approximately 37°C. Routinely used concentrations of isoflurane are 4% for induction and 2% for maintenance of anaesthesia.

2.3.1.2 *Surgical procedure*

As soon as the animal has been anaesthetised by the assistant, it is placed onto a heating pad with an even temperature of 37°C to minimise the risk of hypothermia. Fluids are administered via the femoral catheter after catheterisation to maintain adequate hydration.

The animal is prepared for surgery by clipping the inguinal region overlying the femoral vein and removing all loose fur with a small vacuum cleaner. The clipped area and the animal's tail are scrubbed up for surgery using chlorhexidine in detergent (Vetasept Chlorhexidine Surgical Scrub, Hibiscrub) and disinfected using chlorhexidine in surgical spirit (Vetasept Clorhexidine). The animal is then transferred to the surgery room.

A high standard of asepsis is maintained during the surgical procedure (including handling of equipment). Asepsis is crucial for surgical success and also increases the long-term patency rates of the inserted catheter (Popp and Brennan 1981). Maintenance of the sterile field is ensured by the surgeon wearing sterile gloves and using a new set of sterile instruments and consumables for each animal.

The animal is placed on its back on a sterile drape, and the hind limbs are secured to the drape using surgical tape. Where possible, a sterile self-adhesive drape is placed over the body of the animal during the surgery.

An initial incision is made in the skin of the inguinal region overlying the femoral vein. The underlying fascia is carefully dissected to expose the femoral vein, which lies just caudal to the femur and next to the femoral artery and sciatic nerve. Approximately 8 mm of the vein is carefully separated from the surrounding tissue. Two lengths of sterile non-absorbent suture material are then placed around the vein, about 3 mm apart. To assist the visualisation of the blood vessels and nerves during surgery, a microscope with ×10, ×16, or ×24 objective is used (with sterilised hand pieces).

A narrow-gauge catheter is inserted into the vein, and this is joined to a larger-gauge catheter external to the vein.

The size of the catheter for insertion is limited both by the size of the mouse's femoral vein and by the inguinal canal, where the catheter may be obstructed as it is being fed up towards the vena cava. During catheter insertion, care must be taken not to use excessive force that would cause perforation of the vessel. The wider bore 3-Fr catheter is used external to the vein as it is less fragile and is therefore less likely to be damaged when passing through the tail-cuff and tether. In addition, this gauge of catheter is wide enough to allow connection to a 22-gauge swivel above the cage.

Passing the catheter through the inguinal canal is less problematic in mice weighing in excess of 35 g. Even using the smallest size of catheter commercially available, it may not always be possible to catheterise the femoral veins of small mice (25 g or less). If difficulty is experienced in successfully inserting the catheter, further alteration of the catheter can be attempted. Previous experience has shown that the tip of the catheter can be stretched and made more flexible with hot water to enable insertion into the vein. Care should be taken to maintain sterility if such adjustment is necessary. The advantage of softening the catheter in terms of reduced potential tissue damage is offset by an increased tendency for it to be damaged and leak when exteriorised.

The narrow-gauge catheter is pre-filled with heparin saline (25 IU/mL) (Monoparin 1000 IU/mL), and the syringe and needle are left attached to the distal end in order to prevent leakage of fluid from the catheter. The pre-placed suture lengths are held under tension, and the vein is punctured between them with a 30-gauge needle. The tip of the catheter is then advanced approximately 20 mm into the vein, through the inguinal canal, so that the tip is lying in the caudal vena cava. The cranial suture is then tied around the vein and catheter. The sutures are glued to the surrounding tissues with the surgical glue Vetbond™.

Patency of the catheter is checked at this point in order to ensure that it has not been occluded by the sutures.

The distal end of the catheter is tunnelled towards the tail. A trocar needle, bevelled upwards, is inserted into the ventrolateral area of the tail approximately 10 mm from the base and advanced carefully towards the inguinal surgical site. The trocar allows passage of the catheter through it to exteriorise the catheter. As the catheter is drawn back through the exit point, care is taken to prevent it from kinking. It is advisable to anchor the catheter to the muscular tissues close to the insertion point to prevent accidental removal of the catheter. Immediately after the catheter is drawn through the tail, the syringe and needle are reconnected and the catheter flushed with approximately 0.02 mL of heparinised saline to remove any blood within the catheter.

The needle and syringe are left attached and the inguinal region is then closed with a continuous subcutaneous suture using 7-0 prolene monofilament sutures. The skin is closed with sutures and/or disposable IN011 clips.

A surgical data sheet detailing the anaesthesia and surgical steps is completed for each animal.

There are two major alternatives for the exteriorisation and securing of the catheter: the tail-cuff, and the skin button.

2.3.1.2.1 Tail-cuff attachment. Following wound closure, the animal is repositioned and the tail-cuff attached. A length of sterile surgical stainless steel wire (USP 00 monofilament, Ethicon, USA), is inserted through the skin laterally from one side of the tail to the other, on the dorsal surface of the tail, approximately 3 mm proximal to the exit point of the catheter. The animal is then placed onto its back again. The syringe and needle are removed from the free end of the catheter, which is passed through the tail-cuff and tether. Then the ends of the surgical wire are passed through the fixation holes in the tail-cuff and twisted.

Following completion of attachment of the tail-cuff, it is essential to ensure that the wire suture is suitably tight. If this is too loose, excess movement and irritation tends to occur. If too tight, it may impair the circulation of the tail.

The average surgery time for each mouse with an average body weight of 24 g for femoral vein catheterisation and tail-cuff attachment is 22 minutes. This includes 6 minutes for induction of anaesthesia and preparation time (including injections). Femoral catheterisation and tail-cuff attachment takes 16 minutes on average (Figures 2.2 and 2.3).

2.3.1.2.2 Skin button implantation and harness attachment. After successful catheterisation of the femoral vein, the distal portion of the catheter is tunnelled to a skin incision in the scapular region to allow the introduction of a trocar. The trocar is passed subcutaneously to exit at a point just above the catheterised vein within the original incision site. The free end of the catheter is passed through the trocar, which is then removed, leaving the exteriorised catheter in position. The inguinal surgical site is then sutured closed.

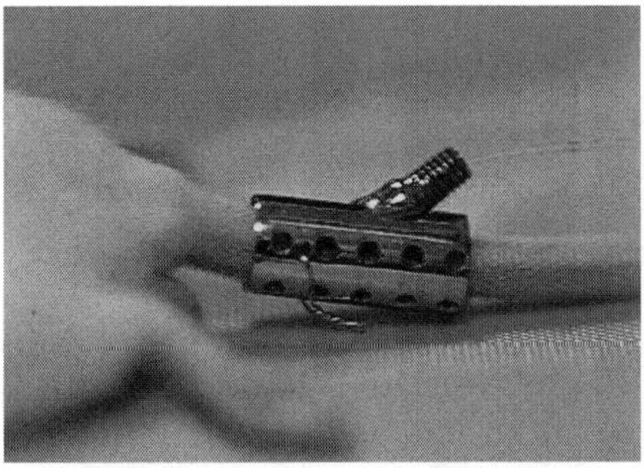

***Figure* 2.2** A tail-cuff attached to tail with catheter coming out of tail-cuff.

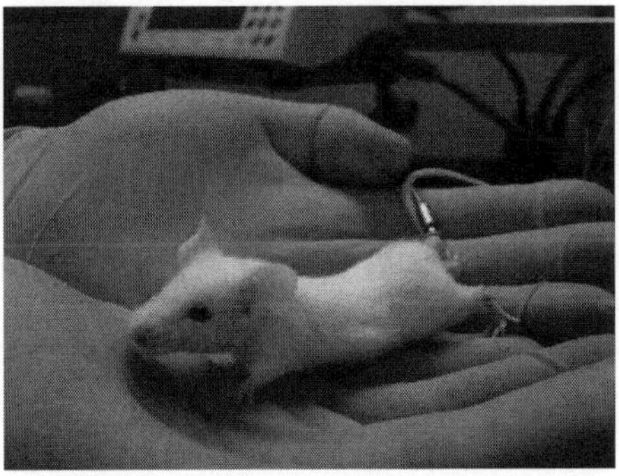

***Figure* 2.3** Surgically prepared continuous i.v. infusion tail-cuff model.

The free end of the catheter is passed through the opening of the skin button and attached to the skin in the scapular region, and the dorsal incision site is closed with a suture.

The free end of the catheter is then passed through the tether and connected to the swivel/infusion line. The harness is fitted to the animal, and a small part of the tether is located through the harness and fixed into the skin button (Figure 2.4).

The average surgery time for a femoral vein catheterisation with skin button implantation and subsequent connection of the tether to the skin button and the harness with catheter exteriorisation in a mouse with an average body weight of 29 grams was 36 minutes. This includes an induction and preparation time (including injections) of 6 minutes and femoral vein catheterisation and harness attachment taking 30 minutes on average. Currently there is less experience with this technique (at this laboratory) compared to the femoral vein catheterisation with tail-cuff exteriorisation method.

2.3.1.2.3 Connecting the infusion line. The animal is returned to its individual home cage for connection of the free end of the catheter to a syringe assembled on an infusion pump. The tether and catheter are passed through the top of the cage lid and the swivel fixed into the swivel holder. The top of the swivel is connected to the syringe via an appropriate tubing, and an inline filter can be included if necessary. The tubing, syringe, and filter (if used) are pre-filled with saline prior to attachment to the implanted catheter. A saline (0.9%) infusion is started immediately after connection, initially at a rate of 0.1 mL/hr. A minimum post-surgical recovery period of 5 days is then allowed. During this time the animals and their surgical wounds are checked regularly.

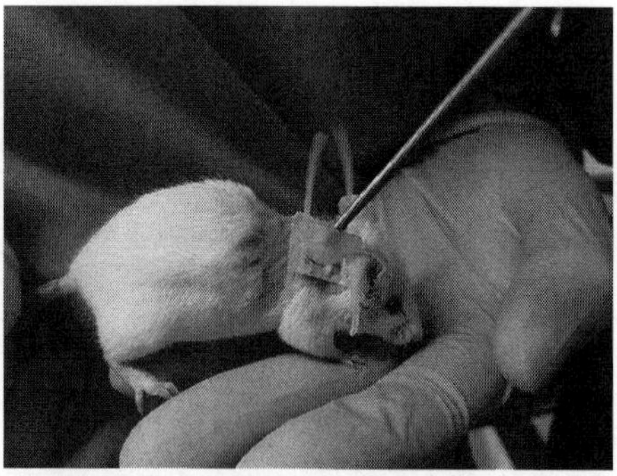

Figure 2.4 Harness was fitted to the animal and a small part of the tether located through the harness and fixed into the skin button.

2.3.1.3 Maintenance

2.3.1.3.1 Recovery procedures. After the animals are returned to their respective home cages, the heating pad remains under each cage until they are fully recovered.

During the post-operative period, the animals are observed at regular intervals and any post-operative clinical signs recorded for a minimum period of 4 hours.

2.3.1.3.2 Housing and husbandry. Each animal fitted with a tail-cuff is tethered to the infusion apparatus post-surgery, using a swivel and stainless steel spring tether.

The animals are individually housed in clear MT 1 polycarbonate cages (33 × 15 × 13 cm). The cage tops are of a standard design, with a linear grid space for a food hopper and water bottle. They are modified in-house so that the mouse and its infusion line can easily be removed from its home cage without interruption of the infusion in order to measure parameters such as body weight and to carry out blood sampling, urine sampling, and ophthalmoscopy in the room (Figures 2.5 and 2.6).

In order to enrich the environment for the welfare of the animals, they are provided with clean paper wool (Beta-Shred FDA paper nesting materials, supplied by Datesand Ltd, Manchester UK) as bedding, which is changed twice weekly.

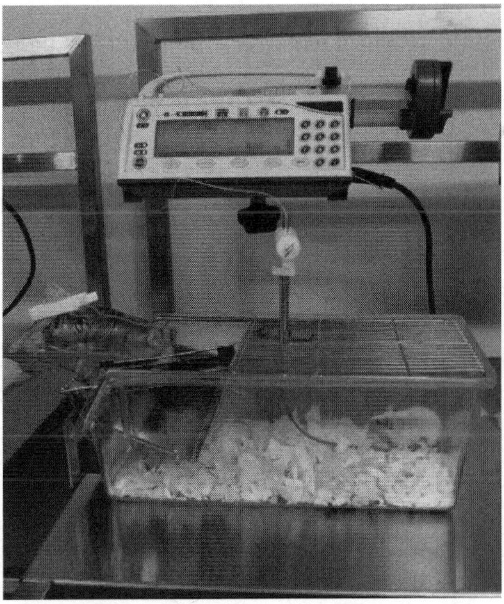

Figure 2.5 Mouse tail-cuff infusion cage set-up.

Figure 2.6 Mouse tail-cuff continuous intravenous infusion room setup.

For the harness or skin button model, a counterbalance lever arm can also be integrated into the system.

The use of a counterbalance can be an improvement of the tethered system since the mouse may be more comfortable, benefiting from the relief from the pressure of a harness or skin button (Figure 2.7).

The use of nude mice necessitates filter-top caging in order to maintain the health status of these immuno-compromised strains. Because of equipment availability, a standard filter-top cage is used. However, the requirement to use a standard filter lid means that the swivel holder has to be lowered from a height of 10 cm to 5 cm above the cage lid. This adjustment causes the swivel to be heavier to rotate, which may contribute to a slightly increased incidence of adverse effects in the tail caused by the tail-cuff.

The availability of a taller filter-top cage would be advantageous. There are special mouse micro-isolator swivel holders that have been developed for the jacket/harness (Figure 2.8). In this system, there is no cage lid, and the feed and water bottle are available separately (not shown in photo).

For tail-cuff exteriorisation animals, it is necessary to have a cage lid, as the mouse has so much freedom of movement that it is able to climb its tether. The same behaviour occurred when measuring body weight, whereby the mouse climbed along the tether and lifted its feet off the balance pan. If the cage lid is absent, the mouse can reach the catheter at

Figure 2.7 Use of a counterbalance arm.

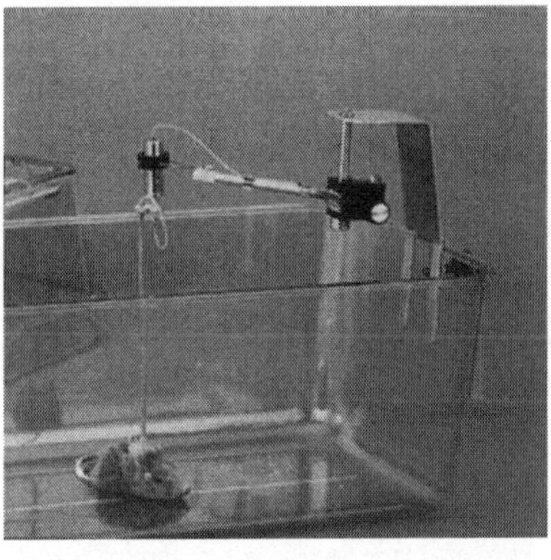

Figure 2.8 Jacket/harness system with infusion swivel arm (multi-axis), specially designed for infusion animals housed in a filter-top cage.

the swivel and chew the infusion line, or climb the swivel arm and escape through the filter top.

Therefore, it is not advisable to use the multi-axis lever arm swivel in tail-cuff exteriorised mice without the use of a cage lid. Higher filter-top cages would be better all round for the tail-cuff and harness systems. Presence of the filter top prevents exposure to unfiltered air in order to protect the immuno-compromised animals from possible infections. This approach would also be useful in mouse infusion studies evaluating human stem cell products; animals in these studies must be given immuno-suppressants to prevent rejection of stem cells, and this leaves animals susceptible to infection.

2.3.1.3.3 Checks. The infusion system, including the patency of the catheter, of each animal allocated to a study is checked twice weekly during and at the end of the study. The infusion system and the exteriorisation site of the catheter, the tail-cuff attachment, and the animal itself are checked at least once daily. At necropsy the location of the tip of the catheter is checked to confirm correct placement. The reliability of the tail-cuff infusion method in toxicokinetic (TK) study animals is also assessed by the number of animals presented for TK blood sampling at the end of the infusion period. The results of the patency check of the TK animals is compared with the number of successfully analysed blood samples obtained at the end of infusion period. Animals that were killed for causes other than those related to the continuous intravenous infusion procedures, such as toxic effect of test article and the orbital sinus sampling procedure, are excluded from the total number of animals infused (see Section 5 for these results).

2.3.1.4 Record keeping

The condition and function of all components of the dosing system are checked daily. The number of animals lost during the study phase as a result of failure of the infusion equipment is recorded and calculated as a percentage of the study population. Occasions when the catheters are blocked and the dose is not received completely are also recorded. Control animals are used for generating background data such as body weight, haematology, clinical chemistry, urine analysis, ophthalmoscopy, and histopathology. Detailed surgery records for each animal are kept using in-house specific forms.

2.3.2 Non-surgical models

2.3.2.1 Methods of restraint

The mice are placed in a restraining tube for the duration of the i.v. bolus injection. Prior to conducting the experiment the animals are acclimatised to the restraint process.

2.3.2.2 Methods of vascular access

In order to facilitate insertion of the needle of the butterfly infusion set into the tail vein of small rodents, the vein can be vasodilated by placing the animal in a warming cabinet (at a temperature of 38°C). Small rodents such as rats and even more so mice are particularly vulnerable to prolonged warming because of their large surface area in relation to their size. Typically mice are placed for no longer than 15 minutes in the warming cabinet. The animal is then placed in a restrainer for the duration of the slow i.v. bolus infusion.

For small infusion volumes of up to 0.03 mL, insulin syringes are used. For larger volumes, 1 mL syringes with 30-gauge needles are used.

2.3.2.3 Housing

Cages that conform with the Code of Practice for the housing and care of animals in scientific procedures are used. Female groups of up to 3 and singly housed males are housed in cages measuring 29 × 11 × 12 cm; these cages are maintained in racks, M2, which can be double- or single-sided.

Aspen bedding (Tapvei®, manufactured by Estonia) or Beta-bed grade 6 (European softwood bedding) is used. Wooden aspen chew blocks and sizzle-nest are used for environmental enrichment.

2.3.2.4 Record keeping

The following records are maintained for in-study animals:

- Dose volume administered to each animal
- Problems associated with dosing
- Daily checks of mice, including condition of tail vein
- Replacement animals due to poor condition of tail (early in study, Days 1 or 2)
- In-life data (including BW, ophthalmoscopy, clinical pathology, clinical observations)
- Blood sampling for TK analysis

2.4 Equipment

2.4.1 Surgical models

All equipment and procedures used were designed to be compatible with continuous intravenous administration without any discontinuation of infusion occurring during measurement of any of the routine parameters normally obtained during a toxicity study.

2.4.1.1 Surgical facilities

First, a dedicated surgical facility is essential. Separate areas for storage, animal preparation, surgery, and recovery are required. This surgical suite consists of a surgery room with a second similar animal recovery room equipped with hot water and a large sink directly attached to it. The surgical suite can be accessed only through the recovery room and not directly from the corridor.

In both rooms, there is an independent air-flow system making it suitable for mouse surgery in an open-plan room. The airflow goes from the surgery room into the recovery room and then into the return corridor. All the rooms are ventilated with fresh, non-circulated air.

Each room has independent temperature control, set at 21°C with a tolerance of 2°C.

2.4.1.2 *Mini-osmotic pumps/vascular access ports/catheters*

2.4.1.2.1 Osmotic pumps. Osmotic pumps can be used for systemic administration of test articles when implanted subcutaneously or intraperitoneally. They can be attached to a catheter for intravenous or intra-arterial delivery.

Alzet® osmotic pumps are miniature implantable pumps used for research in mice, rats, and other laboratory animals. These infusion pumps continuously deliver test articles at controlled rates from 1 day to 6 weeks without the need for external connections or frequent handling. These pumps work by the osmotic pressure difference between a compartment within the pump (the salt sleeve) and the tissue environment in which the pump is implanted.

The administration of the test material starts as soon as the pump is implanted, allowing no recovery period from surgery prior to start of the toxicity study. The stress of surgery and the interference of analgesic administered at surgery could be a complicating factor in determining the response of animals to test article.

In addition, when the osmotic pump is empty, it needs to be surgically replaced with a new one when the complete dose is delivered or be surgically removed (to prevent reverse osmosis). The dose delivery cannot be monitored in real time, and there is the possibility that the dose delivery could be changed by the hydration status of the animal.

In comparison with the use of osmotic pumps, the surgical vascular infusion technique allows a post-surgical recovery period before administration of test article and flexibility in the dosing regime (e.g. variation in infusion rates). In addition, the surgical vascular infusion technique allows the infusion of relatively high volumes of test article, including viscous test articles, by syringe driver pumps for extended periods of time with no additional direct interference with the mouse.

2.4.1.2.2 Vascular access ports (VAPs). The Penny MousePort™ (Figure 2.9) for single site placement is an all-silicone VAP small and light enough for mice. It is used mainly for long-term peritoneal access

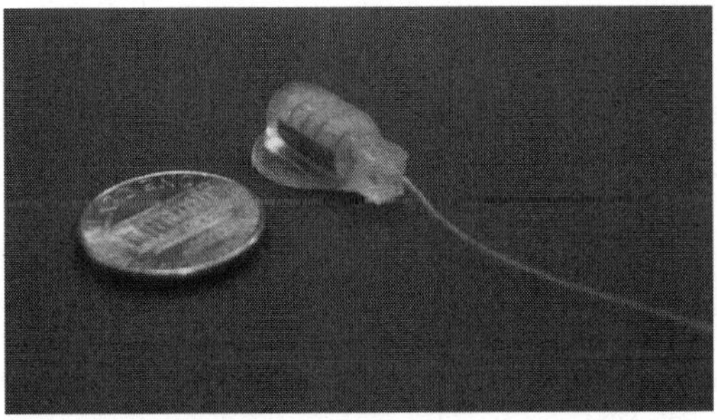

Figure 2.9 The Penny-MousePort™ by Access Technologies with attached catheter.

for peritoneal dialysis studies in mice (González-Mateo et al. 2008) and (Bögels 2006) where the attached catheter is implanted into the peritoneum and the Penny MousePort is implanted into the subcutis. It can also be used for long-term repeated vascular access. The main advantage of the port is that since it is implanted sealed under the skin and does not come through the skin, no tether devices are necessary for restraint.

To the authors' knowledge this device has not been developed to be connected to a tether and swivel for use for continuous i.v. infusion studies. The advantages of this VAP are that no heating is necessary to vasodilate the tail vein, and repeat access of the vein via the catheter is possible.

The vascular access port concept can be used for bolus delivery of infusion materials with extremes of pH or osmolarity, with the Penny MousePort being connected to a catheter inserted into one of the larger blood vessels, which can protect and prevent damage to the vessels.

With the port lying subcutaneously, it is difficult to access the port and to restrain the mouse in a restrainer, but alternative restrainers such as an extra-small rodent restraint bag are available.

2.4.1.2.3 Catheters. Commercially available catheters are suitable for vascular catheterisation in mice. Catheters can be manufactured in-house, but producing high-quality catheters with a rounded tip is time consuming and they are difficult to sterilise. It is therefore recommended that they be obtained from a commercial supplier. Where possible, a catheter with a rounded tip and a smooth anti-thrombogenic film should be used, which will reduce the subsequent thrombosis around the catheter. In general the material of the catheter should be soft and flexible (e.g. silicone) in the blood vessel, and where the catheter exits the body the catheter should be strong (e.g. polyurethane) and robust to withstand wear and tear.

Catheters exist in one size (3-French) suitable for rat infusion studies. For mouse vascular infusion studies a 2-French catheter can be inserted into the jugular vein. For the femoral vein, which is narrower than the jugular vein, 1-French catheter is required. To the authors' knowledge, soft-tip catheters are unfortunately not available at present in 2 or 1 French size.

2.4.1.2.3.1 Femoral vein catheters used in tail-cuff infusion model. For a long time only a limited number of commercially available catheters were suitable for vascular catheterisation of the femoral vein in mice.

For the catheterisation of the femoral vein with the tail-cuff exteriorisation infusion method, the authors have used a small polyurethane catheter (1.2-Fr) with an anti-thrombogenic bevelled tip, with an inner diameter (ID) of 0.18 mm and an outer diameter (OD) of 0.41 mm). The narrow-gauge catheter is supplied sealed to a larger catheter (0.6 mm ID x 0.9 mm OD, 3-Fr), both of which are sterile.

Data presented in this chapter derive from the use of a catheter, the tip of which had been modified in-house (to fit into the femoral vein of approximately 25-gram mice) by stretching the tip of the catheter, although this might have unfortunately destroyed the carefully prepared anti-thrombogenic film.

Recently, 1-French catheters have become commercially available that do fit, without modifications, into the femoral vein of a mouse around 25 grams, but these are normally not used in long lengths (as this increases the risk of becoming clotted) and are connected to a larger-bore catheter, as above. These are being investigated at present.

The disadvantage of those types of catheters is that the flow is less fluent and residues of blood and/or test article can accumulate at the junction of the smaller 1.2-French catheter and the larger 3-French catheter.

2.4.1.2.3.2 Femoral vein catheters used in the harness method. Catheters that were used for the harness infusion model validation study and later for the tail-cuff infusion model are of a slightly different design, consisting of a tapered polyurethane catheter (PU 3-Fr to 1.5-Fr). The catheter tip might need modifying in-house (to fit into the femoral vein of 25-gram mice) by stretching the tip of the catheter, although this does unfortunately destroy the carefully prepared anti-thrombogenic film. The authors believe that when performing a patency check, this catheter provides less back pressure on the plunger of the syringe compared with two different sizes of catheters joined together.

2.4.1.2.3.3 Jugular vein used in tail-cuff. A 2-French catheter can be used for the jugular vein catheterisation and has the advantages that it can be connected in one piece without connections straight to the vascular harness or straight to a 25-gauge swivel using the tail-cuff infusion model. The blood samples are manually obtained via a 2-French catheter connected to an insulin syringe (van Wijk et al. 2007).

The 2-French silicone catheters are frequently used in short-term automated blood sample studies (from 24 hours up to 10 days). The advantage of softer catheters in terms of reduced potential tissue damage has to be countered by the increased tendency for them to be damaged and leak when exteriorised. In addition, when small silicone catheters are used for vascular infusion studies, care has to be taken when they are clotted or blocked. Unblocking silicone catheters can cause the silicone catheter to balloon and burst its wall. The smaller the catheter, the greater the risk. For this reason polyurethane catheters might be preferred when using this jugular vascular catheterisation technique in vascular toxicity infusion studies.

When possible it is prudent to check the compatibility of the test article to the catheter and infusion system prior to performing an infusion study. For example, polyethylene infusion lines might bind less than polyurethane to biological-derived test articles.

2.4.1.3 External equipment

2.4.1.3.1 Tail-cuff. The tail-cuff consists of a stainless steel cylinder with an exit channel leaving the upper surface of the cylinder at a 45° angle. The cuffs are 15 mm in length and are made in two diameters: 5 mm for mice less than 30 g, and 6 mm for mice greater than 30 g. Tail-cuffs weighing over 1.0 g should not be used, as this would compromise the freedom of movement of the mouse. Working with a local manufacturer, we have minimised the weight of the tail-cuff to just less than 1.0 g. The tail-cuff is attached to the tail of the mouse using a sterile stainless steel suture passed through small fixation holes at the proximal end of the tail-cuff.

2.4.1.3.2 Harness. The mouse harness (Figure 2.10) is a small version of the rat harness. The area of harness for the rat is 4.5 cm^2, and for the mouse it is 1.25 cm^2.

The weight of the mouse harness is 0.574 g without the spring, which seems acceptable.

The harness offers a range of benefits over jackets, including adjustability for growth, maintenance of normal body temperatures, and improved air circulation for faster healing.

The Vascular Access Harness™ (Figure 2.11) has been recently developed and permits quick aseptic connection and disconnection of a catheterised mouse from its catheter infusion line and tether. This harness was not available at the time that the harness infusion validation study was done at this laboratory. The possibility that the on-and-off connection

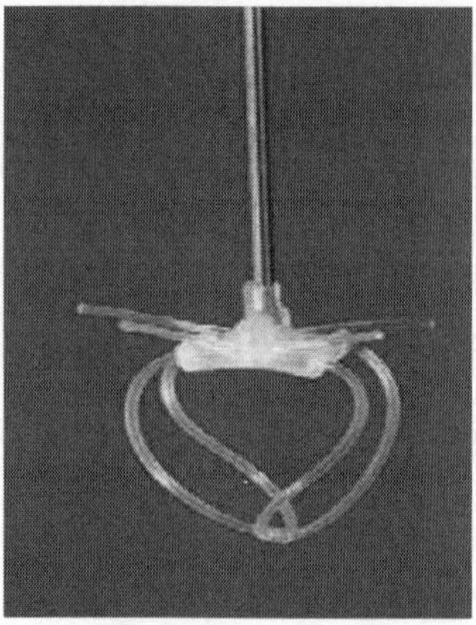

Figure 2.10 Mouse harness system.

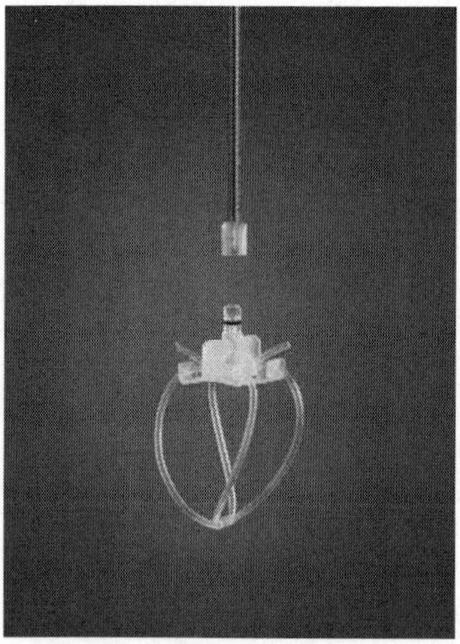

Figure 2.11 Vascular Access Harness™.

can increase clotting of the infusion line requires consideration. Also, daily handling is necessary for daily intermittent infusions studies. The main advantage is that the animals are not connected all the time to the tether during intermittent infusion studies. This makes repair of tethers and infusion lines more practical. The Vascular Access Harness system is also used by those animal suppliers who are providing surgically pre-catheterised jugular vein mice (although pre-catheterised femoral vein mice are not supplied by animal suppliers in Europe).

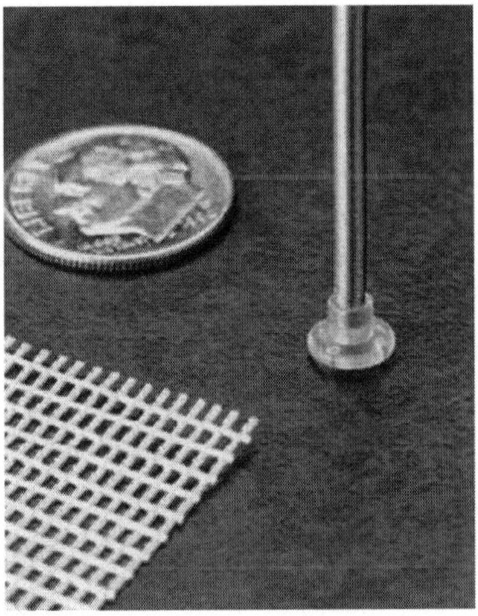

Figure 2.12 Mouse skin button.

2.4.1.3.3 Skin button. A mouse skin button (Figure 2.12) made of polysulfone (LW62S) implanted in the scapular region combined with the harness is thought to be acceptable for providing adequate fixation of the tether, and it is anticipated that a 13 weeks' duration would be feasible.

2.4.1.3.4 Tether system. A flexible stainless-steel spring tether is attached to the exit channel of the tail-cuff or skin button harness or jacket. The spring is then attached to a swivel holder, and the catheter, exiting through the tether, is attached to a 22-gauge swivel. The swivel holder fixed to the cage lid on a metal stand holds the swivel in place above the cage (Figure 2.5). Tethers can be removed and replaced if they become damaged during the dosing period.

2.4.1.3.5 Other exteriorisation methods. There are several other methods described in the literature. Jugular vein catheterisation (Bardelmeijer et al. 2003) with exteriorisation of a catheter from the dorsal neck region using a harness exteriorisation has been performed for serial blood sample studies in the mouse for up to 7 days (Holmberg and Pelletier 2009). The same technique with a jacket exteriorisation method was used for a 10-day infusion period (Inomata et al. 1999; Horii 2000). Other exteriorisation methods, for example the skin button, have also been used (Desjardins et al. 1986).

Most of these systems are detachable and can be disconnected between infusions. Use of appropriate locking solutions (such as heparin solution) can increase the risk of infection and cause blocking (clotting) of infusion lines compared with using a continuous infusion with saline substituted for test article when not being dosed. The use of anticoagulant in locking solutions can also interfere with the test article in toxicity studies.

Another publication described continuous long-term intravenous infusion in unrestrained mice (Lemmel et al. 1971). There, mice were catheterised via the jugular vein and exteriorised by a jacket method that was secured with tape.

Occasional difficulties occurred especially after a prolonged infusion time, with evidence of clotting in spite of being infused with a heparinised 30–40 IU/mL solution at an infusion rate of 0.2mL/h for up to 35 days. Lemmel et al. claimed a success/survival rate greater than 95% (from a total of 200 mice). No other parameters were published, with the exception of a rare instance of otitis or cerebral thrombosis at pathology.

2.4.1.4 *Infusion pumps*

The Medfusion 3500 syringe pump (Smiths Medical Ltd, Lancashire UK) is used in our laboratory. This pump can dose at a flow rate as low as 0.01 mL/hr, with a stated accuracy of ± 2%. It also contains a backup battery, which will allow disconnection from the mains for up to 10 hours to carry out procedures. This pump has a pressure alarm triggered at pressure greater than 16.0 psi, or 35 psi when using a 1-mL syringe.

Immediately after surgery, infusion with physiological saline at an infusion rate of 0.1 mL/h is started to keep the catheter patent. This infusion rate is usually sufficient to maintain catheter patency and avoid clotting within the catheter. This equals an infusion rate of 4 mL/kg/h, optimal for the mouse (see Section 2).

Prior to the start of any study all infusion pumps undergo a performance check to ensure they are fit for use on study.

The Medfusion 3500 pumps are run for 24 hours at either 0.1 mL/hr (mice) or 0.5 mL/hr (rats) using a BD Plastipak syringe. The dispensed volume is weighed to 2 decimal places, and pumps are considered fit for use if the actual dispensed volume is within 10% (mice) or 5% (rats) of the expected dispensed volume. In this case 1 mL of water or saline is assumed to be 1 gram in weight.

The Medfusion pumps are currently serviced annually or earlier if required.

2.4.2 *Non-surgical models*

2.4.2.1 *Restraining devices*

Animals should always be approached and handled firmly, quietly and sympathetically, reassured before placing in a restraining device. The restraining device used by the authors is a restraining tube, with the tail exposed.

The restrainer must be used correctly to avoid the possibilities of physical injury to the tail, limbs, or snout as the animal is manoeuvered into the confined space of the restrainer tube. Extra small rodent restrainer bags are available as an alternative.

See Section 2.2.1.

2.4.2.2 *Vascular access*

Warm and prepare the animal per SOP and place the animal in a restraint cone. Prime the catheter with 0.9% sodium chloride or equivalent by placing a syringe over the stylette onto the attached hub and inject. This must be done slowly because of the resistance of the stylette within a 28G catheter.

Starting in the lower half of the tail, place the bevel of a 23G needle into the tail vein. The needle is used only to make a small hole in the vein in order to advance the catheter. Ensuring that only the bevel of the needle is placed in the tail will reduce trauma to the vein. Once insertion into the vein is confirmed by the presence of blood, insert the catheter into the hole and advance slowly until approximately 2 cm into the tail vein. The catheter should advance easily if it is correctly positioned in the vein. Pressure should immediately be applied to the catheter where it exits the tail while the stylette is pulled out. A small drop of cyanoacrylic adhesive (superglue) or a piece of tape can be placed over the catheter and exit site to provide stability and prevent a backflow of blood or test article. The test article can then be injected slowly. Once test article has been administered, flush the catheter and tail vein with 0.9% sodium chloride or desired solution. The catheter can then be removed and pressure applied to the tail. Return the animal to its home cage.

2.4.2.3 *Infusion pumps*

The Harvard PhD 22/2000 infusion pump has been historically used for intermittent infusions. This pump is dependent on mains power owing to the absence of a battery backup system and uses the lead screw principle to deliver the required flow rate. It has a minimum flow rate of 0.0001 μL/min and a maximum of 220.82 mL/min with an accuracy of 0.35%.

2.5 *Background data*

2.5.1 *Surgical models*

Continuous i.v. infusion models using the mouse in safety assessment studies are still being developed, and background data are scarce compared to data derived from continuous i.v. infusion models using well-established species such as rat, dog, minipig, and non-human primates.

The background data illustrated in this section were generated from two in-house validation studies (one using the tail-cuff exteriorisation model and one using the harness exteriorisation model), together with data from three commercial toxicology continuous infusion studies using the tail-cuff method.

2.5.1.1 *Complications associated with the surgical process and study design*

2.5.1.1.1 Tail-cuff model. The stages during which failure of the model can occur can be categorised, in order to assess percentage failure rates and calculate number of animals required to be surgically prepared:

Stage 1: During surgery
Stage 2: During the post-surgical recovery period until Day 1 of study (minimum recovery period 5 days)
Stage 3: From Day 1 of the study until the end of the infusion period

Overall failure rates of the tail-cuff infusion method during Stages 1 and 2 (from surgery until Day 1 of study, minimum recovery period 5 days) with joined polyurethane catheters (a 1.2-French catheter glued into a 3-French catheter) were 9.4% for males and 10.2% for females. Reasons for deaths occurring in association with surgery included death during surgery (no specific cause determined), air in the implanted catheter, and anaesthesia extract problems. The total percentage of failures during surgery was 1.1% in males and 1.2% in females.

It is thus recommended that a minimum of 10% additional animals be surgically prepared to provide sufficient animals to commence continuous i.v. infusion on Study Day 1, when the joined polyurethane 1.2-French catheter glued into a 3-French catheter is used.

The percentage failure of the tail-cuff infusion method for Stage 3, calculated as the percentage of the number males and females together that did not complete the study because of infusion failures, is presented in Table 2.2.

The femoral vein catheterisation with the tail-cuff exteriorisation method showed no significant adverse effects on the parameters obtained in toxicology studies (ophthalmoscopy, clinical signs, clinical pathology, and histopathology). However, the reliability of this method decreased over time, with success rates of 94%, 88%, and 74% at 7, 14, and 28 days on study, respectively. Results showed that for the tail-cuff infusion method, a minimum of 10% extra animals have to be surgically prepared (minimum recovery period 5 days) in order to have a sufficient number of working, healthy infusion animals for use in studies on Study Day 1.

The use of the FunnelCath™ catheter showed that the reliability of the infusion method increased by using a one-piece catheter instead of 2 pieces joined together. There was no catheter failure during Stage 1 and 2

Table 2.2 Success Expressed as a Percentage of Failure of the Tail-cuff Infusion Method over 7, 14, and 28 Days of Continuous Vascular Infusion

Duration (days)	Study type	Sex	Number of failures	Total animals	Number of failures (%)
7	Main study	Both	2	48	4.17%
7	TK study	Male	4	69	5.7%
7	Main and TK studies combined	Both	6	117	5.1%
14	Main study	Both	18	160	11.3%
14	TK study	Both	24	178	13.5%
14	Main and TK studies combined	Both	42	338	12.4%
28	Main and Recovery (no TK samples)	Both	13	50	26.0%

when the FunnelCath catheter was used with the tail-cuff exteriorisation method. In comparison, the failure rate with the previous 1.2-French catheter glued together with a 3-French catheter was around 3.7% in males and around 3.9% in females.

The diameter of the tail-cuff used on study is an important factor for consideration for the success rate of the infusion model. It is important to use as small a diameter of tail-cuff as is required in order to minimise excess movement, but equally the tail-cuff must be large enough to accommodate growth during the anticipated dosing period. For a chronic study, the tail-cuff is likely to be larger than is ideal, but during the course of the study the wire suture may need to be loosened to allow for growth. A regular check should always be made of tails and tail-cuffs during the study, and the need to loosen or tighten the tail-cuff should be assessed. In cases where the tail lesions become an end-point for an individual animal, the tail-cuff can be removed and the animal retained for post-dose recovery, if appropriate to the experiment (the catheter is cut at the entry of the tail and causes no adverse effects). Ulcerative tail lesions can be expected to heal.

The selection of the optimum design of catheter and diameter of the tail-cuff will affect the outcome of the infusion success rate on study.

2.5.1.1.2 Harness/Skin-button exteriorisation model. The causes of the failures of the harness/skin-button vascular infusion method where the FunnelCath catheter was implanted into the vena cava via the femoral vein were revealed by the harness/skin button 91-day validation study.

Table 2.3 Stage 3 Reasons for Failure Using Harness/Skin-Button Continuous Intravenous Infusion and the Funnel Cath™ Catheter Method from Day 1 until Day 91 Following a 7-Day Recovery Period (10M+10F Animals)

Reason for failure	Total males starting study 10	Total females starting study 10
Catheter related problems	5	4
Other	0	3
Total loss, 91 days	5 (50%)	7 (70%)

Only one animal did not survive surgery to the Study Day 1 period (Stages 1 and 2). This was due to perforation of the catheter during surgery (at the point of exteriorisation of the catheter through the skin button).

No animals were lost during Stage 2. The failure rate during the 91-day infusion period, Stage 3, is presented in Table 2.3. More detailed data of the causes of failures have been published in *The Infusionist* (van Wijk and Fraser 2011).

Four FunnelCath catheters were blocked and 5 were leaking, the majority of those leaking being damaged by the suture needle during unsuccessful re-suturing of the skin button. This action was performed only once on a few animals.

Other issues during infusion study (Stage 3) using the harness/skin-button continuous i.v. infusion method are

- Harness chewed on 66 occasions
- General anaesthesia required to replace the straps on 61 occasions
- Harness straps found caught in animals' mouths on 42 occasions
- Animals escaped/partly escaped from the harness on 31 occasions

For the harness/skin-button infusion validation study, 20 surgically prepared animals were selected on Day 1 of study, with 12 animals failing to complete the 91-day infusion period. Out of the 8 animals that did complete the study, 6 animals showed adverse effects to the eyes, as shown by ophthalmoscopy, clinical signs, and histopathology, the constricting effect of the harness having caused exophthalmos and resultant keratitis.

During the 91 days of a harness/skin-button study, intermittent problems with correct fitting of the harness/skin-button can occur. If too tight, the straps can rub the front limbs, but if the apparatus is then loosened there is more chance of the mouse escaping out of the harness and chewing the infusion line. In this harness/skin-button validation study, mice escaped out of the harness, and they were anaesthetised on several occasions to put it back on. The effect of the isoflurane on the study outcome is not known but could interfere with the test article (especially if it is an anaesthetic itself). The use of a skin button permanently attached to the tether alone was not successful in rats, as it quickly detached from the skin even when softer felt material was used.

2.5.1.2 *In-life changes associated with the model*

2.5.1.2.1 Body weight/growth curves. The surgically catheterised animals showed losses of body weight immediately following surgery. These animals generally recovered within 7 days of surgery. From Study Day 1 to Day 28 normal body weight gain was achieved.

The surgical vascular infusion models had a higher body weight (BW) gain during the study itself (Study Day 1 to Day 28) compared with animals on i.v. bolus injection studies or inhalation studies during which animals are restrained on a daily basis. BW is considered to be a good indicator of animal well-being and degree of stress. For the surgical vascular infusion model, the most stressful period for the animal is recovery from surgery, during which time BW gains were less or there was actual BW loss (females). Higher BW gains during the surgical model studies compared to the BW gains of animals on i.v. bolus studies may be an indicator of increased stress on i.v. bolus studies, which, as previously discussed, may be associated with the daily handling/restraint, warming, and tail vein venepuncture necessary for dosing on the i.v. bolus studies. However, it must be acknowledged that this comparison of stress levels is currently based on one in-life parameter.

2.5.1.2.2 *Clinical observations*

2.5.1.2.2.1 Tail-cuff method. Typical clinical signs associated with the tail-cuff method, such as swollen tail and sores/lesions inside and above the tail-cuff, were observed. The numbers and severity of the sores and lesions increased as the study progressed. However, there were no signs of systemic infection as a result of the tail lesions. In some animals, impaired mobility of the left hind leg was observed immediately after surgery, but these signs diminished during the post-surgical recovery phase.

2.5.1.2.2.2 Harness/skin-button method. Clinical observations associated with the harness method included lesions caused by the straps around the chest, legs, under forearms, and shoulders and lesions around the skin button.

2.5.1.3 *Pathology*

2.5.1.3.1 *Clinical pathology*

2.5.1.3.1.1 Haematology. Group mean haematology data from saline control animals in two 14-day continuous infusion studies (tail-cuff model) are presented in Table 2.4. Individual animal values were usually

Table 2.4 Summary of Haematology Data (Day 14) for Saline Control Animals in Two 14-Day Continuous Infusion Studies (Tail-Cuff Model)

Sex	Statistic	Hb (g/dL)	RBC (10^{12}/L)	PCV (%)	RETA (%)	RABS (10^9/L)	MCV (fL)	MCH (pg)
Male	Mean	14.3	9.00	45.8	3.0	271.2	51.0	15.9
	SD	1.32	0.677	3.04	0.76	67.98	1.83	0.63
	N	12	12	12	12	12	12	12
Female	Mean	13.8	8.85	45.4	3.9	338.2	51.4	15.6
	SD	0.97	0.627	2.99	2.25	167.33	3.44	0.49
	N	7	7	7	7	7	7	7

Sex	Statistic	MCHC (g/dL)	HDW (g/dL)	RDW (%)	PLT (10^9/L)	PCT (%)	MPV (fl)	PDW (%)
Male	Mean	31.2	2.03	12.6	1441	1.28	8.3	53.3
	SD	1.20	0.081	0.62	242.6	0.166	0.09	2.95
	N	12	5	12	12	5	5	5
Female	Mean	30.5	2.48	14.8	1166	0.52	6.7	50.3
	SD	1.58	0.138	3.70	531.5	0.215	0.67	10.87
	N	7	4	7	7	4	4	4

Sex	Statistic	WBC (10^9/L)	N (10^9/L)	L (10^9/L)	M (10^9/L)	E (10^9/L)	B (10^9/L)	LUC (10^9/L)
Male	Mean	7.1	2.3	4.3	0.2	0.2	0.0	0.1
	SD	3.19	2.14	1.96	0.12	0.11	0.00	0.07
	N	12	12	12	12	12	12	12
Female	Mean	8.4	2.4	5.4	0.2	0.2	0.0	0.2
	SD	3.55	1.46	2.71	0.10	0.09	0.00	0.19
	N	7	7	7	7	7	7	7

Sex	Statistic	WBC (10^9/L)	N (% WBC)	L(% WBC)	M(% WBC)	E(% WBC)	B(% WBC)	LUC(% WBC)
Male	Mean	7.1	31	63	2	2	0	2
	SD	3.19	17.5	16.5	1.1	0.7	0.3	0.8
	N	12	12	12	12	12	12	12
Female	Mean	8.4	29	64	2	3	0	3
	SD	3.55	13.7	13.6	1.0	1.3	0.4	3.0
	N	7	7	7	7	7	7	7

within or close to the current 95% reference range for uninfused CD-1 mice of similar age, at the author's laboratory. This indicates that for most parameters, the surgical procedure and infusion of saline did not significantly alter the haematological data. A notable exception to this however was the presence of some variably higher individual neutrophil counts in both sexes, possibly reflecting the inflammatory response seen locally in association with the indwelling catheter.

Group mean haematology data from saline controls in a single 91-day in-house continuous infusion validation study (harness/skin button model) are presented in Table 2.5. Individual animal values were mostly within or close to the current 95% reference range for uninfused CD-1 mice of similar age, at the author's laboratory. The only exception to this was the presence of slightly higher neutrophil counts, particularly in males. The reason for this difference from in-house background data is likely to be the same as that given for the 14-day continuous infusion studies referred to earlier.

2.5.1.3.1.2 Clinical chemistry. Group mean clinical chemistry data from saline control animals in two 14-day continuous infusion studies (tail-cuff model) are presented in Table 2.6. Individual animal values were usually within or close to the current 95% reference range for uninfused CD-1 mice of similar age, at the author's laboratory, again indicating that the surgical procedure and infusion of saline did not significantly influence the clinical chemistry data. Some individual globulin levels were slightly higher in both sexes, resulting in lower albumin globulin ratios (AGR). As for the haematology, the higher globulin values possibly reflect the inflammatory response seen locally in association with the indwelling catheter. In addition, some higher glucose and inorganic phosphorus values were present in males.

Group mean clinical chemistry data from saline controls in a single 91-day continuous infusion validation study (harness/skin button model) are presented in Table 2.7. Individual animal values were mostly within or close to the current 95% reference range for uninfused CD-1 mice of similar age, at the author's laboratory.

2.5.1.3.1.3 Urinalysis. In the author's experience, urine volume is significantly increased and urinary specific gravity significantly decreased in control animals continuously infused with saline (regardless of the surgical model or duration) when compared with uninfused mice. This is a response to the increased fluid input via the indwelling catheter.

2.5.1.3.2 Histopathology. The microscopic findings from control mice from different tail-cuff or harness technique studies have been combined according to length of study in order to obtain a comparative background

Table 2.5 Summary of Haematology Data (Week 13) for Saline Control Animals in an In-House Continuous-Infusion 91-Day Validation Study (Harness/Skin Button Model)

Sex	Animal Number	Hb (g/dL)	RBC (10^{12}/L)	PCV (%)	RETA (%)	RABS (10^9/L)	MCV (fl)	HDW (g/dL)
Male	N	5	5	5	5	5	5	5
	Mean	13.34	9.43	46.42	2.32	214.34	50.1	1.88
	SD	1.05	0.61	4.12	0.29	24.79	1.78	0.06
Female	N	3	3	3	3	3	3	3
	Mean	13.17	9.10	45.1	3.17	285	49.57	1.92
	SD	0.25	0.46	1.59	1.24	99.7	1.07	0.16

Sex	Animal Number	MPV (fl)	MCH (pg)	MCHC (g/dL)	RDW (%)	PLT (10^9/L)	PCT (%)	PDW (%)
Male	N	5	5	5	5	5	5	5
	Mean	5.26	14.44	28.8	14.2	1800	0.948	49.9
	SD	0.19	0.36	0.39	0.54	427.5	0.24	4.39
Female	N	3	3	3	3	3	3	3
	Mean	5.03	14.47	29.2	13.77	1273	0.64	46.9
	SD	0.15	0.45	0.56	1.15	227	0.12	3.14

Sex	Animal Number	WBC (10^9/L)	N (10^9/L)	L (10^9/L)	M (10^9/L)	E (10^9/L)	B (10^9/L)	LUC (10^9/L)
Male	N	5	5	5	5	5	5	5
	Mean	9.04	3.2	5.32	0.24	0.24	0.0	0.1
	SD	3.14	1.69	1.39	0.05	0.17	0.0	0.0

(Continued)

Table 2.5 *(Continued)* Summary of Haematology Data (Week 13) for Saline Control Animals in an In-House Continuous-Infusion 91-Day Validation Study (Harness/Skin Button Model)

Sex	Animal Number	WBC (10^9/L)	N (10^9/L)	L (10^9/L)	M (10^9/L)	E (10^9/L)	B (10^9/L)	LUC (10^9/L)
Female	N	3	3	3	3	3	3	3
	Mean	9.0	3.9	4.37	0.17	0.5	0.0	0.1
	SD	6.2	4.69	1.17	0.15	0.3	0.0	0.1

Sex	Animal Number	WBC (10^9/L)	N (% WBC)	L (% WBC)	M (% WBC)	E (% WBC)	B (% WBC)	LUC (%WBC)
Male	N	5	5	5	5	5	5	5
	Mean	9.04	33.10	61.0	2.8	2.6	0.0	1.0
	SD	3.14	10.65	9.72	0.84	1.14	0.0	0.0
Female	N	3	3	3	3	3	3	3
	Mean	9.0	33.67	58.67	2.0	5.33	0.0	0.67
	SD	6.2	21.08	21.57	1.0	1.53	0.0	0.58

Table 2.6 Summary of Clinical Chemistry Data (Day 14) for Saline Control Animals in Two 14-Day Continuous-Infusion Studies (Tail-Cuff Model)

Sex	Statistic	AST (IU/L)	ALT (IU/L)	ALP (IU/L)	GAMMA GT (IU/L)	Na (mmol/L)	K (mmol/L)
Male	Mean	174	64	70	2	150	6.2
	SD	191.7	45.7	24.5	0.0	1.1	2.37
	N	11	11	11	7	11	11
Female	Mean	73	36	69	2	148	4.7
	SD	10.8	6.9	25.8	0.0	1.3	0.18
	N	8	8	8	3	8	8

Sex	Statistic	Ca (mmol/L)	Cl (mmol/L)	IN PHOS (mmol/L)	TRIGS T (mmol/L)	PROT (g/L)
Male	Mean	2.62	109	3.2	0.85	54
	SD	0.175	2.1	1.58	0.267	4.4
	N	11	11	11	7	11
Female	Mean	2.42	110	2.0	0.60	55
	SD	0.051	0.6	0.55	0.060	4.2
	N	8	8	8	3	8

(Continued)

Table 2.6 *(Continued)* Summary of Clinical Chemistry Data (Day 14) for Saline Control Animals in Two 14-Day Continuous-Infusion Studies (Tail-Cuff Model)

Sex	Statistic	ALBUMIN (g/L)	GLOBULIN (g/L)	AG RATIO ()	TOT CHOL (mmol/L)	GLUC (mmol/L)
Male	Mean	30	24	1.3	3.0	29.7
	SD	4.0	4.6	0.34	0.32	32.4
	N	11	11	11	11	11
Female	Mean	31	24	1.4	2.0	10.7
	SD	3.9	5.9	0.45	0.35	0.76
	N	8	8	8	8	8

Sex	Statistic	BUN (mg/dL)	UREA (mmol/L)	T BILI (umol/L)	HCRE (umol/L)
Male	Mean	21.8	9.3	2.4	12
	SD	4.30	1.05	0.93	2.7
	N	11	4	11	11
Female	Mean	16.7	6.5	2.1	12
	SD	3.19	1.16	0.56	3.5
	N	8	5	8	8

***Table* 2.7** Summary of Clinical Chemistry Data (Week 13) for Saline Control Animals in an In-House Continuous-Infusion 91-Day Validation Study (Harness/Skin Button Model)

Group/ Sex	Animal number	AST (IU/L)	ALT (IU/L)	ALP (IU/L)	Na (mmol/L)	K (mmol/L)	Ca (mmol/L)
Male	Mean	62.75	37.75	35.5	153.75	5.10	2.43
	SD	7.23	9.22	6.45	1.50	0.18	0.08
	N	4	4	4	4	4	4
Female	Mean	53.66	27.33	44.33	151.33	4.5	2.47
	SD	3.21	3.06	9.87	1.53	0.0	0.046
	N	3	3	3	3	3	3

Group/ Sex	Animal number	IN PHOS (mmol/L)	Cl (mmol/L)	T PROT (g/L)	ALBUMIN (g/L)	GLOBULIN (g/L)	AG RATIO ( )
Male	Mean	2.25	114.5	55	32.25	22.75	1.45
	SD	0.13	1.29	3.27	2.22	3.95	0.31
	N	4	4	4	4	4	4

(Continued)

***Table 2.7** (Continued)* Summary of Clinical Chemistry Data (Week 13) for Saline Control Animals in an In-House Continuous-Infusion 91-Day Validation Study (Harness/Skin Button Model)

Group/ Sex	Animal number	IN PHOS (mmol/L)	Cl (mmol/L)	T PROT (g/L)	ALBUMIN (g/L)	GLOBULIN (g/L)	AG RATIO ( )
Female	Mean	2.0	113.67	57.33	33.33	24.0	1.50
	SD	0.0	1.53	2.08	5.51	7.21	0.66
	N	3	3	3	3	3	3

Group/ Sex	Animal number	TOT CHOL (mmol/L)	GLUC (mmol/L)	UREA (mmol/L)	T BILI (umol/L)	HCRE (umol/L)
Male	Mean	2.65	11.2	6.58	2.38	5.0
	SD	0.70	0.93	1.11	0.61	0.0
	N	4	4	4	4	4
Female	Mean	2.57	10.1	5.43	1.8	7.0
	SD	0.38	2.0	0.92	0.17	1.0
	N	3	3	3	3	3

Table 2.8 Microscopic Findings of Vena Cava at Point of Catheter Entry

	Tail-cuff				Harness/skin button			
	14 Days		28 Days		7 Days		14 Weeks	
	n = 32		n = 9		n = 5		n = 8	
Microscopic findings	No.	%	No.	%	No.	%	No.	%
Vena Cava								
Phlebitis/periphlebitis	18	56	3	33	4	80	2	25
Intimal proliferation	10	31	3	33	4	80	6	75
Perivascular haemorrhage	1	3	0	0	0	0	0	0
Pyogranuloma	1	3	0	0	0	0	0	0
Thrombus	0	0	0	0	0	0	2	25

Key: n = number of control animals examined; No. = number of animals with listed microscopic findings; % = percentage of the examined animals with this finding.

Table 2.9 Microscopic Findings of Femoral Vein at Point of Catheter Entry

	Tail-cuff				Harness/skin button			
	14 Days		28 Days		7 Days		14 Weeks	
	n = 32		n = 9		n = 5		n = 8	
Microscopic findings	No.	%	No.	%	No.	%	No.	%
Femoral vein								
Phlebitis/periphlebitis	14	44	0	0	3	60	0	0
Fasciitis/cellulitis	18	56	7	78	5	100	5	63
Myopathy/myositis	4	13	0	0	4	80	0	0
Pyogranuloma/ granuloma	3	9	8	89	5	100	7	88
Mineralisation	9	28	0	0	0	0	0	0
Intimal proliferation	3	9	0	0	0	0	0	0
Thrombus	0	0	0	0	3	60	2	25

Key: n = number of control animals examined; No. = number of animals with listed microscopic findings; % = percentage of the examined animals with this finding.

database. Results are illustrated in Tables 2.8 to 2.11. In Tables 2.8 and 2.9 microscopic findings of the vena cava and femoral vein, respectively, from both tail-cuff and harness methods for different study lengths are displayed. Tables 2.10 and 2.11 illustrate the microscopic findings at the respective catheter exteriorisation sites. Males and females have been combined in the tables as there were no significant differences in results between sexes.

Table 2.10 Microscopic Findings of Point of Catheter Exit via Tail-cuff

	Tail-cuff			
	14 Days n = 32		28 Days n = 9	
Microscopic findings	No.	%	No.	%
Catheter exit (via tail-cuff)				
Fasciitis/cellulitis	31	97	9	100
Dermatitis	29	91	5	56
Granuloma	1	3	0	0
Oedema	13	41	0	0

Key: n = number of control animals examined; No. = number of animals with listed microscopic findings; % = percentage of the examined animals with this finding.

Table 2.11 Microscopic Findings of Point of Catheter Exit via Harness/Skin Button

	Harness/skin button			
	7 Days n = 5		14 Weeks n = 8	
Microscopic findings	No.	%	No.	%
Catheter exteriorisation site via harness				
Dermatitis/folliculitis	4	80	8	100
Fasciitis/cellulitis	5	100	8	100
Granuloma	0	0	4	50
Haemorrhage	1	20	0	0
Oedema	1	20	0	0
Myositis/myopathy	1	20	0	0

Key: n = number of control animals examined; No. = number of animals with listed microscopic findings; % = percentage of the examined animals with this finding.

The microscopic findings of the vena cava in both the tail-cuff and harness methods consisted of tissue changes in response to the presence of the indwelling catheter, including inflammatory changes in and around the vessel wall (phlebitis/periphlebitis, Figure 2.13) and cellular proliferation of the inner vessel cell layer (intimal proliferation). For the harness method, after 14 weeks, 2/8 animals had a thrombus occluding the vessel lumen.

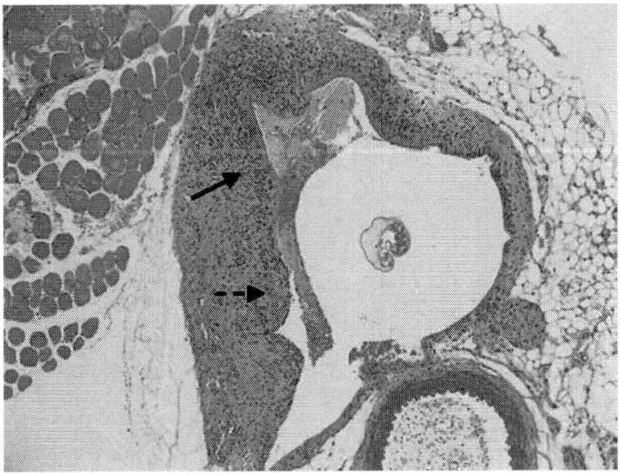

Figure 2.13 (See colour insert.) Vena cava, tail-cuff model, Study Day 14, (H&E ×10 objective). Arrow: Phlebitis/periphlebitis. Dashed arrow: fibrinoid necrosis.

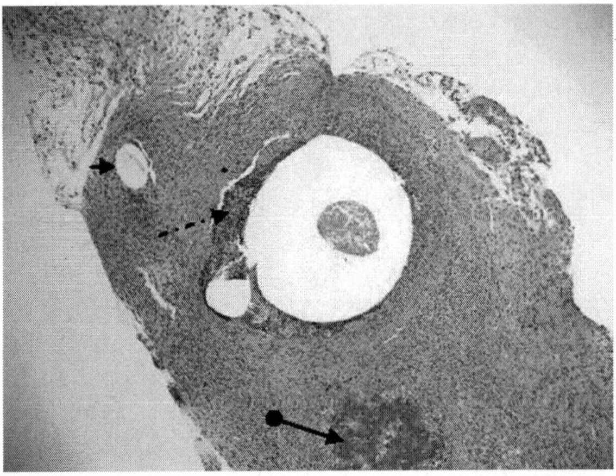

Figure 2.14 (See colour insert.) Femoral vein, tail-cuff model, Study Day 14, (H&E ×4 objective). Short arrow: suture. Dashed arrow: phlebitis/periphlebitis. Arrow with circle: pyogranuloma.

Microscopic findings in and around the femoral vein consisted of phlebitis/periphlebitis in the 14-day tail-cuff and 7-day harness studies. This finding was not seen in the longer-term studies using the tail-cuff or harness. More extensive inflammation of tissues was seen around the femoral vein than the vena cava for both methods (fasciitis/cellulitis, myopathy/myositis, and pyogranuloma/granuloma) (Figure 2.14) correlating with the nature of the surgery, including suture stabilisation of

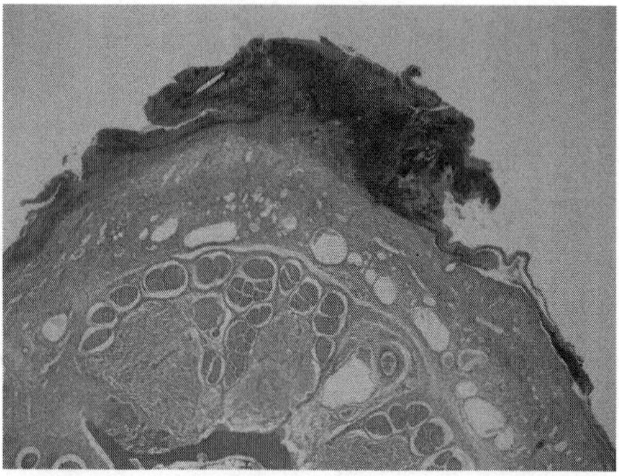

Figure 2.15 (See colour insert.) Catheter exteriorisation site (close to tail-cuff wire), Study Day 8, (H&E ×4 objective). Dermatitis, moderately severe. Fasciitis/cellulitis, moderately severe. Osteomyelitis, slight.

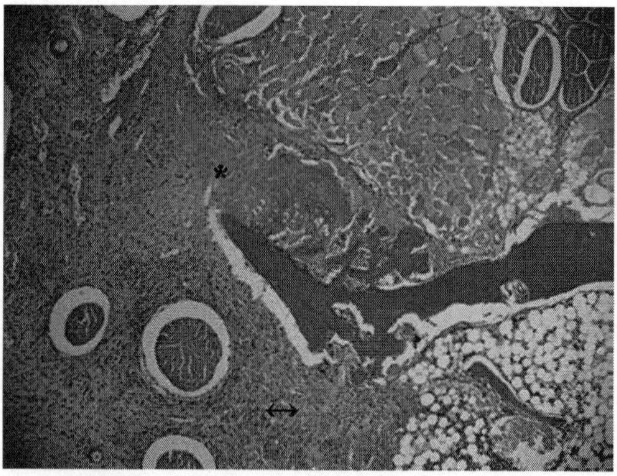

Figure 2.16 (See colour insert.) Catheter exteriorisation site (close to tail-cuff wire), Study Day 8 (H&E ×10 objective). *Osteomyelitis, slight. ↔ Fasciitis/cellulitis, moderately severe.

the catheter, at this site. Thrombus in the vena cava was recorded for the femoral vein in the 14-week harness study.

Microscopic findings at the catheter exteriorisation site in both methods were comparable, consisting of tissue inflammation involving the dermis, subcutis, and deeper fascial planes.

In some cases, where excessive inflammation occurred around the tail-cuff wire and catheter exteriorisation site, animals were removed from study and humanely killed on welfare grounds (Figures 2.15–2.20).

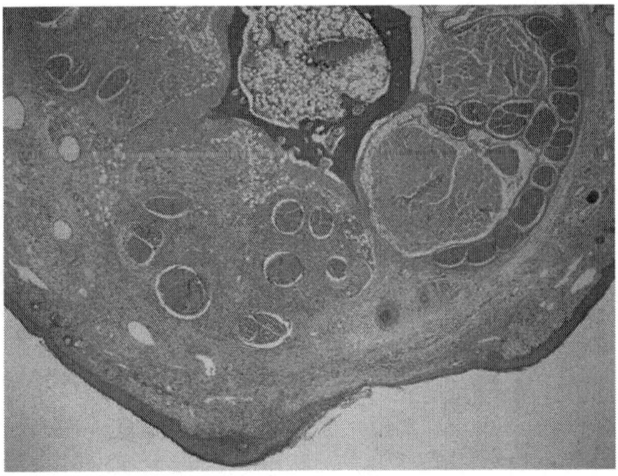

Figure 2.17 (See colour insert.) Tail-cuff wire, tail-cuff model, Study Day 8, (H&E ×4 objective).

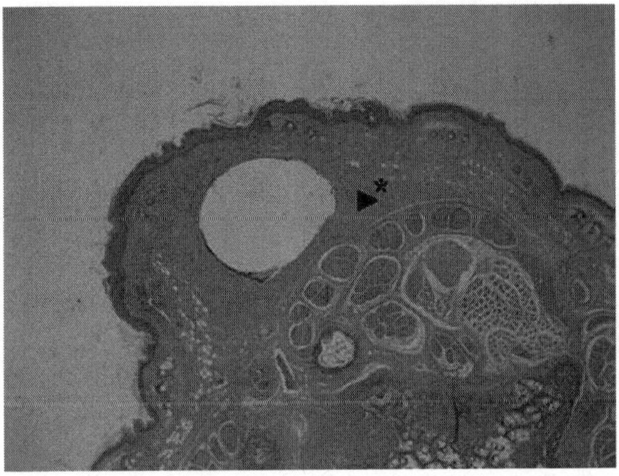

Figure 2.18 (See colour insert.) Catheter exteriorisation site, tail-cuff model, Study Day 15, (H&E ×4 objective). * Fasciitis/cellulitis, moderate. Dermatitis, minimal.

For the 14-week harness validation study, there were microscopic findings of the eye and submaxillary salivary gland. In the eye, 5/8 terminal-kill control mice had keratitis, characterised by focal inflammation and erosion of the cornea. Salivary gland findings in 1/8 animals consisted of oedema.

Microscopic findings in other internal organs of these terminal-kill control animals for both tail-cuff and harness studies were comparable

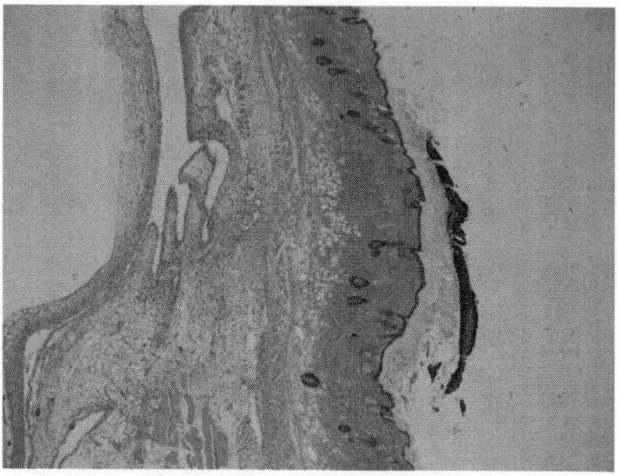

Figure 2.19 (See colour insert.) Catheter exteriorisation site, harness method, 8 days, (H&E ×4 objective).

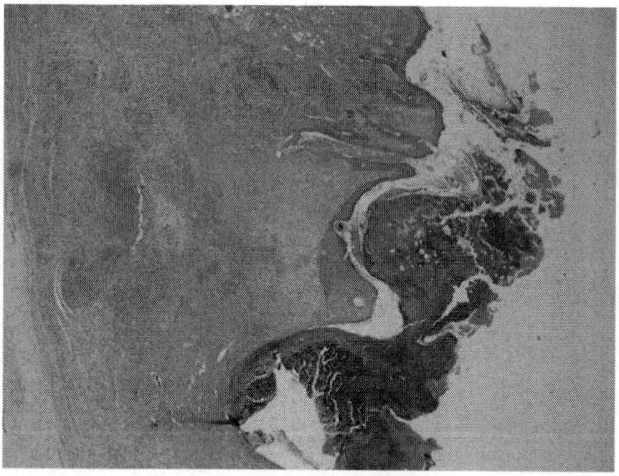

Figure 2.20 (See colour insert.) Catheter exteriorisation site, harness/skin-button model, Study Day 92, (H&E ×10 objective).

and were mostly within the ranges of those seen as background findings in animals of this species, strain, and age. For the 14-day tail-cuff study 3/18 controls had minimal to moderate focal necrosis in the liver. In another 14-day tail-cuff study, 1/11 controls had arteritis of a heart vessel and 1/11 controls had phlebitis/periphlebitis in a perirenal vessel.

2.5.2 Non-surgical models

Studies using non-surgical i.v. bolus mouse models are usually of shorter duration with respect to both study days and actual daily dosing time compared to studies in rats or to studies using surgical i.v. infusion mouse models. This must be taken into account when comparing background data from mouse non-surgical models to these other models.

2.5.2.1 *Issues associated with study design*

- Tail vein damage. Vascular damage can result from needle damage in administering i.v. injections and from the effect of perivascular injection of vehicle and/or test article (and is therefore seen in both controls and treated animals.) Injection problems particularly occur in mice owing to the small diameter of tail vein, which makes i.v. injection technically challenging (see Table 2.8). Additionally, the small volume of blood in this vessel means less dilution of test article and increased direct contact of test article with the vessel wall. Even when mice are restrained, they tend to move their tails more quickly and more frequently than do rats.
- Warming of mice to vasodilate the vein before they enter the restrainer.
- The restraining tube. No access to food or water whilst in the tube for bolus injection.
- Tail vein injection. The actual procedure of sticking the needle into the vein (on a daily basis) causes a pain reaction in the mouse.
- Increased handling procedures. Mice are handled on a daily basis for study procedures (compared to surgical models, which are handled twice weekly only).

The warming, restraint, and amount of handling of the mice are all stress factors that can affect the animals on study. These factors are reduced in the surgical model.

Comparing methods for repeated blood sampling and repeated i.v. dosing (Flynn and Guilloud 1988) suggest that the use of a central catheter (surgical continuous i.v. infusion) would be less stressful for the mice than use of a daily restraint for manual i.v. bolus injection. Stress would be reduced, as the mice would be handled less often and there would be no need for a daily puncture to the tail to access the lateral tail vein for dosing.

It is probable that mice that are warmed to promote vasodilatation and are then subsequently placed in a restraint tube are likely to dehydrate quickly, suffer an increase in body temperature, and lose weight in a similar way to reported data for the rat (Sheehan et al. 2011).

2.5.2.2 *Clinical observations*

In a 7-day mouse study involving i.v. bolus injection (non-surgical method), the effect of repeat i.v. injections caused tail vein damage such that on several days animals could not be dosed. Clinical signs of the tail included sores, lesions, and blue discoloration.

2.5.2.3 *Pathology*

2.5.2.3.1 Clinical pathology. Because tail vein infusion studies featuring test article administration durations longer than 5 minutes on more than 7 consecutive days are not recommended, no clinical pathology data is available for presentation.

2.5.2.3.2 Histopathology. Although one of the issues with the design of i.v. bolus studies includes the condition of the tail vein following technical problems with injection, this usually necessitated replacement of the animal. Hence the affected tail veins were not evaluated microscopically and do not form part of the histopathology data. Histopathology data from mice that did complete the study show fewer pathology findings compared to the surgical models or that seen in rats following i.v. bolus injection studies. This is mainly because mouse i.v. bolus injections studies involve a shorter i.v. dosing duration and shorter studies (by number of days). In rats, i.v. dosing times are often up to 2 hours, whereas in mice the maximum i.v. bolus time is up to 5 minutes (any longer than this and the mice tend to move, causing perforation of the vein).

Histopathology findings at tail vein injection sites of control animals mainly consisted of minimal phlebitis/periphlebitis and minimal haemorrhage in association with the tail vein. The minimal phlebitis/periphlebitis was characterised by scattered, mixed inflammatory cells within and around the tail vein, typical of expected findings following repeated venepuncture.

2.6 *Conclusion*

Experience of continuous infusion in mice using a harness is limited and has mainly been restricted to studies of up to 24 hours in duration. Our experience of attempting to use harnesses on mice suggest there is high risk of either exerting unreasonable pressure on the torso of the animal or failure to achieve security of fit. The consequences of poor fit can lead to the catheter being damaged or perforated by gnawing and exacerbation of injury or infection at the point of emergence of the catheter. When the fit is too tight, adverse effects, such as corneal opacities, impact on measured parameters critical to the outcome of the study.

Continuous infusion involving a site of emergence of the catheter on the tail of the mouse has been used more often. Experienced staff find that the mouse fitted with a tail-cuff remains agile and mobile within the home cage and that it can be removed for procedures such as ophthalmoscopy and recording of body weight during infusion with very little practical difficulty. We can routinely allow a period of post-surgical recovery during which the patency of the catheter is maintained by continuous infusion of physiological saline. By the time the recovery period is complete, the mice show weight gains superior to mice dosed intravenously by methods involving tube restraint. Continuous infusion of mice fitted with tail-cuffs can be maintained with great accuracy for many weeks in regimes involving continuous administration or repeated intermittent intravenous administration of the test article.

As a result of our experience, when continuous infusion of mice is a necessary component of study design, we consider the tail-cuff to be the preferred device for protecting the point of emergence of the catheter. Notwithstanding the anticipated removal of a small proportion of mice from study because of unacceptable tissue damage caused by the cuff, welfare considerations presently dictate that the tail-cuff is preferred over the harness for deployment on definitive toxicology studies intended for regulatory submission.

Acknowledgements

The authors would like to thank the surgical and infusion technicians for their invaluable assistance, suggestions, and ideas for making continous improvements in the technique: D Haida, L Davidson, H Smyth, P Dawson, N Farrington, T Cascone, and D Spencer of the rodent animal facility at Covance Harrogate. We would also like to acknowledge study directors Gemma Gaunt and Sobia Iqbal and the upper management of Covance Laboratories for the support of new techniques.

References

Andes D, Marchillo K, Stamstad T, Conklin R. 2003. *In vivo* pharmacokinetics and pharmacodynamics of a new triazole, voriconazole, in a murine candidiasis model. *Antimicrobial Agents and Chemotherapy* 47(10): 3165–3169.

Ayala JE, Bracy DP, Malabanan C, James FD, Ansari T, Fueger PT, McGuinness OP, Wasserman DH. 2011. Hyperinsulinemic-euglycemic clamps in conscious, unrestrained mice. *J. Vis. Exp.* 57(e3188): 1–8.

Bardelmeijer HA, Buckle T, Ouwehand M, Beijnen JH, Schellens JHM, Tellingen O. 2003. Cannulation of the jugular vein in mice: A method for serial withdrawal of blood samples. *Laboratory Animals* 37: 181–187.

Barr JE, Holmes DB, Ryan LJ, Sharpless SK. 1979. Techniques for chronic cannulation of the jugular vein in mice. *Pharmacology Biochem. Behav.* 11:115–118.

Bögels M. 2006. Intaperitoneal mouse model using mouse-o-ports for treatment of intraperitoneal carcinoma. Presentation at Laboratory Animal Long-Term Access Roundtable, The Netherlands.

Cave DA, Schoenmakers ACM, Van Wijk HJ, Enninga IC, Van der Hoeven JCM. 1995. Continuous intravenous infusion in the unrestrained rat: Procedures and results. *Human and Experimental Toxicology* 14: 192–200.

Clark CT, Haynes JJ, Bayliss MAJ, Burrows L. 2010. Utilization of DBS within drug discovery: Development of a serial microsampling pharmacokinetic study in mice. *Bioanalysis* 2 (8): 1477–1488.

Home Office, London. 1989. Code of Practice for the Housing and Care of Animals Used in Scientific Procedures.

Dainty T, Richmond R, Davies I and Blackwell M. 2012. Dried blood spot bioanalysis: An evaluation of techniques and opportunities for reduction and refinement in mouse and juvenile rat toxicokinetic studies. *International Journal of Toxicology* 31(1): 4–13.

Desjardins C. 1986. Methods of animal experimentation. In Vol. VII, Part A, *Patient Care, Vascular Access and Telemetry*, pp. 166–177. Orlando: Academic Press.

Diehl A, Hull R, Morton D, Pfister R, Rabemampianina Y, Smith D, Vidal JM, Van de Vorstenbosch C. 2001. European Federation of Pharmaceutical Industries Association and European Centre for the Validation of Alternative Methods: Good practise guide to the administration of substances and removal of blood, including routes and volumes. *Journal of Applied Toxicology* 21: 15–23.

Fagin KD, Shinsako, J Dallmann, MF. 1983. Effects of housing and chronic cannulation on plasma ACTH and cortocosterone in the rat. *American Journal of Physiology* 245: E515–E520.

Flynn LA, Guilloud RB. 1988. Vascular catheterisation: Advantages over venipuncture for multiple blood collection. *Laboratory Animals* 17: 39–45.

González-Mateo G, Loureiro-Álvarez J, Rayego-Mateos S, Ruiz-Ortego M, López-Cabrera M, Selgas R, and Aroeira L. 2008. Modelos animales de diálysis peritoneal: Relevancia dificultades y futuro. *Nefrologia Supl.* 6: 17–22.

Hodge DE and Shalev M. 1992. Dual cannulation: A method for continuous intravenous infusion and repeated blood sampling in unrestrained mice. *Laboratory Animal Science* 42: 320–322.

Holmberg A and Pelletier R. 2009. Automated blood sampling and the 3Rs. NC3R 16: April 2009.

Horii I. 2000. Jugular cannulation and efficacy studies in the mouse. In: *Handbook of Preclinical Continuous Intravenous Infusion.* Healing G and Smith D (eds). London and New York: Taylor & Francis.

Inomata A, Shishido, N, Kawashima A and Horii I. 1999. A continuous intravenous infusion technique for pharmacological efficacy study in mouse disease model. Roche Internal Report.

International Conference on Harmonisation (ICH). 1993. Harmonised Tripartite Guideline S5(R2): Detection of Toxicity to Reproduction for Medicinal Products and Toxicity to Male Fertility, 24 June 1993.

Jones PA and Hynd JW. 1981. Continuous long-term intravenous infusion in the unrestrained rat: A novel technique. *Laboratory Animals* 15: 29–33.

Kelley BM, Bandy ALE, Middaugh LD. 1997. A novel chronic and detachable indwelling jugular catheterization procedure for mice. *Physiology and Behaviour* 62: 163–167.

Lemmel E and Good RA. 1971. Continuous long-term intravenous infusion in unrestrained mice method. *Journal of Laboratory and Clinical Medicine* 77: 1011–1014.

MacLeod JN and Shapiro BH. 1988. Repetitive blood sampling in unrestrained and unstressed mice using a chronic indwelling right arterial catheterization apparatus. *Lab. Anim. Sc.* 38: 603–608.

McCarthy A and McGregor J. 1999. Comparison of cannula exteriorisation methods and equipment used on continuous infusion methods. *Animal Technology* 50: 221–224.

Mokhtarian A, Meile MJ and Even PC. 1993. Chronic vascular catheterisation in the mouse. *Physiology and Behaviour* 54: 895–898.

Popovic P, Sybers H and Popovic VP. 1968. Permanent cannulation of blood vessels in mice. *Journal of Applied Physiology* 25: 626–627.

Popp MB and Brennan MF. 1961. Long-term vascular access in the rat: Importance of asepsis. *Am J Physiol* 241: H606–H612.

Sheehan J, Ramaiah L, Willard-Mack C and Vanterpool I. 2011. Intravenous infusing dosing in the rat: Considerations when selecting the optimal model. *The Infusionist* 1 (1): 44–57.

Van Wijk H. 1997. A continuous intravenous infusion technique in the unrestrained mouse. *Animal Technology* 48, 115–128.

Van Wijk H, Anstruther P, Forsyth L, Livingstone J and Lynagh S. 2006. Validation of a continuous intravenous technique in tumour bearing nude mice. *Animal Technology and Welfare* 5(3): 175–176.

Van Wijk H and Fraser A. 2011. Best practice in the development of mouse continuous intravenous infusion models. *The Infusionist* 1(1): 65–70.

Van Wijk H, Harrison R, Brown J, Short T and Pugnaghi F. 2007. Evaluation of a serial blood sampling technique in freely moving mice. *Animal Technology and Welfare* 6(3): 143–146.

Van Wijk H and Robb D. 2000. Femoral cannulation using the tail-cuff model in the mouse. In: *Handbook of Preclinical Continuous Intravenous Infusion*. Healing G and Smith D (eds). p. 61. London and New York: Taylor & Francis.

Van Wijk H and Robb D. 2000. Multidose infusion toxicity studies in the mouse. In: *Handbook of Preclinical Continuous Intravenous Infusion*. Healing G and Smith D (eds). p. 71. London and New York: Taylor & Francis.

Webb AJ. 2011. Best practise in developing infusion models for dogs and minipigs. *The Infusionist* 1(1): 80–85.

Weber K, Mowath V, Hartman E, Razinger T, Chevalier HJ, Blumbach K, Kaiser S, Jackson A, Casedesus A. 2011. Pathology in continuous infusion studies in rodents and non-rodents and ITO (Infusion Technology Organisation)-recommended terminology for tissue sampling and method-related lesions. *J Toxicol Pathol* 24(2): 113–124.

chapter three

Dog

John Cody Resendez, MS, RLATG, SRS, CMAR
MPI Research, USA

David Rehagen, DVM, DACVP
MPI Research, USA

Contents

3.1 Introduction

3.1.1 Choice and relevance of the species

The dog (*Canis familiaris*) has been used as the non-rodent model of choice in experimental research to predict the potential toxicology of new drug entities meant to be administered to humans (FDA 2006). Because of its extensive use in physiological and surgical research over the past century, the canine has amassed a large historical database of anatomic and physiologic endpoints. This vast scientific knowledge has revealed similarities to humans, and enhanced our understanding of anatomy and physiology, making the canine almost exclusively the large animal species for general toxicology testing. Coupled with its extensive use in experimental surgical research, the dog is quite advantageous as a model for long-term infusion (Haggerty et al. 1992). In most toxicology studies involving canines, the breed of choice is the beagle, mainly because of its availability, size, and temperament (Gleason and Chengelis 2000).

In testing investigational drugs, pre-clinical study designs that mimic clinical protocols have become more complex. Intravascular administration is the most common infusion route. Long-term intravascular infusion permits the maintenance of a constant, steady-state blood level concentration of a drug. Successful infusion techniques require the integration of numerous technologies, including catheter implantation, externalised equipment protection, and methods of compound delivery.

From the regulatory guideline viewpoint, despite the technically challenging and logistically demanding nature of continuous intravenous infusion studies, regulatory agencies expect these studies to be comparable to general toxicology designs in regard to overall endpoints, data parameters, and animal numbers. Similarly, there is no regulatory guideline that limits the duration of animal infusion studies (Washer 2001). As a result, the global guidelines, including the United States Food and Drug Administration (FDA) and International Conference on Harmonization (ICH) guidelines appropriate for non-clinical safety studies, are applicable to infusion studies. These guidelines are generic for the testing of chemicals worldwide and generally make no dispensations for any specific route of delivery.

3.1.2 Choice of infusion model

Generally, when evaluating whether to use a surgically prepared or non-surgically prepared (restrained) model, the following must be considered:

- Clinical regimen/dosing regimen
- Steady-state systemic drug levels required
- Study duration
- Length of acclimation (post-surgery recovery)
- Test-article–system compatibility
- Frequency of dosing (single vs. multiple)
- Duration of dose administration (minutes vs. hours vs. days)
- Infusate characteristics
- Infusion rates and volumes
- Surgical/post-surgical care (training)

IV infusion studies should be designed and conducted to mimic clinical dosing protocols. Once the study duration and endpoints have been identified, evaluate the logistics associated with each modality. Identify which regimen is more appropriate, understanding its limitations, equipment and material needs, the compound characteristics, study objectives and anticipated test article effects (if known), practical and regulatory considerations, and overall feasibility. For many study designs, percutaneous IV administration is not appropriate for extended dosing durations and repeated procedures. In these cases, chronic vascular implantation is the more appropriate and efficient regimen to consider. Based on the frequency of dose administration, infusion rates, and infusion volume, a restrained, non-surgical model can be selected, or alternatively either the tethered or the backpack method of delivery, both of which involve surgical implantation of IV access devices. As with any surgical procedure, proper training for pre-operative, intra-operative, and post-operative care is essential for the success of the procedure.

Having the capability for both tethered and ambulatory models allows the researcher to select the model that is more appropriate for the physiochemical properties of the test article and duration and frequency of infusion. The ambulatory infusion model is the more acceptable model. Ambulatory infusion models incorporate a custom-fit infusion jacket, a battery-powered infusion pump with dose solution reservoir, an external infusion access line, and a subcutaneously implanted vascular access port attached to an implanted catheter. Tethered systems may also be used for long-term infusion in the canine when total dose volumes exceed limitations in reservoir size associated with an ambulatory infusion system. In the development of a successful infusion model, the requirement for a system versatile enough to accommodate the activities associated with an IACUC-approved (IACUC in full) and regulatory-compliant study protocol is paramount. The infusion system must be incorporated into the animal caging, must optimise use of space, and must allow standard animal manipulations without interrupting dosing. Enhancements to animal cages and the infusion systems (catheters, catheter material, and vascular access ports) have been essential in meeting these requirements.

Improvements to jacket design, implantation techniques, and maintenance procedures have made both ambulatory and tethered infusion successful and efficient. Novel refinements, adequate planning, proper catheter maintenance, and careful experimental design are crucial to the successful conduct of infusion models in the laboratory canine.

3.1.3 Limitations of available models

Although most procedures and end parameters included in a typical toxicology study can similarly be included as part of an infusion study, the technical and logistical issues surrounding the use of an indwelling vascular catheter, either peripherally or centrally, must be considered and planned for:

- Acclimatization: For the restrained model, animals will need to be acclimatised to the method of restraint, usually in slings over a prolonged period of time and building up to the desired daily restrained duration required in the study. For the surgically cannulated animals, a period of recovery from surgery will be required prior to commencing the dosing.
- Post-surgical treatment: Only required for the surgically implanted models but the pre- and post-surgical antibiotic and analgesic regimen must be considered, as there is the potential for test article interactions.
- Physical characteristics of the test material: These should be known and planned for in all infusion models. Drug-catheter compatibility/ suitability studies should be conducted to evaluate any adsorption

or leaching of test article formulations into the catheters and/or other components of the infusion system. Particular attention should be given to vehicle and test article properties such as pH, osmolality, and stability. The stability of the formulated test material may significantly impact the frequency of required formulation preparations. Extremes in pH and osmolality of the infused formulation may result in local and/or systemic vehicle- or test-article-related irritation or development of infusion phlebitis.

- Volume and rate of delivery: The infusion rate and volume must be considered when designing infusion studies. Low volumes and rates of delivery would be suitable for administration into peripheral vessels in the restrained models, whereas high volumes and rates would benefit from being delivered via a surgically implanted catheter in a central blood vessel.
- Duration of infusion: Animals should not be subjected to long periods of restraint; therefore, where the duration of the infusion period is to be for a prolonged daily period because of infusate volume demands or clinical regime demands, then surgically implanted models are preferred.
- Clinical signs assessment: Full and proper assessment of clinical signs is not possible in immobilised restrained animals, and thus when such assessment is important, as in most safety assessment programmes, the fully mobile surgically prepared animal model is preferred.
- Animal welfare: For large and active animals such as dogs it is important for environmental enrichment issues and general well-being that the animal is able to perform the full range of natural behaviour throughout the duration of a study. Consequently, mobile surgically prepared animals are preferred, where the fully ambulatory, jacketed model provides the most complete freedom of all.

3.2 Best practice

3.2.1 Surgical models

3.2.1.1 Training

Education and training are critical for ensuring that qualified personnel are performing surgery on laboratory animals. Those performing surgery vary in educational backgrounds and experience from having doctorates (physicians, veterinarians, and dentists) to non-doctoral individuals (graduate students, medical students, and technicians). Assessment of the surgical skills and knowledge of persons performing animal surgery is difficult owing to the vast array of procedures that are performed. There are, however, basic tenets of surgical technique to which individuals must adhere and demonstrate competence, including a proficiency in proper sterile technique, tissue handling, haemostasis, and closure techniques in addition to peri-operative and post-operative care. In recent years it has become more common for non-doctoral personnel to perform biomedical research-related animal surgical procedures. These individuals are responsible for the well-being of the animal and must have adequate training and experience. Individuals should be assessed independently based on experience, background, and training before being allowed to perform surgery. Most research institutions performing experimental surgery have, and conform to, regulations, laws, and standards requiring that personnel be trained in performing animal surgery. However, regulations pertaining to surgical training are few, and those that exist lack definition with regard to specifics about surgical training and surgical qualification. For example, in the United States, the Academy of Surgical Research (ASR) is an international organization that promotes the advancement of professional and academic standards, education, and research in experimental surgery. Through the ASR, a certification programme has been established to provide verification of competency in anaesthesia, minor surgical procedures, and more complex experimental surgery, with corresponding certification as Surgical Research Anesthetist (SRA), Surgical Research Technician (SRT), and Surgical Research Specialist (SRS), respectively. Recommendations for surgical research guidelines have been described in a 2006 publication by the ASR, 'Guidelines for Training in Surgical Research with Animals', for a laboratory animal surgical training programme requiring surgeons to demonstrate proficiency in all facets of the procedures being performed. Other training resources for surgical methods are available through hands-on surgical wet labs sponsored by various organizations, including the ASR, the Safety Pharmacology Society (SPS), and the American Association for Laboratory Animal Science (AALAS). Numerous virtual training opportunities are available through webinars and other forms of electronic media. Ultimately, it is incumbent upon the surgeon to demonstrate a working knowledge of the ethical considerations and laws that govern the performance of experimental surgery in animals, and it is the responsibility of an institution to ensure that all personnel performing survival or terminal surgical procedures on these animals be qualified by experience, education, and training to perform such procedures and, ultimately, are appropriately certified/licensed (by a relevant national authority or organization).

3.2.1.2 Applicability of the methodology

For both tethered and untethered (ambulatory) infusion models, the jugular and femoral veins are the two most common sites of catheter implantation for continuous intravenous infusion in the dog. Irrespective of the implantation site, the surgical techniques used for isolating, cannulating, and ligating the target vein are similar. More importantly, established general surgery principles apply to the implantation of intravascular catheters. Asepsis, closure of dead space, haemostasis, minimal tissue trauma, and careful approximation and apposition of the wound/incision are paramount for a successful surgical outcome.

As previously mentioned, pre- and post-surgical antibiotic and analgesic regimens must be considered, as the potential for test article interactions exists. This in no way would preclude the use of antibiotic/analgesic treatment following surgery, but would more specifically identify which drugs are most appropriate (based on the potential for test article interaction) and/or would define an adequate wash-out period that could impact the duration of recovery. Along with anticipating drug contraindications, the anticipated infusion rate and dosing duration and frequency may dictate the infusion model (tethered or ambulatory) and equipment required to accomplish the dose administration. Maximum infusion volumes are dictated by the duration of exposure, the length of the dosing period, chemical characteristics of the formulation, and toxicity of the test material (Walker 2000). During the drug delivery portion of an infusion study, recommended flow rates for test article delivery range from a minimum of 2 mL/hr to approximately 100 mL/kg/day (~4 mL/kg/hr). Infusion rates below 2 mL/hr (typical large animal keep vein open [KVO] rate) may lead to patency issues such as test article precipitation or clotting and occlusions. Very low infusion rates may lead to insufficient mixing of infusate with intravascular fluids (Blacklock et al. 1986). High infusion rates can lead to fluid overload, resulting in pulmonary oedema. However, compensatory responses (such as initial vasodilation and resulting increased renal excretion) enable infusion of much larger volumes over extended infusion durations. The upper limit of the infusion rate is theoretically set by the glomerular filtration rate, which has been calculated to be approximately 5 mL/kg/min (300 mL/kg/hr) in the dog (Gleason and Chengelis 2000). The requirements for maintaining normal fluid balance in the dog have been estimated at 60 mL/kg/day (Maddison 2000).

The use of vascular access ports (VAPs) as an alternative to externalised catheters has become common in research institutions. With advancements in VAP materials and design, incorporation of access port technology into both tethered and ambulatory infusion models is now possible. The use of VAPs has opened up new research opportunities and refined techniques that have improved the intravenous infusion model as a whole. Vascular access port use, advantages, and implications are discussed further in subsequent sections of this chapter.

3.2.1.3 *Risks/advantages/disadvantages*

As mentioned, for both tethered and ambulatory infusion models, the jugular and femoral veins are the two most common sites of catheter implantation for continuous intravenous infusion in the dog. Both vessels have advantages and disadvantages. The jugular vein is easy to access and cannulate, but because of its proximity (the catheter tip's proximity) to the heart, test articles have less chance to mix with the blood, and the catheter must be carefully positioned so as to not lie directly in the left atrium. Additionally, cannulation of the jugular vein prevents use of that vessel for the percutaneous blood sample collections often required for toxicokinetic/pharmacokinetic evaluations during the study; blood collection is thus limited to the contralateral jugular vein and/or limb vessels. The femoral vein approach provides for more complete systemic distribution of the test article and avoids potential catheter interaction with the heart. Additionally, femoral catheter placement allows for the use of both jugular veins for percutaneous blood sample collection.

The VAP model provides unique advantages over the traditional externalised catheter technique. The VAP is not exteriorised. Once the blood vessel is cannulated, the catheter is attached to the port body (usually comprising a small reservoir with an injection septum) that is then positioned subcutaneously. The entire system is therefore completely internalised, which greatly decreases the risk of infection and prevents possible catheter dislodgement, thereby improving overall catheter patency over longer durations.

3.2.2 Non-surgical models

3.2.2.1 Methods of restraint and applicability of the methodology

The jugular vein and peripheral limb veins provide options with regard to percutaneous catheters. Percutaneous placement of limb catheters is considered a non-invasive technique that allows short-term access to peripheral vessels through the skin. A standard over-the-needle intravenous catheter (e.g. the BD Angiocath™) is placed into a saphenous or cephalic vein. The catheter is then secured to the limb with tape, thus allowing percutaneous vascular access to facilitate dose administration. Because the peripheral catheter is positioned on the dog's forelimb or hind limb, the dog must be restrained during the dose administration. Typically, dogs can be restrained in a canine sling apparatus for up to 4–6 hours per dose, although this would be perceived as an atypical circumstance. Short intermittent peripheral infusions will require appropriate acclimation to the restraint sling.

Another option permits percutaneous placement of a central, jugular cannula. A percutaneous indwelling catheter is introduced into the jugular vein through the skin using a break-away catheter introducer while the animal is under light sedation. Once the cannula is threaded into the jugular vein, the introducer is removed from the vessel and then 'split' away, leaving the cannula in place. The cannula is then secured in place to the skin using surface sutures, tape, and overwrap for protection.

A variation of this procedure involves making a slight surface incision at the implantation site to expose the vein. Following placement of the catheter using the catheter introducer, the introducer is removed and the skin is apposed around the proximal, reinforcing strip of the catheter, providing additional catheter protection and stabilization. This model provides a potential alternative to the restrained model and can allow full ambulatory options should the volume of infusion permit.

3.2.2.2 Risks/advantages/disadvantages

Percutaneous peripheral infusion provides an alternative to the more invasive process of surgical implantation of cannulas. Short intermittent infusion study designs may allow the use of temporary peripheral vein catheters. There are, however, considerations that may impact the benefit and utility of this model. The potential for subcutaneous exposure and resultant local reactions are increased by the need for repeated needle sticks to gain vascular access. Consequently, the characteristics of the test article become very important when considering a percutaneous peripheral dose design. The effects of pH and osmolality are often exacerbated depending on the size of the vessel in relation to the catheter size; a smaller vessel/catheter size ratio would suggest the increased likelihood of test material coming into direct contact with the blood vessel wall before any significant dilution could occur. In this situation, local intolerance associated with vehicles/test articles prone to induce irritation (by test article properties, extremes in pH or osmolality, associated volume and rate of infusion, etc.), becomes more problematic, less tolerated, and often a dose-limiting issue that could result in vascular occlusion or extravasation. Specifically related to the percutaneous placement of a central jugular cannula, this technique may limit the repeat accessibility of vasculature used for blood sample collection. Since the jugular vein is the blood collection vessel of choice in the dog, blood sampling is limited to the non-catheterised, contralateral jugular vein (since venipuncture techniques performed on the side of the dog that is catheterised may result in a needle puncture of the catheter). In addition there are the potential stress-related issues in animals restrained for up to 2 hours or more on a frequent basis, as well as the reduced ability to assess a full range of clinical signs in the animal during the infusion of test article.

3.3 Practical techniques

3.3.1 Surgical models (adapted from Resendez and Rehagen, 2012)

3.3.1.1 Preparation

3.3.1.1.1 Pre-medication and anaesthesia. To determine the suitability of animals for surgery, a physical examination and an evaluation of clinical pathology parameters should be conducted within 15 days of surgery. Only animals determined to be suitable for study should undergo surgical procedures. Animals should be fasted overnight prior to surgery to prevent emesis during the surgical or recovery periods and possible aspiration pneumonia after recovery. Prophylactic cefazolin should be administered (25 mg/kg) intravenously (IV) before surgery. The animals should be pre-medicated with an anticholinergic such as atropine (0.05 mg/kg), tranquilised with acepromazine (0.1 mg/kg), and anaesthetised by intravenous administration of propofol (6 mg/kg to effect). An endotracheal tube will be inserted and general anaesthesia will be maintained with isoflurane (0.5–5.0%, to effect) delivered in oxygen via a precision vaporiser and rebreathing anaesthetic circuit. Lactated Ringer's solution (LRS) at approximately 25 mL per hour will be given via a peripheral catheter during surgery. Other pre-operative procedures should be performed as indicated in Table 3.1.

3.3.1.1.2 Pre-surgical preparation. All surgical procedures should be conducted utilizing routine aseptic techniques, and the procedures should be conducted in a dedicated surgical suite. Pre-operative preparation should be conducted in a room separate from the operating room.

Table 3.1 Scheduled Medications and Dosages

	Interval, dose, and route	
Drug	Surgery	Post-surgery
Acepromazine maleate	0.1 mg/kg SC	-
Atropine sulfate	0.05 mg/kg SC	-
Propofol	6 mg/kg IV to effect	-
Isoflurane	To effect by inhalation	-
Buprenorphine	0.01 mg/kg SC TID every 6–9 hours	0.01 mg/kg SC TID every 6–9 hours x 3 days
Meloxicam	0.2 mg/kg SC sid	
Carprofen		4 mg/kg SC SID x 3 days or 25–50 mg PO BID x 3 days
Bupivacaine	2 mg/kg maximum injected into incisions	-
Cefazolin	25 mg/kg IV (pre-operative)	-
Cefazolin	-	25 mg/kg SC
Cephalexin	-	250–500 mg PO bid every 9–12 hours x 7 days
LRS	10–15 mL/kg/hr IV	-

The animals will be prepared for surgery by shaving the area over and adjacent to the surgical sites. The shaved areas should be cleansed with a surgical scrub and solution (e.g., iodine, chlorhexidine). Following the final removal of the surgical scrub, an appropriate antiseptic solution should be applied (e.g., iodine, alcohol). A sterile ocular lubricant should be applied to the corneal surface of each eye.

3.3.1.2 Surgical procedure

Basic tenets of surgical technique have been previously described and should be followed. An incision is made to the inguinal region and the femoral vein is isolated from the surrounding adventitia (Figure 3.1).

A venotomy is made in the vein, and a medical-grade polyurethane catheter is inserted and advanced cranially (Figure 3.2) until the catheter tip is located within the vena cava; the distance the catheter is advanced into the vein is documented. The catheter is secured to the vessel with non-absorbable suture (Figure 3.3) and then is subcutaneously routed to an exteriorization site in the dorsal-lateral thoracic region.

The catheter is attached to a vascular access port placed into a subcutaneous pocket created at the exteriorization site, and is secured to underlying muscle using an appropriately sized non-absorbable suture.

All the subcutaneous tissues are closed with absorbable suture. The skin incisions are closed using a subcuticular suture pattern and skin staples and/or tissue glue.

3.3.1.3 Maintenance

3.3.1.3.1 Recovery procedures. Cefazolin (25 mg/kg) should be administered IM or IV the afternoon following surgery. Analgesics (e.g., Meloxicam) should be administered the afternoon following surgery and for the three days immediately following surgery for control of pain. Other post-operative procedures should be performed as indicated in Table 3.1. For at least 7–10 days following surgery, animals should be observed for pain and/or discomfort and post-operative complications. Generally, animals should be allowed at least 10–14 days post-surgical recovery to ensure that the prophylactic antibiotics have been eliminated from the animal and that the animal has fully recovered prior to study initiation.

3.3.1.3.2 Housing. Applicable national guidelines on appropriate housing should be followed, along with applicable animal welfare regulations. On arrival at the laboratory, the dogs should be acclimated for a period of at least 1–2 weeks. Upon receipt, the dogs should be placed in housing commensurate with the guidelines applicable to the country involved. For example, this would include large, 4-m^2, floor pens with no height restrictions under European legislation, or 1.2-m^2 floored

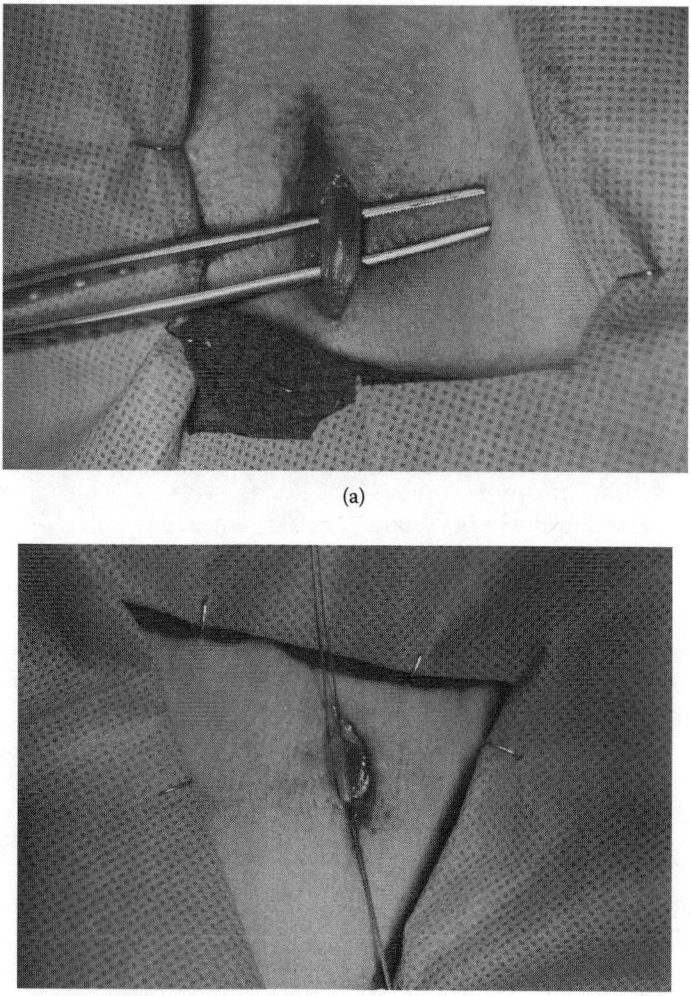

(a)

(b)

Figure 3.1 (a) Femoral vein isolated from surrounding adventitia. (b) Non-absorbable suture used to further isolate the vein and secure the catheter in place.

stainless steel cages with stainless steel or plastic coated flooring under US legislation. These housing variations can significantly affect the type of surgically prepared model that is selected for the infusion study. While dogs may be pair-housed (single sex) in large pens or double-sized cages during the acclimation period, housing options become more restrictive when conducting continuous infusion studies. Depending on the study design and infusion model utilised, pair-housing may or may not

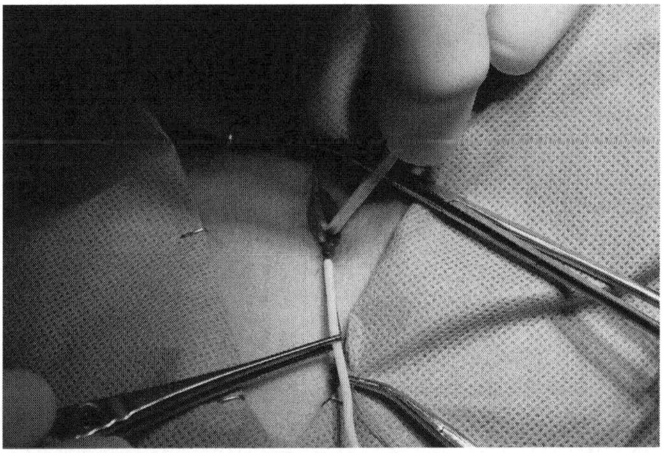

Figure 3.2 A medical grade polyurethane catheter is inserted and advanced cranially until the catheter tip is located within the vena cava.

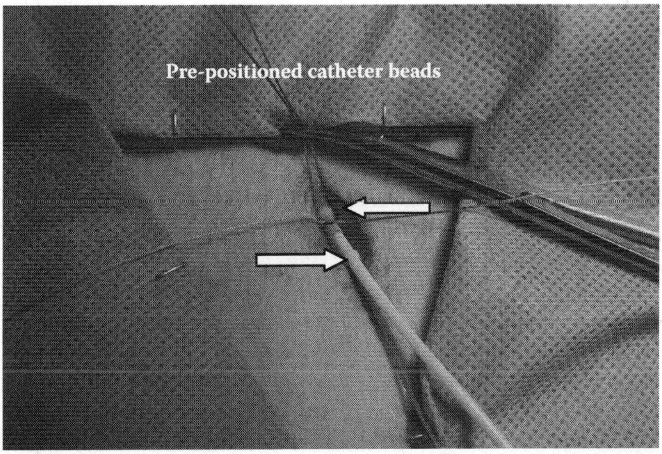

Figure 3.3 Catheter is secured to the vessel by ligating the vein around the catheter, distal to a pre-positioned catheter bead. The distal portion of the vein is then ligated, and the ligating suture is used to further secure the catheter in place by tying around the catheter, proximal to a second pre-positioned catheter bead.

be appropriate. When using the fully ambulatory (non-tethered) model, the infusion equipment (jackets, catheters, etc.) can either remain in place or be removed following a dosing period, thus allowing same-sex pair-housing between test article administrations. However, the dogs will be individually housed when protocol functions require individual data collection, such as food consumption measurement and urine collection.

When using the tethered infusion model, two tethered dogs cannot be kept in the same cage. Physical limitations associated with the equipment prevents pair-housing the dogs, because of the potential for tether entanglement and/or infusion system damage. In all cases, the dogs should be provided the opportunity for exercise unless limited by the tether system used to facilitate a continuous dose administration.

3.3.1.3.3 Checks and troubleshooting. The incision and exteriorization site (if applicable), the jacket fit, and the entire infusion system should be inspected daily for signs of infection (e.g., erythema, oedema, and exudation), equipment damage, leaking, disconnections, etc. The exteriorization site should be shaved as needed. The entire length of the subcutaneous catheter tract should be shaved to assist in observing any possible indications of inflammation, swelling, or infection. A cleansing agent (e.g., Novalsan® solution) can be used to clean the incision/exteriorization site as needed. Care should be taken to ensure the site is kept dry for 7–14 days post-surgery. An appropriate antibiotic (e.g., gentamycin) should be applied to the incision site(s) for at least 14 days post-surgery and daily following catheter exteriorization.

If any portion of the infusion system is disconnected during the maintenance and/or test administration period, any disconnected ends will be aseptically handled and appropriately disinfected (using chlorohexadine, alcohol, etc.) prior to being reconnected.

Once a dog has been jacketed and attached to the infusion apparatus, the infusion system should be maintained with 0.9% sodium chloride for injection USP (saline) at a rate of 2–4 mL/hr. VAPs in animals that are not placed on catheter maintenance should be routinely flushed and locked with an adequate volume of catheter lock solution (e.g., heparinised saline or taurolidine citrate catheter solution) to fill the VAP and implanted catheter.

3.3.1.4 Record keeping

Record keeping is essential to ensure institutional regulatory compliance. Records pertaining to individual dogs and groups/colonies of dogs usually contain such specific information as identification, date of birth, date of receipt, animal vendor, physical examination findings/physical condition, vaccination records, laboratory results, body weight data, and health history. Institutional recordkeeping is usually dictated by international guidelines, including those of the USFDA, USDA, and ICH, that are appropriate for non-clinical safety studies, and which are also applicable to infusion studies. For example, the USDA has set forth recommendations for detailed health records that describe the results of medically related observations, diagnosis, prognosis, and treatment plans along with the specific descriptions of treatment, dosages, administration route, and treatment duration. These requirements are particularly important in regard to the records that must be maintained for animals that are surgically altered for infusion administration in order to document adequate animal care and use.

Procedures that require anaesthesia use drugs that are controlled by a regulatory agency (e.g., the Drug Enforcement Administration in the US). Specific administration of these drugs requires exact records of usage. During surgical procedures, individual animal anaesthesia records are maintained that document the surgical procedure being performed, specific drugs and dosages used, and the monitoring parameters. Typically, dogs undergoing surgical procedures to implant a vascular catheter will be monitored for cardiovascular function (e.g., heart rate, blood pressure, electrocardiographic waveform), respiratory function (e.g., respiratory rate, tidal volume, minute volume, end tidal CO_2, inhaled gas concentration), metabolic status (e.g., oxygenation, body temperature), and haematologic status (e.g. capillary refill time). These parameters should be monitored and recorded (as appropriate while monitoring equipment is still in use) until the animal is conscious. Further post-operative records should include assessment of general behaviour, appetite, hydration, and condition of the surgical incision(s) until the wounds have healed and sutures/staples have been removed (if appropriate).

Dose administration records that pertain to the test material(s) being administered are essential to the researcher in documenting the specifics associated with the infusion and any significant infusion interruptions. Test article/vehicle dose administration records should document information that includes

- Animal number and gender
- Group (associated test material/vehicle)
- Body weight
- Target dose volume
- Infusion rate
- Infusion duration (target and actual)
- Infusion start time
- Infusion end time
- Infusion interruptions (and associated interruption durations)
- Dose site (if applicable, peripheral percutaneous infusions)
- Dose accountability (requires reservoir start/end weight; calculated target/actual dose volume)

Record-keeping and appropriate data documentation for infusion studies is an arduous, usually manual, but necessary process. The researcher must interpret study data based on the information documented in these records. More importantly, any spurious or confounding results may be further explained by the details provided by these records (e.g., toxicokinetic outliers associated with documented infusion interruptions). As infusion equipment evolves and technology improves, more data are being collected using electronic data capture systems. With the advent of remote infusion pump control, more aspects of the dose administration processes will become automated and, as a result, require less manual recordkeeping.

3.3.2 *Non-surgical models*

3.3.2.1 *Methods of restraint and acclimatization*

As previously described, the jugular vein and peripheral limb veins provide options for continuous intravenous dose administration. Percutaneous placement of limb catheters allows short-term access to peripheral vessels through the skin. Typically, dogs can be restrained in a canine sling apparatus for up to 4–6 hours per dose. Slings provide mechanical restraint with minimal stress to the animal while allowing easy and unhindered access to all four limbs. Since the restraint slings are mounted on casters, the animal(s) can easily be moved around within and between study rooms with the animal completely immobilised. Short intermittent peripheral infusions will require appropriate acclimation to the restraint sling, but dogs tend to tolerate short restraint periods well.

3.3.2.2 *Methods of vascular access*

Percutaneous vascular access usually involves a standard over-the-needle intravenous catheter (e.g., BD Angiocath™). The catheter is percutaneously placed into a saphenous or cephalic vein and then secured to the limb with tape, thus allowing vascular access to facilitate dose administration. Another option allows for percutaneous placement of a central, jugular cannula using a break-away catheter introducer while the animal is under light sedation. The cannula is secured in place to the skin using surface sutures, tape, and overwrap for protection. A variation of this procedure involves making a slight surface incision at the implantation site to expose the vein. Following placement of the catheter using the catheter introducer, the introducer is removed and the skin is apposed around the proximal, reinforcing strip of the catheter, providing additional catheter protection and stabilization (Gleason and Chengelis 2000). For both techniques, the area over the catheter site should be cleansed and surgically scrubbed prior to catheter placement.

A third peripheral non-surgical technique actually incorporates a percutaneous catheter and a jacketed infusion model, thus providing an intermediate model between restrained non-surgical and fully ambulatory surgical models. The specific study design requires an infusion model that more appropriately mimics a 24-hour infusion in humans. Of particular importance is the catheter-to-vessel-size ratio. A percutaneous peripheral limb placement in the dog better approximates a 1:1–2 catheter-to-vessel-size ratio (and more appropriately mimics the catheter:vessel-size-ratio in humans). To facilitate continuous infusion dose administration, the peripheral catheter is attached to an extension set routed underneath a canine jacket and then attached to a tether infusion system (note: this model can easily be adapted to the ambulatory jacketed infusion system). Each animal should be catheterised on the day of or the day prior to dose administration. To prevent the animals from accessing or removing the peripherally placed percutaneous catheter, the catheterization site should be bandaged and an Elizabethan collar or padded collar should be placed on the animals for the duration of the treatment (the catheterised period). During the pretreatment period, the venous catheter should be infused continuously at a rate of approximately 2–4 mL/hr with 0.9% sodium chloride for injection, USP to maintain catheter patency.

3.3.2.3 Housing

Typical percutaneous peripheral infusions use short-term indwelling catheters that are removed following dose administration. As a result, standard canine housing is appropriate. Additionally, since there is no external equipment associated with the model following dose administration, the dogs can easily be pair-housed and socialised when appropriate. However, if the peripheral technique incorporating a percutaneous catheter and a jacketed infusion model is employed, housing limitations are similar to those described in Section 3.1.3.2 above.

3.3.2.4 Checks and troubleshooting

During the infusion period, the entire infusion system should be checked for equipment damage, leaking, disconnections, etc. The injection site should be continuously monitored for signs of vehicle or test material extravasation. Following removal of the peripheral catheter, the catheterization site should be inspected daily for signs of infection, such as erythema, oedema, and exudation.

3.4 Equipment

3.4.1 Surgical models

3.4.1.1 Surgical facilities

Regulations require that survival surgery performed in dogs occur in a dedicated surgical facility. Animals should always arrive for surgery via a separate preparation room. Following surgery, animals should leave via a separate recovery room; however, in many facilities, the preparation room also serves as the recovery room. Before each surgery or session of surgeries, the surgical suite should be thoroughly cleaned. Between each animal, the surgical tables, instrument trays, and all work surfaces should be cleaned and disinfected. Separate, sterile surgical packs and instruments will be used for each animal. Each set of instruments should be cleaned, disinfected, and then autoclaved prior to subsequent use. Surgeons and surgical support staff should wear single-use, disposable, or autoclavable gowns, hats, masks, and sterile disposable gloves between each animal.

3.4.1.2 Catheters/vascular access ports

The use of VAPs has significantly improved the conduct and success of infusion models and more broadly, infusion programmes. The VAP has in many ways revolutionised modern infusion techniques, resulting in improved catheter patency, reduced the incidence of catheter related infection, and all but eliminated the potential for catheter dislodgement. As a result, the use of VAPs has furthered improved research capabilities, refined infusion techniques, and improved animal welfare.

The VAP system is composed of three basic parts: port, septum, and catheter. The port (or port body) allows percutaneous access into a small hollow port reservoir through a compressed silicone septum. To maintain the functional integrity of the VAP, a non-coring Huber-point needle must be used to puncture the silicone septum. The non-coring needle allows for repeated VAP access without damaging the silicone rubber. The catheter is implanted into a vessel and connects to the port. The connected port and catheter system make up the typical VAP. The port body is generally designed as either a top-accessed 'dome' or a side-accessed lateral port that lies parallel to the skin and muscle. The dome port is accessed perpendicular to the body wall, while the lateral port is accessed parallel to the body wall. Although these are the two basic configurations of a VAP, there are numerous iterations of each that differ in size, shape, and port materials ranging from polysulfone plastic to titanium (Figure 3.4).

Similarly, the catheter design and catheter material also come in a variety of options. When selecting an appropriate vascular catheter, one must consider the catheter size (both inner and outer diameters), catheter material and pliability, and intravascular tip shape and configuration. Catheter materials available include silicone, Teflon/PTFE, polyethylene, polyvinyl chloride (PVC/Tygon), and polyurethane. Most catheters used for VAP systems are either silicone or polyurethane. Each of these materials has obvious benefits and disadvantages. It is often either personal preference of the researcher or suitability/compatibility issues with a specific test formulation that will dictate the catheter material.

Maintenance of VAPs has almost exclusively involved 'locking' the port and attached catheter with heparin-containing solutions designed

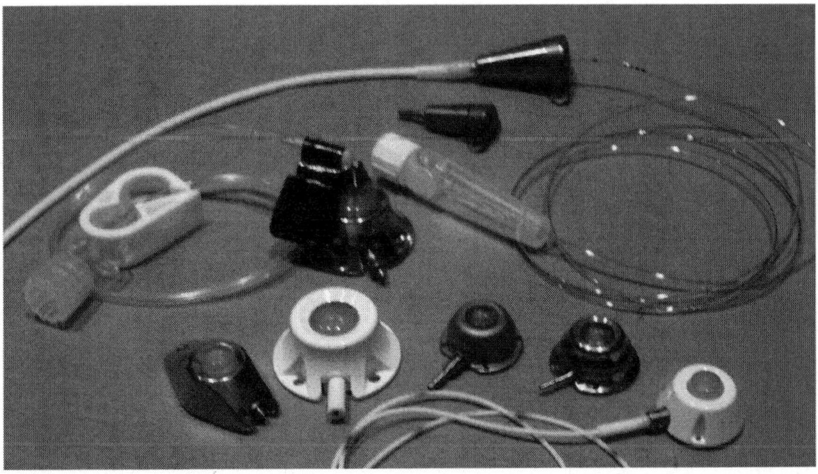

Figure 3.4 Typical vascular access ports and catheters.

to aid in reducing thrombotic-related occlusions. A non-heparin-containing lock solution, Taurolidine Citrate Catheter Solution™ (TCS), can also be used to maintain VAP and catheter patency in laboratory-housed canines. Extensive evaluation at the author's laboratory has concluded that repeated systemic administration of TCS, often a consequence of port and catheter maintenance, had no effect on clinical pathology parameters. More importantly, the desired anticoagulant properties of TCS did not cause a systemic prolongation of prothrombin time or activated partial thromboplastin time (Resendez, unpublished data). These results support the use of TCS as an alternative to heparin-containing lock solutions.

3.4.1.3 External equipment

Recent advancements to the side-accessed vascular access port have provided a refined approach to long-term infusion studies in dogs both for the preferred ambulatory model and also for the tethered model. Also, modifications to caging to support dog tethered infusion systems have resulted in a refined, standardised infusion programme in this species. Refinements to the dog infusion model have made it possible for technicians to access most of the infusion system without direct interaction with the animal. This greatly reduces the time necessary for most repairs and inspections of the infusion system. The lateral vascular access port (L-VAP) has been paired with an externalised Cath-in-Cath™ (CIC) nylon infusion system (Figure 3.5) that can accommodate various types of infusions in the dog. The CIC infusion system has been adopted for its ease of use and to aid in the potential reduction of infusion-system-related infections. When the L-VAP and CIC system are utilised together, the exteriorization

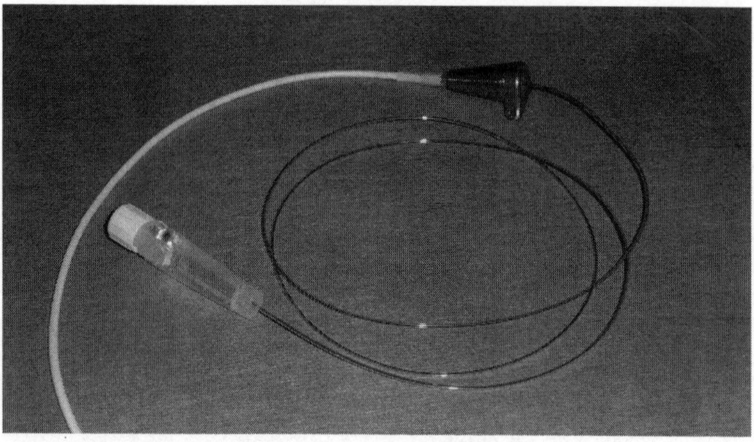

Figure 3.5 Assembled L-VAP and CIC infusion system.

site is reduced in size considerably to that of an 18-gauge needle puncture. When the CIC is not required, it can be removed from the L-VAP system, the port can be locked, and it can remain a closed VAP system requiring minimal maintenance.

Prior to the initiation of an infusion, the CIC system is inserted into the L-VAP. This can be accomplished with either a sedated or appropriately restrained dog. The area over the port is aseptically prepared, and a sterile drape is placed over the animal, leaving only the port area exposed. Sterile gloves and equipment are used to access the port. A needle and inducer are used to advance the catheter through the dermal layer directly cranial to the L-VAP. Using depth markings on the CIC, the nylon catheter is advanced approximately 15–20 cm through the L-VAP into the surgically implanted 6-French polyurethane femoral catheter.

The introducer is then removed over the CIC, and the CIC is left positioned inside the implanted femoral catheter, held securely in place by the L-VAP silicone septum. Once the CIC is in place and exteriorised, the exteriorization site will be inspected daily for signs of infection. Topical antiseptics and antibiotics may be used to reduce the potential of infection.

The CIC system is perfectly designed for the ambulatory infusion model. In this model, the exteriorised end of the CIC nylon catheter is connected directly to a programmable battery-operated ambulatory infusion pump positioned in a 'back-pack' canine infusion jacket. This model is ideal for intermittent infusions with extended periods of recovery between treatments, thus allowing the externalised CIC to be removed.

Under the legislation of the United States, where tethered dogs are permissible for this mode of delivery, the CIC system can be readily adapted for this model also. In this case, additional external equipment such as tether, swivel, and specialised extension lines are required.

In both cases the animal is required to wear a canine infusion jacket, and the proper sizing of this jacket to the animal is paramount to ensure proper function of the infusion system and comfort for the animal. In the event the externalised catheter becomes compromised, minor repairs may be made by technical staff without the need for the animal to return to surgery. In most cases, the damaged catheter can simply be removed and a new CIC can be quickly replaced, minimizing the amount of down time during a continuous infusion.

3.4.1.4 Infusion pumps

There is a wide selection of infusion pumps available for infusion studies in canines and, depending on a variety of factors, of which rate and total infusion volume are often the most limiting, there will be an appropriate infusion device to accommodate almost any study design (Figure 3.6). Syringe pumps offer an exceptional level of precision and a wide range of flow rates and dosing libraries. They are, however, limited by total infusion volume and the size of the syringe (usually a maximum volume of either 60 or 120 mL). Larger total infusion volumes would require more frequent syringe changes. Ambulatory pumps offer a larger reservoir capacity limited only by the size of IV bag or reservoir connected to the pump. Battery-operated ambulatory pumps have become compact in size, allowing for true ambulatory administration. Most pumps used in the pre-clinical setting are designed for laboratory use or human clinical dosing and have been used in some variation in clinical medicine

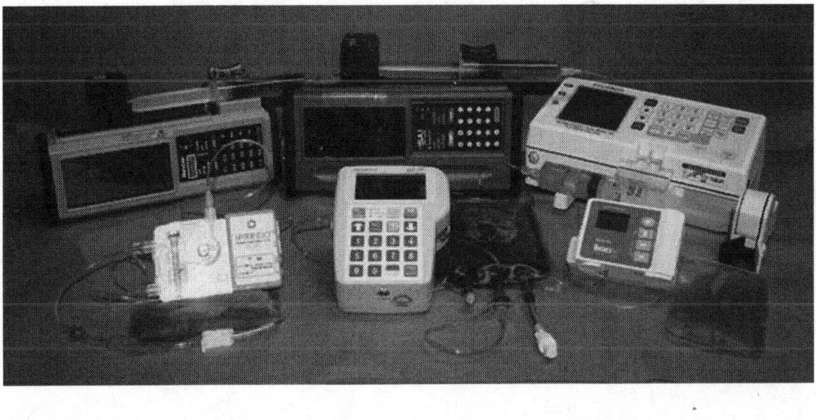

Figure 3.6 Various types of infusion pumps allow the researcher to accommodate different study designs. Newly designed infusion pumps are capable of wireless and/or Ethernet connectivity, allowing for remote pump monitoring and/or control.

for many years. However, these infusion pumps are not designed for the rigorous demands of large-scale GLP pre-clinical animal studies. It is only recently that manufacturers have targeted the pre-clinical market with infusion pumps developed specifically for the needs of the pre-clinical end user.

Recent advancements in infusion pump technology have led to improved functionality and state-of-the-art communication and connectivity capabilities. Infusion pumps are now being developed with unique features that allow for complete remote monitoring and control while still maintaining GLP Part 11 compliance. Through the use of wireless and/or Ethernet connectivity, newly designed smart-pumps are providing researchers with a level of system integration that, to date, has not been possible. Software platforms are being developed that allow multiple infusion design profiles to be used concurrently, across multiple infusion pumps, with minimal manual 'at the pump' interaction. This automation process will reduce human error while still maintaining the user-control required to react as necessary to study issues. By incorporating this technology into infusion projects and programmes, new efficiencies will be realised, new and innovative infusion solutions will emerge, and substantial time and effort will be saved.

At our laboratory, all infusion pumps are bench validated quarterly to confirm that the unit is functioning normally and to confirm accuracy of dose delivery. All pumps are tested at two infusion rates (low-accuracy rate and high-accuracy rate determined by manufacturer specifications). The pump accuracy must be ±10% of the target volume. If the pump fails the accuracy test, it is returned to the manufacturer for repair. All infusion pumps are uniquely identified. All validation and accuracy data are maintained in the study files and are collected either by written record or using an electronic data-capture system.

For intermittent infusion studies, dose accuracy is determined daily during the study. Dose accuracy is calculated as a percentage of the target dose volume based on pre- and post-dose reservoir weights. Most dose accuracy calculations assume a specific gravity of 1 (unless identified differently by protocol). For continuous infusion dosing, dose accuracy is determined as a function of time and dose delivery. The target volume is calculated based on the actual time of delivery. This value is then factored into the actual dose delivery based on the pre- and post-dose reservoir weight as described above.

3.4.2 *Non-surgical models*

3.4.2.1 *Restraining devices*

The most commonly used restraint device used for canines while undergoing infusion procedures is the sling. The mobile restraint unit (the sling) provides the most proficient and comfortable way to immobilise a dog for short periods. It also provides an efficient and safe working platform for the handler.

3.4.2.2 *Vascular access*

All appropriate details are described in Section 3.3.2.2.

3.4.2.3 *Infusion pumps*

All appropriate details are described in Section 3.4.1.4.

3.5 *Background data*

3.5.1 *Surgical models*

3.5.1.1 *Complications associated with the surgical process*

Implantation of a catheter into the jugular or femoral vein in the dog has become a relatively standard surgical procedure. However, as with any surgery, there is always the possibility of complication during the procedure. Vascular access ports and associated catheters are foreign bodies. Although rare, there is always the possibility that the implant will be rejected. More common but still rare are the occasional localised reactions to suture. Animal surgery involving implanted materials requires strict attention to aseptic technique and should be performed with a minimalist approach to the procedure. The surgical technique should include gentle tissue handling, effective haemostasis, strict asepsis, accurate tissue apposition, and expeditious performance of the surgical procedure (Mendenhall 1999). Complications associated with improper tissue handling can be significant, often requiring correction and use of the contralateral vessel. A wrong-size catheter or improper vessel dissection can easily lead to a torn vein. Similarly, improper isolation of the vessel could result in trauma to the adjacent nerve and subsequent limb impairment. It is essential that surgeons understand the appropriate anatomy and aim to leave the animal as normal as possible following the procedure.

3.5.1.2 *Issues associated with study design*

Appropriately designing an infusion study requires consideration of many technical, scientific, and logistical issues that must be understood in order to successfully initiate and complete the project.

The surgical component of most infusion studies is likely the first challenge. Enough animals must be surgically prepared to ensure that the appropriate number of animals per group are available at study initiation. Potential surgical/post-surgical issues must be considered along with VAP patency. Additionally, the post-surgery recovery period must be planned for. Adequate recovery should be allowed for incisions to heal and to ensure that post-surgical medications have been eliminated (when considering potential drug interaction with the test compound).

Target dose volume and infusion rates may limit the equipment available for study use (e.g., the planned rate is outside the functional range of the pumps). Similarly, the duration and frequency of exposure, the length of the dosing period, chemical characteristics of the dose formulation, and toxicity of the test material will all dictate limitations for infusion volumes. The required total dose volume required for a particular study often determines the infusion model that can be used: either the fully ambulatory or tethered model, or the percutaneous peripheral vein model.

There are also the issues of formulation components and consequent design of control groups for infusion studies, much of which is outside the scope of this chapter.

Infusion studies should also include additional pathological assessments over and above routine regulatory study protocols. As a minimum, the catheterised vessel near the catheter tip should be collected at necropsy. Some labs also collect the entire catheter tract, including the vascular insertion point, the subcutaneous catheter tract, the vascular access port, and the exteriorization site. For percutaneous administration using temporary catheters, the collection only of the infusion site and the surrounding tissues may be necessary. Some laboratories prefer to incise the catheterised vessel lengthwise to identify thrombi and document other macroscopic changes, while others may refrain from incising the vessel at necropsy, keeping the vessel and perivascular tissues intact in order to decrease tissue artefacts that result from handling. The most important tissue for interpretation of infusion-related pathology is the infusion site consisting of the vessel and perivascular tissues near the catheter tip. The location of the tip of the catheter should be identified and marked, using dye if necessary, at necropsy and at tissue trimming. If the catheter tip is palpable in life, the location of the catheter tip should be marked on the skin to help facilitate identification at necropsy, especially if the catheter is removed prior to necropsy. Some labs sample the infusion site longitudinally, in cross section, or use a combination thereof. If cross sections are collected, sections approximately 1 cm proximal to the tip of the catheter (closer to the heart), at the tip of the catheter, and approximately 1 cm distal to the tip of the catheter (farther from the heart) are recommended. For venous catheters, cross sections at the tip of the catheter and proximal to the tip of the catheter will typically capture changes associated with the infusate, while cross sections at the tip of the catheter and distal to the tip of the catheter (along the path of the catheter) will typically demonstrate vascular changes associated with the catheter. A comparison of the vessel in these areas will help distinguish changes associated with catheterization from changes associated with the test material. Consistent and complete collection of the infusion site at necropsy should include a method for identifying tissue orientation, especially if the catheter is removed and cannot be used for orientation at trimming.

3.5.1.3 In-life changes associated with the model

Following surgical implantation of a dosing catheter, there is no appreciable change in body weight compared to orally treated animals (Pickersgill and Burnett 2000; Walker 2000). Slight decreases in weight gain during the pre-treatment period following surgery are expected and are consequences of anaesthesia and the surgical procedure. Clinical findings related to the infusion procedure are primarily the result of often extended periods of wearing the jacket. Dermal ulcerations, scabbing, and sores in the cervical, axillary, and dorsal thoracic regions are primarily caused by an improperly fitted jacket or simple jacket movement over time. Jacket fit should be monitored daily as part of the infusion system checks. Jackets should be changed at least every two weeks or as needed if excessive wear or soiling is apparent.

3.5.1.4 Pathology

3.5.1.4.1 Clinical pathology. Presented below is historical control clinical pathology data from a population of surgically altered infusion canines 5–9 months old (Table 3.2; adapted from Resendez 2012). All samples were obtained via a percutaneous jugular collection. The surgically altered canines were sampled following surgery pretest (prior to dosing) during the post-surgical recovery period. There were no appreciable changes in clinical pathology parameters in the surgically altered animals compared to the historical controls.

3.5.1.4.2 Macroscopic and microscopic pathology. When test materials are administered to dogs via intravenous infusion, distinguishing between test-system-related and test-article-related pathological changes can be challenging. Even catheterised animals administered physiologic saline solution via infusion can exhibit a range of localised and systemic effects. Familiarity with the spectrum of possible background pathological changes in the canine test system is important in order to accurately interpret potential test-article-related findings. The development of pathologic changes in these studies may be impacted by variations in techniques and materials utilised, unexpected complications, and the inherent biological variability in the animal model. Therefore, this section will attempt to describe the pathogenesis of infusion-related macroscopic and microscopic changes that may be observed in catheterised dogs.

Table 3.2a Reference Canine Haematology Parameters

		Males			Females		
Interval		Colony historical control	Pre-surgery	Post-surgery	Colony historical control	Pre-surgery	Post-surgery
Animal age		7–9 Months	5–9 Months	5–9 Months	7–9 Months	5–9 Months	5–9 Months
Leukocyte count ($x10^3/\mu L$)	Mean	10.84	13.22	11.39	10.66	12.84	11.57
	SD	*2.553*	*2.934*	*2.191*	*2.778*	*2.744*	*2.702*
	N	*1073*	*65*	*80*	*1019*	*65*	*73*
Erythrocyte count ($x10^6/\mu L$)	Mean	6.894	6.716	6.756	6.992	6.874	6.814
	SD	*0.5501*	*0.7317*	*0.6493*	*0.5784*	*0.7132*	*0.6548*
	N	*1080*	*65*	*81*	*1028*	*65*	*73*
Haemoglobin (g/dL)	Mean	15.18	14.65	14.58	15.44	15.17	14.78
	SD	*1.192*	*1.653*	*1.342*	*1.273*	*1.655*	*1.347*
	N	*1080*	*65*	*81*	*1031*	*65*	*73*
Haematocrit (%)	Mean	44.61	43.47	43.56	45.23	45.01	44.32
	SD	*3.602*	*3.962*	*3.553*	*3.847*	*4.055*	*3.444*
	N	*1081*	*65*	*81*	*1031*	*65*	*73*
Platelet count ($x10^3/\mu L$)	Mean	349.0	391.8	239.5	353.4	384.2	234.8
	SD	*74.89*	*68.83*	*50.69*	*79.08*	*74.14*	*56.80*
	N	*1070*	*65*	*81*	*1026*	*65*	*72*

Absolute Reticulocytes (x10³/µL)	Mean	48.15	—	47.17	42.25	—	36.26
	SD	*25.386*	—	*24.791*	*23.337*	—	*18.886*
	N	*1027*	—	*77*	*991*	—	*73*
Reticulocytes (%)	Mean	0.70	—	—	0.60	—	—
	SD	*0.359*	—	—	*0.327*	—	—
	N	*1027*	—	—	*990*	—	—
Neutrophils (x10³/µL)	Mean	6.961	8.176	7.184	6.809	7.683	7.138
	SD	*2.1444*	*2.4346*	*1.8591*	*2.2935*	*2.0424*	*2.3466*
	N	*1075*	*64*	*80*	*1019*	*65*	*73*
Lymphocytes (x10³/µL)	Mean	2.938	3.800	3.232	2.977	4.038	3.508
	SD	*0.8380*	*0.9659*	*0.7724*	*0.7669*	*1.2374*	*0.9066*
	N	*1082*	*64*	*81*	*1028*	*65*	*73*
Monocytes (x10³/µL)	Mean	0.622	0.811	0.555	0.557	0.720	0.525
	SD	*0.2406*	*0.2748*	*0.2090*	*0.2453*	*0.2690*	*0.1896*
	N	*1082*	*64*	*81*	*1028*	*65*	*73*
Eosinophils (x10³/µL)	Mean	0.220	0.256	0.340	0.209	0.240	0.283
	SD	*0.1285*	*0.1282*	*0.1870*	*0.1317*	*0.1354*	*0.1628*

(*Continued*)

***Table 3.2a** (Continued)* Reference Canine Haematology Parameters

		Males			Females		
Interval		Colony historical control	Pre-surgery	Post-surgery	Colony historical control	Pre-surgery	Post-surgery
Animal age		7–9 Months	5–9 Months	5–9 Months	7–9 Months	5–9 Months	5–9 Months
	N	*1081*	*63*	*81*	*1029*	*65*	*73*
Basophils (x10³/µL)	Mean	0.060	0.067	0.043	0.062	0.074	0.052
	SD	*0.0326*	*0.0290*	*0.0198*	*0.0314*	*0.0311*	*0.0185*
	N	*1083*	*64*	*81*	*1030*	*65*	*73*
Other Cells (x10³/µL)	Mean	0.047	0.080	0.046	0.047	0.087	0.056
	SD	*0.0323*	*0.0350*	*0.0266*	*0.0315*	*0.0472*	*0.0249*
	N	*1084*	*63*	*81*	*1031*	*65*	*73*

Table 3.2b Reference Canine Clinical Chemistry Parameters

		Males			Females		
Interval		Colony historical control	Pre-surgery	Post-surgery	Colony historical control	Pre-surgery	Post-surgery
Animal age		7–9 Months	5–9 Months	5–9 Months	7–9 Months	5–9 Months	5–9 Months
Sodium (mEq/L)	Mean	147.2	145.7	146.5	147.2	146.0	146.8
	SD	1.97	1.57	1.89	1.91	1.69	1.63
	N	1081	65	75	1026	65	75
Potassium (mEq/L)	Mean	4.75	4.95	4.58	4.68	4.85	4.48
	SD	0.335	0.294	0.245	0.332	0.251	0.241
	N	1082	65	75	1026	64	75
Chloride (mEq/L)	Mean	110.5	109.8	110.4	110.7	110.2	110.7
	SD	2.05	1.43	1.50	2.01	1.88	1.71
	N	1080	65	75	1024	65	75
Calcium (mg/dL)	Mean	10.76	—	10.63	10.77	—	10.64
	SD	0.492	—	0.295	0.502	—	0.292
	N	1079	—	75	1023	—	75
Phosphorus (mg/dL)	Mean	6.10	—	6.47	5.88	—	6.30
	SD	0.859	—	0.655	0.867	—	0.781
	N	1076	—	75	1022	—	75

(*Continued*)

Table 3.2b *(Continued)* Reference Canine Clinical Chemistry Parameters

		Males			Females		
Interval		Colony historical control	Pre-surgery	Post-surgery	Colony historical control	Pre-surgery	Post-surgery
Animal Age		7–9 Months	5–9 Months	5–9 Months	7–9 Months	5–9 Months	5–9 Months
Alkaline phosphatase (U/L)	Mean	83.0	—	89.3	83.4	—	84.5
	SD	*24.61*	—	*24.83*	*27.11*	—	*30.16*
	N	*1077*	—	*75*	*1021*	—	*75*
Total bilirubin (mg/dL)	Mean	0.21	—	0.14	0.23	—	0.16
	SD	*0.068*	—	*0.052*	*0.082*	—	*0.063*
	N	*1077*	—	*75*	*1023*	—	*74*
Gamma glutamyl transferase (U/L)	Mean	**3.3**	—	3.6	3.1	—	3.6
	SD	*1.11*	—	*1.32*	*0.89*	—	*1.26*
	N	*971*	—	*75*	*887*	—	*75*
Aspartate aminotransferase (U/L)	Mean	29.1	—	29.8	28.0	—	30.4
	SD	*6.82*	—	*5.07*	*6.01*	—	*5.08*
	N	*1078*	—	*75*	*1020*	—	*75*
Alanine aminotransferase (U/L)	Mean	31.0	30.2	24.0	29.5	29.4	24.7

	SD	8.91	8.01	4.59	7.42	6.91	5.77
	N	1087	65	75	1024	65	75
Sorbitol dehydrogese (U/L)	Mean	7.07	—	5.20	7.49	—	5.34
	SD	2.657	—	1.264	2.803	—	1.324
	N	849	—	18	744	—	18
Creatine kinase (U/L)	Mean	211.9	—	223.0	186.2	—	214.3
	SD	86.57	—	56.01	72.93	—	62.30
	N	1086	—	57	203	—	57
Urea nitrogen (mg/dL)	Mean	13.2	12.4	12.8	13.9	12.7	12.3
	SD	2.79	2.44	2.61	2.72	2.11	2.04
	N	1080	65	75	1027	64	75
Creatinine (mg/dL)	Mean	0.61	0.55	0.56	0.60	0.55	0.55
	SD	0.100	0.089	0.110	0.094	0.089	0.088
	N	1083	65	75	1027	65	75
Total protein (g/dL)	Mean	6.03	5.72	5.57	5.88	5.63	5.49
	SD	0.399	0.460	0.312	0.378	0.464	0.287
	N	1082	65	75	1027	65	75
Albumin (g/dL)	Mean	3.11	—	3.05	3.18	—	3.06
	SD	0.251	—	0.202	0.253	—	0.204
	N	1079	—	75	1023	—	75

(Continued)

Table 3.2b *(Continued)* Reference Canine Clinical Chemistry Parameters

		Males			Females		
Interval		Colony historical control	Pre-surgery	Post-surgery	Colony historical control	Pre-surgery	Post-surgery
Animal age		7–9 Months	5–9 Months	5–9 Months	7–9 Months	5–9 Months	5–9 Months
Globulin (g/dL)	Mean	2.92	—	2.52	2.70	—	2.43
	SD	0.362	—	0.199	0.350	—	0.195
	N	1079	—	75	1020	—	75
Albumin/globulin ratio	Mean	1.08	—	1.21	1.20	—	1.27
	SD	0.174	—	0.113	0.207	—	0.128
	N	1067	—	75	1001	—	75
Triglycerides (mg/dL)	Mean	31.0	—	37.7	31.7	—	37.4
	SD	7.64	—	10.37	7.31	—	9.33
	N	238	—	59	258	—	59
Cholesterol (mg/dL)	Mean	167.8	—	155.2	157.9	—	151.5
	SD	29.69	—	23.04	29.20	—	26.86
	N	1076	—	75	1019	—	75
Glucose (mg/dL)	Mean	92.2	93.1	96.0	91.3	94.1	93.6
	SD	9.10	12.44	9.91	9.19	9.05	8.90
	N	1079	65	75	1027	65	75

Table 3.2c Reference Canine Coagulation Parameters

Interval		Males			Females		
		Colony historical control	Pre-surgery	Post-surgery	Colony historical control	Pre-surgery	Post-surgery
Animal age		7–9 Months	5–9 Months	5–9 Months	7–9 Months	5–9 Months	5–9 Months
APTT (sec)	Mean	9.61	—	10.89	9.81	—	11.11
	SD	0.899	—	0.853	0.978	—	0.795
	N	352	—	52	376	—	52
Prothrombin time (sec)	Mean	7.05	—	7.27	7.16	—	7.27
	SD	0.383	—	0.437	0.678	—	0.526
	N	349	—	52	376	—	52
Fibrinogen (mg/dL)	Mean	275.9	—	181.1	248.8	—	167.2
	SD	43.69	—	32.95	50.16	—	20.93
	N	78	—	16	104	—	16

Table 3.2d Reference Canine Urinalysis Parameters

		Males			Females		
Interval		Colony historical control	Pre-surgery	Post-surgery	Colony historical control	Pre-surgery	Post-surgery
Animal age		7–9 Months	5–9 Months	5–9 Months	7–9 Months	5–9 Months	5–9 Months
Volume (mL)	Mean	155.66	—	122.6	130.78	—	136.4
	SD	153.940	—	98.40	120.403	—	235.03
	N	853	—	52	803	—	34
Specific gravity	Mean	1.0199	—	1.0375	1.0212	—	1.0390
	SD	0.01060	—	0.01648	0.01054	—	0.01873
	N	844	—	52	778	—	34
pH	Mean	7.12	—	7.03	7.02	—	7.16
	SD	0.690	—	0.614	0.656	—	0.636
	N	853	—	52	796	—	34

The specific procedures may vary from laboratory to laboratory based on study requirements and previous laboratory experience. These are described in Section 5.1.2.

3.5.1.4.2.1 External lesions. During the in-life portion of the study and during the necropsy, external lesions can occur in the skin associated with the jacket and at the externalization site. The jacket can be abrasive to the skin, especially if it is not fitted properly, resulting in hair loss, epithelial hyperplasia, abrasions, surface exudation (scabbing), and/or ulceration. The externalization site can become reddened, swollen, and/or scabbed with inflammation and/or infection. Redness and swelling may also be present macroscopically around the VAP or along the catheter tract. Acute microscopic lesions may include haemorrhage, oedema, and acute inflammation with predominantly neutrophilic infiltrates. With longer periods of catheterization, thick capsules of fibrous tissue and chronic inflammation may form around the VAP or catheter.

The externalization site is a portal of entry through the skin, bypassing the normal defence mechanisms against contamination and infection. Even with impeccable sterile catheter handling techniques, bacteria may gain entry through this site. Medical devices such as the catheter, suture material, and VAP can be a favourable environment for the growth and spread of bacteria if contamination occurs during implantation or during a study. Bacteria may form a biofilm on the surfaces of these materials, which can provide protection against the host's natural defences and antibiotics. Morphologic changes at the externalization site or around the VAP and catheter may be severely exacerbated when bacteria are present, including the formation of abscesses.

Catheterization-related inflammation may cause changes in the draining lymph nodes, which may become enlarged from reactive hyperplasia, sinus histiocytosis, inflammatory cell infiltrates, and/or sinus dilatation from fluid drainage. If inflammation is severe or the catheterised vessel becomes occluded, poor venous or lymphatic drainage can lead to dependent oedema of the extremity. With jugular catheterization, oedema of the face and head is possible.

3.5.1.4.2.2 Internal lesions—Infusion site findings. Blood vessels, including veins, are generally composed of three layers: an inner layer adjacent to the lumen called the *tunica intima,* a middle layer called the *tunica media,* and an outer layer called the *tunica adventitia.* The tunica intima is composed of a flattened, smooth layer of cells, called endothelial cells, that line the lumen. The endothelial cells are supported by subendothelial stroma composed of low numbers of fibrocytes and smooth muscles, elastic fibers, and a small amount of collagen. The tunica media is the thickest layer, primarily composed of smooth muscle and elastic fibers with fewer fibrocytes and collagen. The tunica adventitia blends somewhat with the surrounding tissue and is composed of fibrocytes, collagen, elastic fibers, and nerve fibers. The tunica adventitia of larger vessels also includes the *vasa vasorum*, which is composed of small vessels and capillaries that provide a blood supply to the tunica adventitia and tunica media.

An intact endothelial layer is critical for normal blood flow through a vein or artery. Biochemical interactions between the endothelium and the blood maintain an environment that inhibits the formation of clots (*thrombi*). Disruption of the endothelium and exposure of the underlying subendothelial stroma triggers a localised release of chemical mediators and cellular responses that lead to thrombus formation and haemostasis.

Three triggers of thrombosis that can act concurrently, known as *Virchow's triad*, include endothelial injury, changes in blood flow (haemodynamics), and hypercoagulability. All three of these can be present in a catheterised vessel, leading to the disruption of vascular integrity. At the insertion point, the catheter must breach the vessel wall to gain access to the intravascular space. The presence of the catheter maintains the exposure of subendothelial stroma, which sustains the localised chemical environment conducive to thrombogenesis. The presence of the catheter within the vessel can lead to changes in blood flow, stasis, and turbulence. Furthermore, the catheter wall and tip can cause mechanical damage to the endothelium and vessel wall. Starting at the insertion point, thrombi can form along the catheter, forming a fibrin sheath around the catheter. Thrombi may also form at any point along the intravascular portion of the catheter, including the tip of the catheter. The edge of the catheter tip may abrade the endothelium, but there can also be increased turbulence in the blood flow around the catheter tip. Changes in haemodynamics due to the infusion of fluids can contribute to turbulence at the catheter tip. Hypercoagulability may also contribute to thrombus formation when bacteria are present.

Acute thrombi are composed primarily of layers of fibrin, platelets, erythrocytes, and leukocytes. Thrombi can propagate as layers are added, extending in length within the vessel lumen. With chronicity, fibrin thrombi break down (fibrinolysis), or they can become organised, undergoing fibrosis and contraction. The fate of thrombi depends on the severity of the thrombus, the degree of vascular damage, and whether the inciting causes persist. A thrombus may resolve completely, restoring normal vessel structure and function. Remodelling may take place, incorporating the thrombus into the vessel wall. Recanalisation of the thrombus may lead to the formation of a new lumen or multiple smaller lumina. With complete blockage of the affected vessel, neovascularisation may lead to new anastomotic vessels that bypass the area of thrombosis. If the catheterised vessel is incised at necropsy, larger thrombi may be visualised macroscopically and documented.

Catheter-induced acute damage to the vessel wall can cause a loss of endothelium, degeneration and necrosis of the endothelium, and, depending on the severity of the insult, degeneration and necrosis of other layers of the vessel. Microscopically, with acute injury there may be fibrin deposition, haemorrhage, and inflammation within the vessel wall or perivascular tissues. Perivascular necrosis can also occur. Acute injury from catheterisation can be difficult to distinguish from test-article-related vascular toxicity.

Over time, the layers of the vessels have a remarkable ability to adapt to injury and/or regenerate. The endothelium can undergo hypertrophy and hyperplasia, restoring the endothelial cell layer where it was lost. The persistent presence of the catheter frequently induces intimal proliferation, a thickening of the endothelium and subendothelial stroma often associated with the formation of frond-like projections of endothelial-lined stroma into the vessel lumen. The smooth muscle of the tunica media can undergo regenerative hypertrophy and hyperplasia, giving this layer a basophilic appearance. This is generally accompanied by some degree of fibroplasia. Fibroplasia is also a common change in the tunica adventitia and perivascular tissues. Subacute to chronic inflammatory cell infiltrates, haemosiderin-laden macrophages, and mineralization may be present in any of the layers of the vessel wall. Occasionally, foreign body granulomas are present at the infusion site surrounding hair fragments, keratin, suture material, or other foreign material. Remodelling and incorporation of a chronic thrombus into a vessel wall can obscure the distinctions between the thrombus, the tunica intima, and the tunica media. With severe damage from catheterization, inflammation and fibrosis may completely obscure or replace the entire vessel wall. Chronic vascular changes can also be exacerbated by test materials.

With prolonged catheterization, acute and chronic changes may be present concurrently. During the necropsy, acute and/or chronic vascular changes may appear as redness and swelling at the insertion point into the vein and in tissues around the tip of the catheter. Consistent documentation of acute and chronic microscopic and macroscopic vascular changes at the infusion site is necessary in order to distinguish between catheter-induced/infusion-related changes and test-article-related vascular damage. Owing to the temporary nature of percutaneous catheters, microscopic changes in the vessel wall at the infusion site are generally less severe and more conducive to regeneration, and may lack chronic changes observed with long-term catheterization because of a lack of a persistent insult to the vessel wall.

Bacterial contamination of the infusion site can also exacerbate acute and chronic vascular changes at the infusion site of catheterised animals, often resulting in severe morphologic changes. Bacteria may gain entry during surgical placement of the catheter or from the spread infection along the catheter line. Contamination may also occur through the lumen of the catheter during infusion or from flushing the catheter during maintenance procedures. With bacterial colonization, an increased severity of thrombosis, acute inflammation, and degeneration/necrosis are the most common findings at the infusion site; however, with prolonged infection, chronic-active changes and abscess formation can also occur.

3.5.1.4.2.3 Internal lesions—Systemic findings. Catheterised dogs on infusion studies can develop systemic lesions that are often sequelae to the morphologic changes described previously. Systemic lesions are generally associated with thromboembolism, bacterial infection, inflammation, and/or stress. On rare occasion, foreign body granulomas may develop in tissues such as the lung, surrounding hair or keratin fragments, which are considered to be present as a result of embolisation during surgery, during catheter placement, or perhaps during other injection procedures.

Fibrin thrombi are often friable, and portions may break off in the bloodstream, resulting in embolism. Embolisation may occur spontaneously, but vigorous flushing of the catheter may also dislodge fragments of thrombi. Emboli generally pass freely through the right atrium and right ventricle, lodging in the first capillary bed that they encounter in the lungs. The lungs may have red foci in multiple lobes from haemorrhage and necrosis within the alveoli and pulmonary parenchyma. Occasionally, emboli may pass through the lung and occlude a vessel in other organs. Red foci of haemorrhage and foci of necrosis (acute infarcts) may be present in the heart, kidneys, or other tissues. Chronic infarcts characterised by areas of tissue loss, chronic inflammation, and fibrosis may appear as shrunken tan scars at necropsy.

With bacterial infection of the infusion sites, embolisation of bacteria or bacteria-laden thrombi can occur. The heart valves are particularly susceptible to infection in bactaeremic animals and may develop vegetative endocarditis. Affected heart valves become thickened by inflammation, oedema, and proliferation of the endothelium or subendocardial stroma. Fibrin and thrombi generally form on the surface of affected valves, which may fragment, further predisposing the animal to systemic thromboembolism. Foci of inflammation, haemorrhage, necrosis, or abscessation may occur in any tissue associated with bacteraemia; however, these lesions may be more common when heart valves are also affected. Highly vascularised tissues such as the adrenal gland and pituitary gland, or other capillary beds such as the meninges of the brain or the synovium of joints, can be particularly sensitive to systemic infections. In addition to infarction and other focal inflammatory lesions, the kidneys can develop glomerulitis from the deposition of bacteria, fibrin, or other inflammatory mediators within the capillaries of the glomerular tufts. The lungs may develop multifocal to diffuse interstitial inflammation. Body cavities such as the abdominal cavity, thoracic cavity, and pericardium can also become infected and develop peritonitis, pleuritis, and pericarditis, respectively.

In addition to changes described above, septicaemia may cause a systemic drop in blood pressure as a result of widespread vasodilation. Leukocytosis and disseminated intravascular coagulation can also occur, resulting in haematology changes and lengthening of clotting times. Organs that filter pathogens from the blood may exhibit microscopic changes, including Kupffer cell hypertrophy and hyperplasia in the liver, generalised neutrophilic infiltration, leukocytosis, or reticuloendothelial hyperplasia in the red pulp of the spleen, and lymphoid hyperplasia in the white pulp of the spleen.

Bacterial infections of the catheter and infusion site, along with secondary systemic changes, can be more common with immunomodulatory compounds and with compounds that induce immunosuppression such as chemotherapeutic drugs and steroids. It is important to be familiar with the range of findings that can be associated with bacterial infections in catheterised dogs because dose-related increases in bacterial infections and associated morphologic changes may be present. Distinguishing dose-related secondary findings from direct test-article-related findings can be difficult.

Erythrocytes can be damaged by passing through areas with thrombi or through areas of turbulence. Erythrophagocytosis and haemosiderin may be present in tissues such as the liver and spleen as damaged erythrocytes are removed from the circulating blood. Localised inflammation, systemic inflammation, and/or decreases in red cell numbers can trigger bone marrow hyperplasia and extramedullary proliferation in tissues such as the liver and spleen.

Interpretation of macroscopic splenic enlargement can be difficult in the dog. The spleen may become enlarged through septic shock, leukocytosis secondary to inflammation, extramedullary haematopoiesis, and/or erythrophagocytosis. However, in dogs, barbiturate euthanasia solutions typically cause splenic congestion and enlargement. The degree of splenic enlargement due to barbiturate administration can vary from animal to animal; therefore, histopathology is necessary to accurately interpret splenic enlargement in the dog.

With stress, adrenal enlargement due to cortical vacuolation and/or hypertrophy may be present. The thymus is particularly susceptible to stress and may decrease in size as a result of lymphocyte apoptosis, markedly decreasing the numbers of lymphocytes within the thymic cortex and medulla. Lymphoid depletion due to stress can be difficult to distinguish from physiologic thymic involution in dogs. Infusion studies using dogs frequently have a low number of animals in each group. By chance, spontaneous changes such as physiologic thymic involution can be more common in animals treated with test article, artifactually creating a dose-related distribution of findings. On these occasions, historical control data should be utilised if available. Interpretation of stress-related pathologic changes can be difficult and should utilise all of the available information from haematology results and other examined tissues such as the infusion site, the adrenal glands, and any other lesions in the animal.

3.6 Conclusion

Successfully conducting an infusion study in canines requires specific attention to study design and operational planning while understanding the limitations of not only the infusion model(s), but also the limitations and experiences of the laboratory. Various factors will influence the selection of an appropriate infusion model or system, and having the flexibility to adapt a model to the unique challenges that often arise because of the complexity of these studies can greatly maximise a successful outcome. Infusion studies are quite involved, and similarly require the involvement of a collaborative group of talented individuals. From study design and model selection to daily operational planning and accurate data collection, all aspects of a project require acute attention to detail. Presented in this chapter are general guidelines and best practices to follow. Various infusion models, both fully ambulatory and tethered, are discussed, and the advantages and disadvantages of each are presented.

With the advent of remote infusion pump monitoring and control, more aspects of routine study processes will become automated. Improved wireless technologies are allowing multiple (and different) pumps to communicate with a single automation platform. This will ultimately allow the standardization of operational process across projects and, more importantly, between functional groups. Infusion pumps are becoming more compact with added features and technology, all of which will soon revolutionise how infusion studies are conducted. With continued emphasis on the basic principles of infusion techniques and technology, strict adherence to the basic tenets of surgery and asepsis, and proper study design, study conduct, and data collection, the outcome of an infusion study will be valuable.

References

Academy of Surgical Research. 1996. Guidelines for training in surgical research with animals. *Journal of Investigative Surgery* 22(3): 218–225.

Blacklock JB, Wright DC, Dedrick RL, Blasberg RG, Lutz RJ, Doppman JL and Oldfield EH. 1986. Drug streaming during intra-arterial chemotherapy. *Journal of Neurosurgery* 64(2): 284–291.

CDER. 1996. Guidance for Industry, Investigators, and Reviewers: Single dose acute toxicity testing for pharmaceuticals. U.S. FDA Center for Drug Evaluation and Research August 26, 1996.

CDER. 2006. Guidance for Industry, Investigators, and Reviewers: Exploratory IND studies. U.S. FDA Center for Drug Evaluation and Research. January 2006.

Evans JG and Kerry PJ. 2000. Common pathological findings in continuous infusion studies. In *Handbook of Pre-Clinical Continuous Intravenous Infusion*. Healing G and Smith D (eds.), pp. 253–264. New York: Taylor & Francis.

Federal Register. 1985. Principles for the Utilization and Care of Vertebrate Animals Used in Testing, Research, and Training. Federal Register, May 20, 1985: 50(97).

Gleason TR and Chengelis CP. 2000. The ambulatory model in dog multidose infusion toxicity studies. In *Handbook of Pre-Clinical Continuous Intravenous Infusion*. Healing G and Smith D (eds.), pp. 148–160. New York: Taylor & Francis.

Haggerty G, Thomassen S and Chengellis C. 1992. The dog. In *Animal Models in Toxicology*. Gad S and Chengelis C (eds.), pp. 567–674. New York: Marcel Dekker.

ICH (International Conference on Harmonisation). 2009. S6(R1): *Preclinical Safety Evaluation of Biotechnology-derived Pharmaceuticals*. October 2009.

ICH. 2010. M3(R2): *Guidance for Industry on Non-Clinical Safety Studies for the Conduct of Human Clinical Trials and Marketing Authorization for Pharmaceuticals*. January 2010.

Lilbert J and Burnett R. 2003. Main vascular changes seen in the saline controls of continuous infusion studies in the cynomolgus monkey over an eight-year period. *Toxicologic Pathology* 31: 273–280.

Lilbert J and Vasanthi M. 2004. Common vascular changes in the jugular vein of saline controls in continuous infusion studies in the beagle dog. *Toxicologic Pathology* 32: 694–700.

Maddison J. 2000. Adverse drug reactions. In *Textbook of Veterinary Internal Medicine*, 5th edition. Ettinger JT and Feldman EC (eds.), pp. 321–347. Philadelphia: W.B. Saunders.

Mesfin GM, Higgins MG, Brown WP and Rosnick D. 1988. Cardiovascular complications of chronic catheterization of the jugular vein in the dog. *Veterinary Pathology* 25: 492–502.

Mitchell RN. 2010. Hemodynamic disorders, thromboembolic disease, and shock. In *Robbins and Cotran Pathologic Basis of Disease*. Kumar V, Abbas AK, Fausto N, and Aster JC (eds.), pp. 111–134. Philadelphia: Saunders Elsevier.

Morton D, Safron JA, Glosson J, Rice DW, Wilson DM, and White RD. 1997. Histologic lesions associated with intravenous infusions of large volumes of isotonic saline solution in rats for 30 days. *Toxicologic Pathology* 25: 390–394.

NIH. 1993. Position Statement on the Use of Animals in Research. February 26, 1993, NIH Guide 22(8).

Pickersgill N and Burnett R. 2000. The non-ambulatory model in dog multidose infusion toxicity studies. In *Handbook of Pre-Clinical Continuous Intravenous Infusion*. Healing G and Smith D (eds.), pp. 135–147. New York: Taylor & Francis.

Plendl J. 2006. Cardiovascular system. In *Dellmann's Textbook of Veterinary Histology*. Eurell JA and Frappier BL (eds.), pp. 117–133. Ames, Iowa: Blackwell Publishing.

Resendez JC and Rehagen D. 2012. Infusion toxicology and techniques. In *A Comprehensive Guide to Toxicology in Preclinical Drug Development*. Faqi A (ed.), pp. 277–307. Waltham, MA: Academic Press/Elsevier.

Walker MD. 2000. Multidose infusion toxicity studies in the large primate. In *Handbook of Pre-Clinical Continuous Intravenous Infusion*. Healing G and Smith D (eds.), pp. 181–209. New York: Taylor & Francis.

chapter four

Primate

Christine Copeman, BSc, Dip Ecotox
Charles River Laboratories, Canada

Stephanie Clubb, BSc
Charles River Laboratories, UK

Contents

4.1 *Introduction*

4.1.1 *Choice and relevance of the species*

Non-human primates (NHPs) are routinely used to conduct preclinical studies and often represent the sole pharmacologically relevant species for a given drug under development. As such, clinically relevant routes of administration are required to allow assessment under comparable dosing regimens in this species. Vascular infusion, a route of administration required clinically for both small and large molecules, has also been used in preclinical models to provide more-constant blood levels of the test material, whereas the clinical route provides variable levels in the test species (Clarke 1993) or may also be used to overcome problems of local irritation or dosing restrictions (e.g. limits of solubility).

4.1.2 *Regulatory guidelines*

There are no guidelines specific to the use of non-human primates or the use of the intravenous infusion route of administration. The non-human primate represents one of the non-rodent species options that may be used during the conduct of preclinical studies. The International Conference on Harmonization, which focuses on biotechnology-derived products in its S6 document, includes guidelines on the selection of species, making reference to the use of relevant species for conduct of preclinical safety assessment of these entities. These define a relevant species for conduct of the preclinical safety assessment as 'one in which the test material is pharmacologically active due to the expression of the receptor or an epitope (in the case of monoclonal antibodies).' In many cases the relevant non-rodent species, or the only relevant species, as defined by these criteria, will be the non-human primate. The intravenous route of administration is not required by regulatory guidelines specific to this mode of administration, but rather is expected to be used in preclinical safety assessment studies when this is the anticipated clinical route or where sufficient multiples of the anticipated systemic exposure could not be achieved by the use of the planned clinical route of administration. The dose volumes, duration of the administration, and frequency would be defined, as it would be for most routes of administration, to mimic the clinical regimen and to be performed at a frequency and level to minimally be comparable or be a multiple of the clinical plan. Consideration for these various factors may necessitate the dose to be administered as an infusion instead of a bolus injection as may be planned in the clinic.

4.1.3 Available infusion models

Administration of the dose formulations by intravenous infusion can be conducted using the cannulated approach, where animals will have a catheter implanted in an appropriate central vein, or by using a non-cannulated approach, where a peripheral vein is temporarily accessed for the given period of infusion.

Techniques and different approaches for vascular infusion in this species are described in this chapter along with considerations for the various approaches described.

4.2 Best practice

Techniques for vascular infusion can require cannulated or non-cannulated administrations depending on parameters surrounding dose administration such as duration and frequency of the infusion as well as the dose volume and formulation properties. For cannulated models, most common techniques in NHPs consist of the use of tethered or ambulatory set-ups. For non-cannulated models, NHPs need to be restrained via the use of a chair or sling restraint device for the period of infusion.

4.2.1 Surgical models

4.2.1.1 Training

Staff undertaking surgical implantation in non-human primates is restricted to those previously trained and proficient with these and similar techniques and procedures in rodents and other non-rodents, such as dogs, in order to ensure that the surgeon already has experience with the various techniques involved.

Training of the technical staff in dosing procedures needs to be undertaken to allow them to learn how to do the various steps involved in the dosing procedures but also to understand each aspect of the steps, as they may be called upon to trouble-shoot problems that may arise with the set-up or parts of the apparatus used at any given time during the dosing procedures. The technicians need to be trained on how to place the syringes and infusion bag/cassette on the infusion pump, and to understand the significance of the alarms and the steps to take when a given alarm appears, as well as how to document the incidents when they occur, in order to permit a clear reconstruction of the events from the data. Along with use of the infusion pumps, there is also a need for training on the assembly of the infusion lines, the various components and connections, and inspection of the infusion lines prior to and during use, all the while maintaining an awareness for sterility of components. Training technicians on mock set-ups within a training colony to permit them to gain confidence with the apparatus and familiarity with the various alarm codes, resolutions, and pump configurations and the various steps involved generally provides satisfactory results.

4.2.1.2 Applicability of the methodology

Considerations of the limitation of each technique and its impact on the animal's housing conditions need to be balanced with the optimal conditions for the dosing scheme required and risks presented by each approach. Group housing of animals has been recognised in research and industry as the preferred housing condition for most animal species. Co-housing of cannulated animals is feasible with the use of ambulatory systems but presents many challenges. Some laboratories have adopted modified approaches to their set-ups or housing conditions to maximise animal contact and environmental enrichment within the limitations of the model(s) utilised when co-housing is not feasible, as would be the case for the jacket/tether system. For such set-ups, modifications to the home cages have been adopted in order to increase tactile contact with a cohort and provide additional environmental enrichment, such as access to foraging toys, perches, and auditory and visual stimuli, rotated over the course of the housing study period. One of the major benefits with ambulatory set-ups is the housing of animals in social groups and in enclosures in which they can exhibit normal behavioural patterns, such as escape to height. The benefits, including the reduction of stress, for these animals needs to be assessed in regard to the risk of the cage mates interfering with the infusion system, and many consider the benefits to outweigh the risks. There are a number of solutions that can be put in place to minimise the risks of interference with the system and overcome the challenges posed by inquisitive primates with both great strength and dexterity.

A vest may be used under the jackets, and fixtures and fittings placed to improve jacket security. Animals suited with an ambulatory system can also simply be housed in single-occupancy cages.

4.2.2 Non-surgical models

Non-surgical approaches for intravenous infusion can be used for relatively shorter infusion periods. Animals generally need to be restrained during these procedures to prevent them from interfering with the temporary indwelling needles or cannulas inserted in a peripheral vein.

4.2.2.1 Methods of restraint

Various models of restraint devices are commercially available. The size and weight of the non-human primate need to be taken into consideration in the selection of the restraint device. Generally, chair restraint is preferable to a sling restraint device for larger NHPs.

4.2.2.2 Applicability of the methodology

This approach can generally be considered when the duration of the infusion is relatively short, when dosing frequency may be less than daily, and when the formulation is not anticipated to elicit irritation at the administration site.

4.2.3 Risks/advantages/disadvantages

Non-cannulated set-ups are anchored using quick connect systems and bandages or medical tape to secure the set-ups to the animals. However, these systems are still at risk of displacement through movement of the animal, which could lead to damage of the vasculature and possible leakage of the dose in the extravascular space. These approaches, however, do present options of rotation of infusion sites across at least 4 relatively accessible peripheral veins (both saphenous and brachial veins); there are no maintenance requirements between dose administrations. Animals can be co-housed outside of dosing procedures without risks to the dosing procedures or set-up.

4.3 Practical techniques

4.3.1 Surgical models

4.3.1.1 Preparation

4.3.1.1.1 Presurgical preparation. Animals will be instrumented for infusion once they are considered acclimated to their laboratory environment. The preparation of the animals for surgery as described below is consistent with those of various laboratories and in accordance with appropriate animal care and use recommendations. Animals are food-deprived overnight prior to preparation for surgery.

4.3.1.1.2 Premedication. The standard approach for surgical procedures in non-human primates consists of pre-anaesthesia by an intramuscular injection (0.25 mL/kg) of a cocktail of glycopyrrolate, xylazine, and ketamine HCl injection USP, or alternatively with medetomidine, ketamine, atropine, and potentiated amoxicillin (all intramuscularly), to achieve sufficient sedation prior to intubation and shaving. The analgesic carprofen (Rimadyl™), a non-steroidal anti-inflammatory, is also injected subcutaneously shortly prior to surgery.

4.3.1.1.3 Anaesthesia. Anaesthesia over the course of the surgical procedures is maintained by isoflurane/oxygen. The animals are intubated (lidocaine spray applied prior to intubation) to allow maintenance of anaesthesia.

4.3.1.2 Surgical procedure

The catheter is implanted via the femoral vein and exteriorised dorsally at the scapular level. The dorsal exteriorization site and femoral sites are shaved and washed with chlorhexidine gluconate 4% followed by a liberal application of 70% isopropanol and povidone-iodine 10%. A bland lubricating ophthalmic agent is applied to each eye to minimise eye dryness.

A small incision is made typically in the right or left groin region, and the femoral vein isolated. A medical grade silicone-based catheter (or other appropriate catheter material considered compatible with the formulation) is inserted and the tip of the catheter is positioned in the vena cava at approximately the level of the kidneys. The catheter is secured in place with an appropriate suture material and anchoring bulbs located along the catheter. The catheter is looped to reduce risks of catheter blockage associated with the animal's normal activity and movements, which could result in a folding or kinking of the catheter. With the aid of a trocar, a tunnel is created, permitting the catheter to be brought subcutaneously to the exteriorization point on the animal's back. The patency of the catheter is verified over the course of the surgery using a normal saline-filled (0.9% Sodium Chloride Injection, USP) syringe connected to the end of the catheter. The femoral site is irrigated with warm normal saline (approximately 37°C). The femoral site is closed with sutures.

From this position the animal can either be equipped as a fully ambulatory model in a pair or group-housed scenario, or individually housed as a tethered model. In both cases, the exteriorised catheter is protected by a jacket system. In some models, the catheter is connected to a subcutaneous vascular access port with the infusion pump connected to the port via a specialised connecting tubing and needle system.

4.3.1.2.1 Ambulatory model. Animals intended for continuous infusion surgery are acclimated to their jackets as well as the weight of the pump for at least 5 days prior to surgery. Once acclimated or preconditioned, catheter implantation can be undertaken.

Following pre-anaesthesia (as described in Section 4.2.1.1) and once sufficient sedation is achieved, a mask is placed to allow anaesthesia to be maintained with isoflurane/oxygen, and thereafter the animal is intubated following application of lidocaine spray to the larynx. A paraffin-based ocular lubricant is applied to both eyes and the animal monitored using a pulse oximeter applied normally to the lip.

In preparation for the surgery, the animal is clipped over the inguinal area, the lateral thigh, and the dorsum. The skin is then cleaned with chlorhexidine 4% and disinfected with chlorhexidine 5% in 70% alcohol; the disinfected surgical site is then delineated by sterile drapes. Depending on the vessel to be cannulated (jugular or femoral), the animal will be placed in a lateral or dorso-lateral recumbency, and an incision is made over the targeted vessel to be cannulated. The vein is isolated and the catheter inserted using a polyurethane cannula or alternative compatible catheter. The catheter is secured into the vein and surrounding tissue using anchoring bulbs, as described in Section 4.3.1.2. An exteriorization cannula is attached to the catheter and then connected to a saline-filled subcutaneous access port. The edge of the latissimus dorsi muscle is freed from underlying musculature and the port implanted between the muscle layers.

The port and catheter system is filled with sterile lock solution generally consisting of heparin and glucose. A catheter coated with an anticoagulant can also be used for implantation based on compatibility of the test item/material formulation with the catheter coating. For infusion administrations, the exteriorised infusion line is inserted via the skin into the port using the 2-piece non-coring needle, to ~17 cm and attached to the luer lock catheter connector. All surgical wounds are closed using absorbable suture materials, and levobupivacaine is injected subcutaneously around the surgical wounds. A vest and jacket are placed on the animal while still anaesthetised, and the catheter connector is attached to an infusor to provide a background infusion of saline.

4.3.1.2.2 Tethered model. The tethered approach utilises a jacket (adjusted for a comfort fit around susceptible areas such as the neck and underarms) placed over the torso of the animal, to which a tether has been anchored (via a leather plate sewn onto the jacket). The implanted catheter, exteriorised at the scapular level, is passed through the jacket and tether, attached to a swivel system at the front, side, or top of the animal's home cage, and is then connected to the infusion apparatus (a syringe pump or larger volumetric infusion pump) external to the cage. The infusion pump is encased in a plexiglass box to safely secure the pump and prevent interference of the infusion set-up by the animal. The jacket/tether system permits the animal to have freedom of movement within its cage while dosing is occurring. The connection of the catheter to a swivel prevents the catheter from being twisted or tangled by the animal's movement, which would otherwise result in blockages of the catheter interfering with the delivery and possibly resulting in loss of catheter patency. Since the infusion pump is external to the cage, the set-up allows modifications to the infusion parameters, including the infusate, rate, and status of the infusion, without restraint or removal of the animal from its cage. This also permits the use of larger infusion reservoirs without concern for the weight-bear on an animal, and it reduces the technical requirements for multiple infusate changes needed for infusion of larger volumes or longer infusion durations.

The catheter, prefilled with normal saline, is fed through the tether system and attached to a swivel secured to the outside of the cage. The infusion line is attached to the outer portion of the swivel, and all animals are continuously infused with normal saline at an appropriate rate. This rate can vary based on the size (inner bore/diameter) of the catheter implanted, but it generally ranges between 1 to 4 mL/hour.

4.3.1.3 Maintenance

4.3.1.3.1 Recovery procedures. The external sutures at the femoral site are removed depending on the progression of healing (generally approximately 10 days following surgery). A topical antibiotic (a formulated mixture of polymyxin B and bacitracin) is administered to the catheter exteriorisation site daily until infusion termination, and to the femoral site until considered unnecessary.

A post-operative period of 7 days is typically allowed and may be extended as considered appropriate based on the properties of the dose formulation to be administered. For example, for a test item/material known to have anticoagulant properties, the post-operative period will be extended to 10 days.

4.3.1.3.2 Checks and troubleshooting. Common concerns with any type of implantation or instrumentation are the risks of bacterial infections. The catheter presents an easy port of entry for bacteria, and following good practices is critical in preventing infections and septicaemia.

For the jacket/tether system, the catheter is continuous from the tip of the catheter in the vein to the point of connection at the swivel (a stainless steel pin/cannula with rotating discs that is attached to the cage) and protected by a stainless steel coil (the tether), as such leaving the vulnerable points of bacterial entry outside of the animals' environment. Effective control of these vulnerable points is obtained by applying good technical practices and using materials that allow the system to be closed where possible. On the outside of the cage, the catheter is connected to the external portion of the swivel, which is continuous with an injection site when attached to the syringe for smaller volumes of delivery, or to tubing used for infusion when using larger flow/volume peristaltic pumps for delivery. Control of bacterial entry at connection points can therefore be easily maintained by use of alcohol wipes. The use of in-line filters (low-protein-binding 0.22-µm pore-size filter) to further reduce the chances of bacterial introduction is also highly recommended, in particular for longer-term regimens.

As the catheter in the vein is susceptible to blockage by the formation of a blood clot at the tip or by back flow of blood up the catheter, patency is maintained through continuous infusion of a physiological or close to physiological solution at an appropriate rate, with or without the use of an anticoagulant. Alternative options have been the use of anticoagulant-coated catheters. The preference in our laboratories has generally been to utilise continuous infusion of normal physiological saline at appropriate rates, for the given species, following implantation of the catheter post-operatively. The infusion of normal saline is maintained until initiation of dose formulation in continuous infusion dosing regimens, and in between dose administrations for intermittent dosing regimens. Based on these procedures, the use of anticoagulants is not required and as such eliminates any concerns of potential interference or introduction of a confounding variable with the planned test item/material evaluated. This practice requires consideration for assessment of the compatibility of the test formulation with physiological saline. In instances of non-compatibility, the use of an interface or alternate maintenance infusate can be adopted.

The basic principle in considering rates and dose volumes of intravenous infusion is to remain within a range that at a minimal flow will still prevent blood from flowing back up the catheter and clots from forming within the catheter at the tip, thus impeding the flow, while at the upper end of the range not exceeding a rate or dose volume that will overburden the kidneys. The duration of the infusion over the course of 24 hours needs to also be considered, as well as characteristics of the dose formulation. From data gathered postmortem in our laboratories over the recent years from rats, non-human primates, and dogs of average size (mean body weights of 350 g for rats, 2.8 kg for non-human primates, and approximately 9 kg for dogs), the vena cava of rats is 7 times smaller, of primates 3 times smaller, and of dogs approximately ½ the diameter of that reported in humans. The femoral vessel diameters differ slightly less, with the femoral vein in rats on average being 4 times smaller than the average femoral vein in humans, while in non-human primates it is approximately 1.5 times smaller, and in the population of dogs evaluated slightly larger (5%) than that reported in humans.

Based on the considerations discussed above, we have established guidelines of use for continuous (24-hour) intravenous infusion rates of 2–3 mL/kg/hr in non-human primates of 2 to 3 kg that are considered overall well tolerated. Infusion rates and dose volumes above and below ones stated above can be used, but as indicated previously the frequency of dosing, duration of the infusion, and dose formulation properties all need to be taken into consideration.

4.3.1.4 Record keeping

During the post-operative period, the delivery and system integrity is closely monitored by performing accountability of the volume of infusate delivered on a daily basis. This provides the opportunity to address the set-up for any animal where the delivery is expected to meet the anticipated accuracy of delivery. The infusion pumps alone are calibrated to deliver within an accuracy of ±5%, and once these are connected to the implanted animal, because of the variables a live organism introduces to the set-up, the overall accuracy of delivery is expected to be within ±15% from target volume. This accuracy of delivery may be modified for the period of dose formulation delivery if the intended rate of delivery for the conduct of the infusion work will utilise rates at the lower end of the range of the recommended infusion rates, as the loss of even a drop of infusate through the manipulation of the syringes and purging of the system may represent a large proportion of the dose volume to be administered. An accuracy of ±20% is generally recommended to be targeted in these circumstances.

4.3.2 Non-surgical models

4.3.2.1 Methods of restraint and acclimatisation

Non-surgical models generally require restraint of the animal during the period of infusion. Some alternatives that allow infusion via a peripheral vessel whilst still enabling the animal to be unrestrained are feasible with other species, but these have not thus far been explored in non-human primates in our laboratories. Most common modes of restraint for non-surgical infusion include the use of a chair or sling. For either mode of restraint, acclimation to the restraint device by gradual increments of time over periods of several days or weeks, to meet the targeted period of restraint for dosing or monitoring, is required. This method of performing an intravenous infusion study is normally restricted to shorter dosing periods, preferably no more than 30 minutes.

4.3.2.2 Methods of vascular access

Temporary indwelling cannulas such as Abbocath™ (24 gauge) or butterfly (21 gauge) are recommended for use in these species. When an Abbocath is used, it is connected to an injection port and flushed with saline (1 to 2 mL) following insertion into the vessel (a saphenous or brachial vein). An infusion extension set or a Butterfly is used between the injection port and syringe pump. Typically the set-up consists of the use of one pump per animal; however, if co-administration of dose formulations is required, the set-up can allow more than one pump to be used for a given animal.

4.3.3 Best practice

4.3.3.1 Surgical models

In general, the use of implanted models is recommended when longer administration periods (greater than 3 hours) and larger dose volumes are required, or dose formulations are anticipated to have some irritant potential; the decision may be based on a combination of factors, including the likely clinical set-up or study design. An example of study design would be a study monitoring animals for physiological changes such as in cardiovascular telemetry. Typically it is preferable to conduct such studies using a cannulated model for dosing by infusion, since the restraint of the animals during these relatively longer dosing periods may artificially increase heart and respiratory rates during and shortly after the restraint period. Though there are ways to reduce this interference, using a cannulated model obviates this issue. In addition, when dealing with larger study population sizes, the use of a cannulated model to conduct an infusion for longer periods (e.g., more than 3 hours) will reduce the time researchers spend on dosing, may allow greater feasibility for conducting various timed investigations, and, depending on the frequency of dosing, may represent some economic benefits.

As a cannulated model will utilise a central vessel of greater diameter than that of a peripheral vessel, it can provide a greater dilution factor upon first contact with the blood, and as such may allow an irritant potential to be dampened. This is particularly beneficial in species where the number of sites for vascular access is more limited (e.g. rats and pigs). In dogs, where the inferior vena cava may be close in diameter to that of humans, a cannulated model can serve as a good indicator of the potential clinical outcome regarding susceptibility to local irritancy. For non-human primates, where the inferior vena cava is generally smaller than that in an adult human, the cannulated model may possibly not be as predictive as the dog of the clinical scenario, but the use of this model may allow better tolerance of a higher frequency dosing regimen (such as daily or every-other-day dosing) of a potentially irritant dose formulation when compared to the use of peripheral veins with a rotation of sites. Typically four sites (two saphenous and two brachial) are used in non-human primates, and therefore a frequent dosing scheme may not allow sufficient time between scheduled use of a given vessel when using a rotation of peripheral sites for dosing.

4.3.3.1.1 Tethered versus ambulatory set-ups. The disadvantage of the jacket/tether approach is that it does not permit co-housing of the animals. The concerns over single-housing animals with a social nature such as non-human primates have been increasingly heightened over the past decades, and consideration for the use of alternate approaches that may permit social housing of the animals and inclusion of additional enrichment approaches during the conduct of the study have taken high priority. Therefore, use of less limiting set-ups may be sought, although co-housing of instrumented primates with ambulatory systems would still present some challenges. Interference with the infusion system by co-housed animals remains a possibility and may result in significant complications beyond just the interruption of dosing, so this aspect should be weighed in deciding which approach to use.

Ambulatory systems typically include a pocket on the back of the jacket that houses the infusion system, including a pump; the pump and reservoir can be attended to without removing the jacket. The use of ambulatory set-ups can allow adherence to the European legislations for co-housing of animals, including movement across two-storey enclosures. These set-ups are often used for intermittent regimens but can also be used with continuous infusion, with consideration for the frequency of refilling the reservoir; the weight of the pump and filled reservoir that also need to be taken into account, as stress may be induced by the extra weight. Refilling the reservoir may or may not be required, depending on the targeted dose volume. Generally, animals at least two years of age are used. Ambulatory systems are being developed with these limitations in mind and are being optimised where possible to reduce the size and weight of the pumps themselves and ease the exchange or refilling of the reservoirs.

Surgery with percutaneous exit is not suggested for group-housed animals: should cage mates gain access to the catheter, it could be moved from its original location in the vena cava. It is suggested that a vascular access port be used, the port of choice being a side-entry port into which the cannula is directly placed. If the cannula is removed, it can be replaced by a non-coring needle up to 10 to 15 times. There are a number of ports on the market, including ones with a grid lock base, and these may also be suitable for use.

There are ongoing discussions as to whether it is more beneficial to leave the jackets on rather than taking them off and putting them back on intermittently; but there is also the welfare concern that suggests they should be removed whenever possible. Yet our experience thus far would suggest that removal may be problematic and that whenever possible they should be left on. Jackets, whether in an ambulatory or jacket/tether system, will regardless need to be removed periodically to allow change of soiled jackets with clean ones.

4.3.3.2 Non-surgical models

4.3.3.2.1 Chair versus sling restraint. The necessity of restraining the animals on a daily or every second day basis to perform dosing in periods longer than 30 minutes may present fewer benefits and be less practical than a cannulation. When considering a study utilizing a larger population (e.g. more than 32 animals), the time required to perform the dose-staggered infusion in all animals may be more than that for a cannulated approach, and overall may cause increased stress to the animals compared to an unrestrained approach.

4.3.3.2.2 Acclimation considerations. Acclimation to the restraint device should be done gradually, with an incremental increase in duration only when the animal appears to have adjusted to the interval of restraint just completed. Our standard practice is to acclimate the animals to the restraint device generally for 3 to 5 days at each given length of time, with initial increments of 15 minutes being used. A calm environment and close monitoring of the animals and their behaviour during restraint assist in evaluating the ability of the individual animals to acclimate and in further adapting the schedule as required.

4.3.3.2.3 Infusion frequency/duration considerations. The use of a sling restraint is best suited for dosing regimens requiring shorter, less frequent periods and/or cyclical dosing (e.g. dosing once weekly) of intravenous infusion. This approach would not be considered under most circumstances with dose administration of longer than 3 hours. Generally, periods of infusion utilizing these procedures are no longer than 30 minutes, but may exceptionally be for periods of up to 1 hour. The use of this approach for periods of infusion of 1 to 3 hours also requires assessment of the suitability of this set-up. For studies where dose administration is performed daily or every second day, rotation of the vessel selected for dosing is critical and, depending on the length of the study (especially beyond 28 days), requires careful evaluation of the dose formulation properties (the known potential of irritability) and the dose volume.

4.4 Equipment

4.4.1 Surgical models

4.4.1.1 Catheters/vascular access ports

The size and length of catheters will vary depending on the vein to be cannulated. For non-human primates, a silicone or polyurethane-based catheter to be implanted via the jugular or femoral vein with an outer diameter of 0.065 inches will be well tolerated. The inner diameter of the catheter can vary based on the needs for the study (standard versus lower infusion rates, as previously discussed). For standard recommended infusion rates, a catheter with an inner diameter of 0.030 inches has been shown to be appropriate. The selection of the catheter tubing is dependent on the compatibility of the test item/material formulation with this tubing material; however, background tissue reaction to less malleable tubing material should be carefully considered. Although the catheter material used in the preclinical studies does not have to mimic exactly that planned to be used clinically, using similar materials may help minimise additional compatibility assessments.

Vascular access ports are commercially available. A side-entry port in which the cannula is directly placed has been shown to provide good results. With this set-up, if the cannula is removed it can be replaced by a non-coring needle. There are a number of ports on the market, including ones with a grid-lock base and minimum volume ports.

4.4.1.2 External equipment

Purpose-made infusion jackets should be verified on a regular basis to ensure proper fit on the animal and minimise risks of chaffing in vulnerable areas such as the armpit. Inadequate fit and insufficient monitoring can quickly result in the formation of lesions in this area, which can easily become infected. The infusion jackets should be changed on a regular basis, or as they become soiled or damaged. For ambulatory set-ups, there will be a pocket on the outside of the jacket to house the infusion pump. The pocket should be designed to allow access by the technicians to perform changes of reservoirs and to verify the connections, while still being sufficiently secure to reduce the risks of a cage mate gaining access to the pump and infusate. For jacket/tether systems, a tether plate will be present on the jacket onto which the metal tether is secured and through which the catheter will be passed. The tether and catheter will be connected to a swivel system at the top or front of the inside of the animal's cage, while an additional portion of catheter will be connected from the swivel pin on the outside of the cage to the infusion pump.

4.4.1.3 Infusion pumps

Various models of infusion pumps utilizing a syringe set-up for smaller volumes of administrations requiring 60-cc syringes or less, ambulatory pumps with a reservoir of variable upper volume limits, and peristaltic infusion pumps for larger volumes using infusion bags of 1 litre or less are available on the market. Use of syringe pumps for dose administration volumes of more than 60 cc over a given day can be managed with a change of syringe over the course of the planned infusion period on that day. The limits of the rates of infusion for the corresponding size of syringe need to be taken into account as per operation manuals. As with syringe pumps, points for consideration when dealing with ambulatory or peristaltic pumps are the possible interactions with the components of the housing reservoir and/or tubing. The potential of the formulation foaming as a result of the movement of the animal carrying the ambulatory pump and reservoir or of the pumping mechanism itself, with air bubbles becoming trapped in it or in an in-line filter, needs to be monitored closely to reduce the risk of interference with the infusion.

4.4.2 Non-surgical models

4.4.2.1 Restraint device

The two restraint devices, as described previously, are the chair and the sling. Animals can also be manually restrained for short infusion periods. The use of chair or sling restraint can allow more than one animal to be infused at one time.

4.4.2.2 Vascular access

A temporary indwelling catheter, such as an Abbocath, can be placed and connected to an injection site with a septum. An infusion extension set or Butterfly can be inserted through the septum and connected to an infusion pump. If a temporary indwelling catheter is used, it should be filled with saline, and an occasional flush with saline may be required to minimise risks of blockage prior to the start of infusion, depending on the time lapse between placement of the catheter and the start of the infusion. Animals should be restrained throughout the infusion period, and close monitoring of the animals is needed at all times to ensure the cannula does not exit the vasculature. The site can be clipped to better visualise the selected peripheral vein and access signs of potential irritation at the site of entry/infusion.

4.4.2.3 Infusion pumps

A syringe pump or peristaltic pump can be used depending on the volume and rate of infusion required.

4.5 Background data

4.5.1 Surgical models

4.5.1.1 Complications associated with the surgical process

The presence of the catheter in the vessel will knowingly elicit a given level of tissue response, which is subject to being exacerbated by the parameters described above (rate, frequency, and duration of infusion with a given formulation).

4.5.1.2 Issues associated with study design

It is important to thoroughly evaluate the selection of vehicle and the nature of the final formulations prior to undertaking safety assessment studies, in particular in primates, as the successful outcome of the studies may otherwise be jeopardised. Even when a proposed formulation is considered likely to be tolerated in humans, the influence of the test item/material on the formulation properties needs to be evaluated carefully during the design of the preclinical studies, as higher concentrations are likely to be used over the course of the safety assessment studies compared to the target concentration(s) in clinic.

4.5.1.3 In-life changes

In our experience, the surgical procedures and maintenance of the animals and infusion systems as described have not been associated with any marked changes in body weights, food consumption, or ophthalmological changes. Slight transient decrease in both body weights and food consumption may be observed in the post-operative week, but animals have generally been shown to rebound to pre-surgery values by the start of treatment. Changes in behaviour associated with the presence of the jacket and tether may be seen following the initial introduction to these apparatus, but these soon resolve within a few days. Clinical signs associated with the cannulation procedure, consisting of occasional slight swelling at the infusion site and thinning of the fur cover in areas in contact with the infusion jacket, have been noted.

A graph (Figure 4.1) of mean body weights comparing cannulated animals on continuous infusion to non-cannulated animals over a period of more than 28 weeks does not show any marked effects or differences in body weight trends that would be considered related to the surgically cannulated model.

4.5.1.4 Pathology

4.5.1.4.1 Clinical pathology. Clinical pathology parameters from samples collected during the post-operative period have been shown not to differ significantly from that of non-cannulated animals. Red

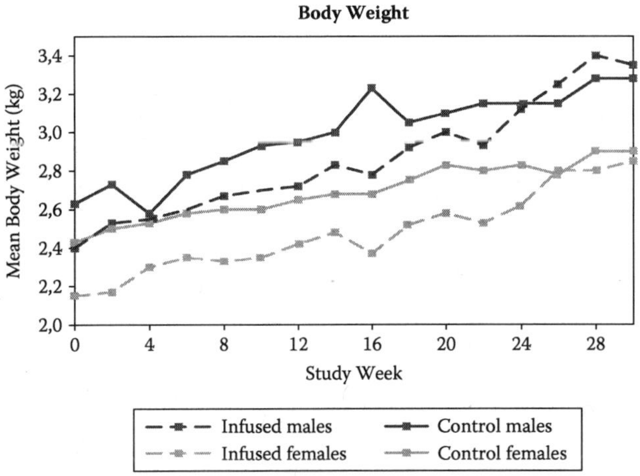

Figure 4.1 ***(See colour insert.)*** Mean body weights comparing cannulated animals on continuous infusion to non-cannulated animals over a period of more than 28 weeks.

and white blood cell mass remains within normal physiological range for species of NHPs routinely used (cynomolgus monkeys and rhesus macaques). Comparative data of cannulated versus non-cannulated cynomolgus monkey are presented in Figures 4.2–4.5 below (red blood cell counts, haemoglobin, haematocrit, and white blood cell count, respectively).

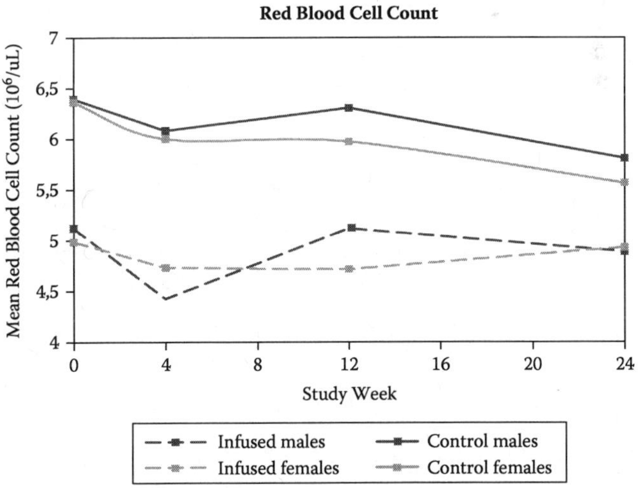

Figure 4.2 (See colour insert.) Group mean red blood cell counts.

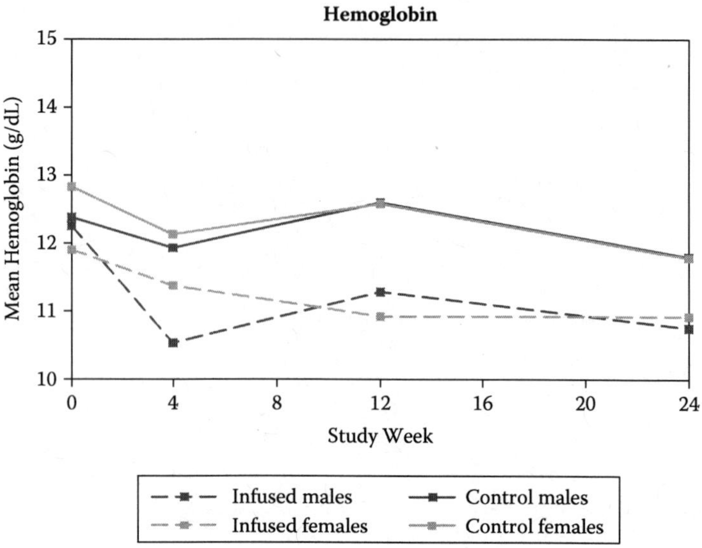

Figure 4. 3 (See colour insert.) Group mean haemoglobin.

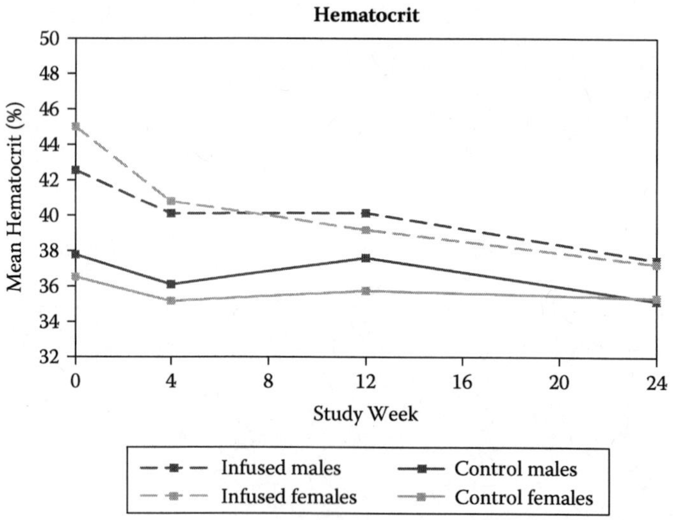

Figure 4.4 (See colour insert.) Group mean haematocrit.

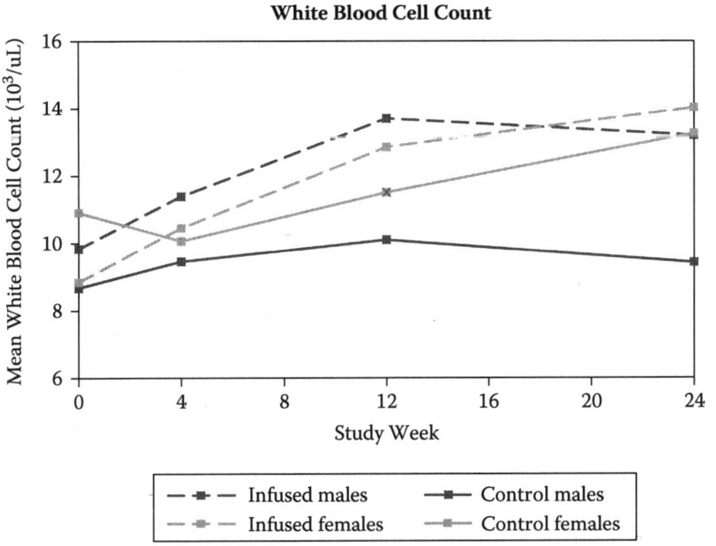

Figure 4.5 (See colour insert.) Group mean white blood cell count.

4.5.1.4.2 Histopathology. Background histopathological changes associated with the cannulation of the vena cava via the femoral vein include intimal proliferation, thrombosis, and inflammation at the infusion site (Table 4.1). The severity of these background changes are generally slight and do not result in significant health complications or interfere with or mask findings associated with the safety assessments of a compound. However, maintaining good aseptic procedures during the conduct of the study is critical for limiting introduction of bacteria. As the exteriorization site presents an easy point of entry for opportunistic bacteria, using prophylactic antibacterial agents at this site can help provide an effective barrier. The above described background changes can be exacerbated by formulations with an irritation potential. As well, the tip of the catheter can oscillate at high infusion rates, causing turbulence at the infusion site,

Table 4.1 Incidence (%) of Background Infusion Site Lesions in Cannulated Saline/Physiological Vehicle Infused Cynomolgus Monkeys

Infusion site histopathological change	Male	Female
Inflammation (perivascular, vascular, subacute)	14%	16%
Intimal proliferation	19%	16%
Thrombosis	33%	32%

Table 4.2 Incidence (%) of Background Infusion Site Lesions in Non-Cannulated Peripherally Infused (Saline/Physiological Vehicle) in Cynomolgus Monkeys

Infusion site histopathological change	Male		Female	
	Saphenous vein	Brachial vein	Saphenous vein	Brachial vein
Haemorrhage	6 to 9%	3 to 5%	12%	2%
Inflammation	6%	2%	7%	0%

which creates an environment that can catalyze or increase the incidences of precipitation of formulations at the infusion site, especially when infusing a formulation that is close to its limit of solubility. Again these aspects should be considered in the interpretation of histopathological changes as well as the toxicokinetic profile of individual animals.

4.5.2 *Non-surgical models*

Histopathological background changes associated with peripheral infusion are generally associated with the trauma resulting from repeat insertion of a needle or temporary indwelling cannula. These are described as perivascular inflammation and haemorrhage (Table 4.2). The extent of the changes depends on the frequency of dosing within a given vessel and the elapsed time between doses, as well as the proficiency of the attending staff at the time of insertion. Rotation of the vessels used for the dosing procedures is recommended in order to allow time for background changes to partially resolve between doses. Therefore, both saphenous and brachial veins, as well as tail veins, should ideally be considered as options. Evaluation of the injection/infusion sites can be monitored over the course of the study and can be scored to allow a semi-quantitative measurement of erythema and oedema.

4.6 *Conclusion*

In conclusion, the appropriateness of the methods of infusion in non-human primates, as in other species used for preclinical safety assessment, should be assessed using a case-by-case approach. For formulations close to physiological and anticipated to present relatively minimal potential for irritation at the site of administration, the choice will be based on the duration, frequency, and dose volume needed to be administered in order to define whether using a cannulated or non-cannulated approach is more appropriate. For formulations with a greater irritation potential, the risks for local reaction and vessel size need to be included with these same factors in defining the best approach. When following these evaluations, should a cannulated approach be defined as the method of choice, the use of ambulatory versus jacket/tether systems may need to be investigated. In order to meet European housing requirements, ambulatory systems are required. Where housing condition expectations may differ (with both single- and co-housing permitted), consideration of the benefits of each approach and successful conduct of infusion studies under the given conditions, as well as the experience with infusion procedures within a laboratory, should be carefully evaluated to ensure appropriate guidance and problem resolution.

References

Clarke DO. 1993. Pharmacokinetic studies in developmental toxicology: Practical considerations and approaches. *Toxicology Methods* 3: 223–251.

Hertzberg BS, Kliewer MA, DeLong DM, Lalouche KJ, Paulson EK, Frederick MG and Carroll BA. 1997. Sonographic assessment of lower limb vein diameters: Implications for the diagnosis and characterization of deep venous thrombosis. *American Journal of Roentgenology* 168(5): 1253–1257.

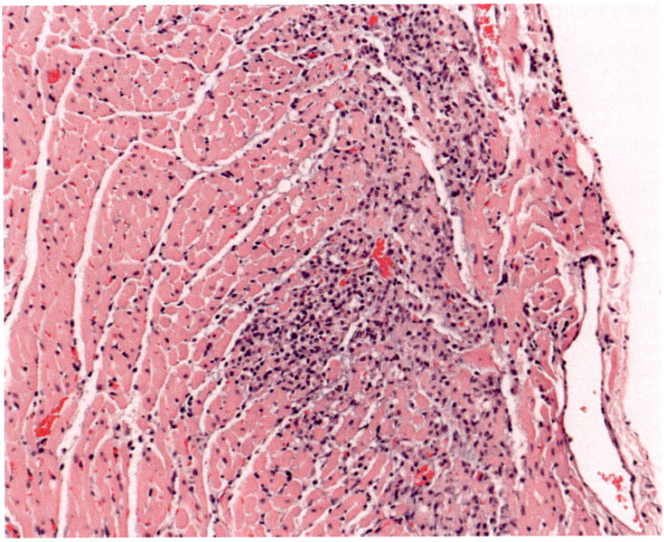

Figure 1.25 Right ventricle: Myocardial degeneration.

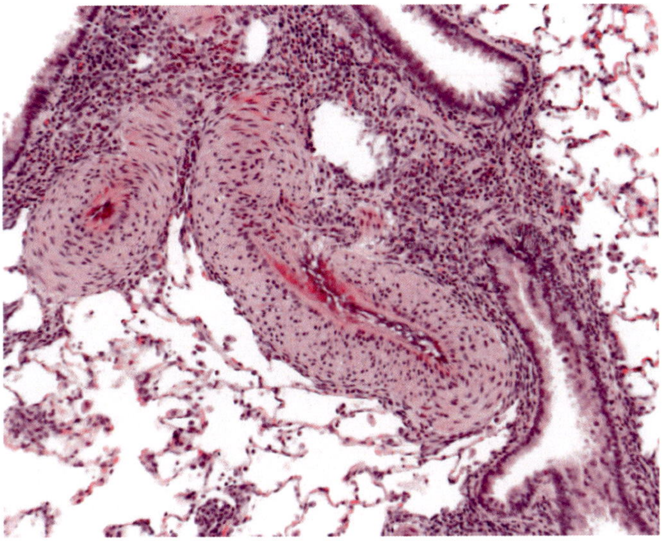

Figure 1.26 Medial hypertrophy of pulmonary arteries.

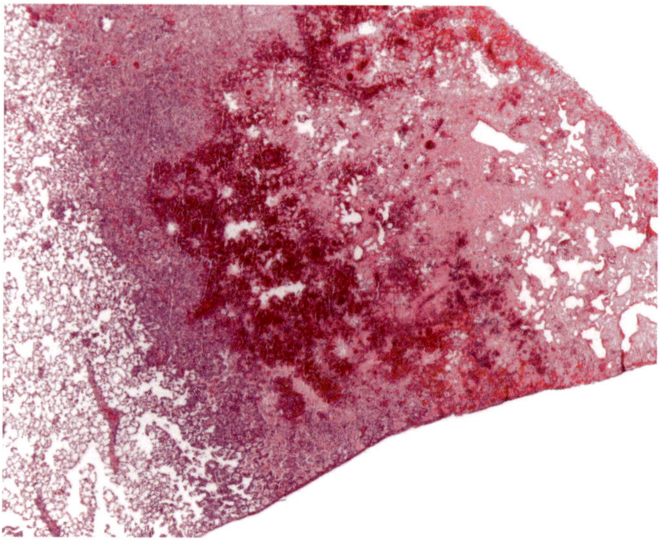

Figure 1.27 Infarction of rodent lung.

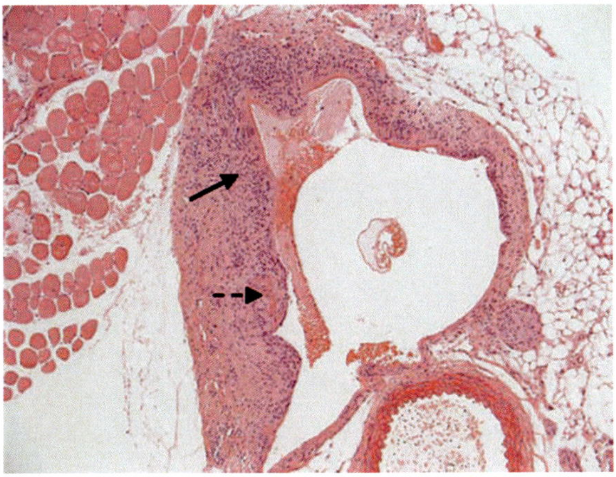

Figure 2.13 Vena cava, tail-cuff model, Study Day 14, (H&E ×10 objective). Arrow: Phlebitis/periphlebitis. Dashed arrow: fibrinoid necrosis.

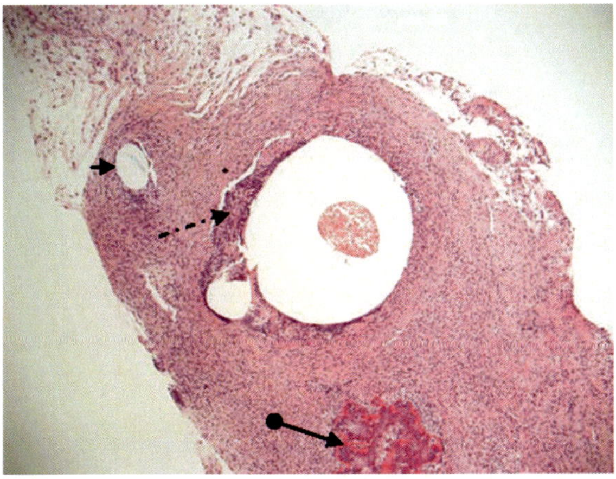

Figure 2.14 Femoral vein, tail-cuff model, Study Day 14, (H&E ×4 objective). Short arrow: suture. Dashed arrow: phlebitis/periphlebitis. Arrow with circle: pyogranuloma.

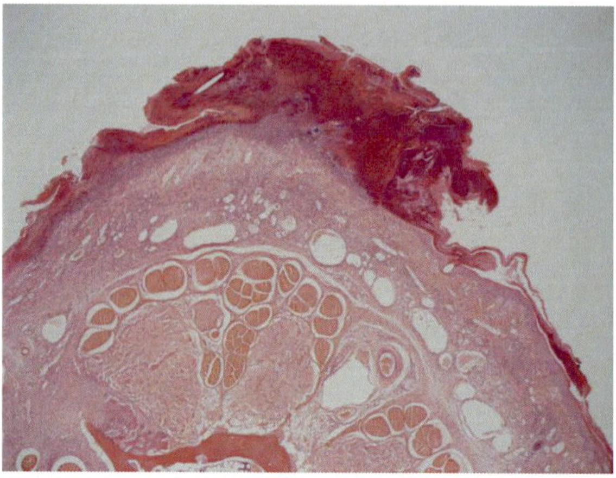

Figure 2.15 Catheter exteriorisation site (close to tail-cuff wire), Study Day 8, (H&E ×4 objective). Dermatitis, moderately severe. Fasciitis/cellulitis, moderately severe. Osteomyelitis, slight.

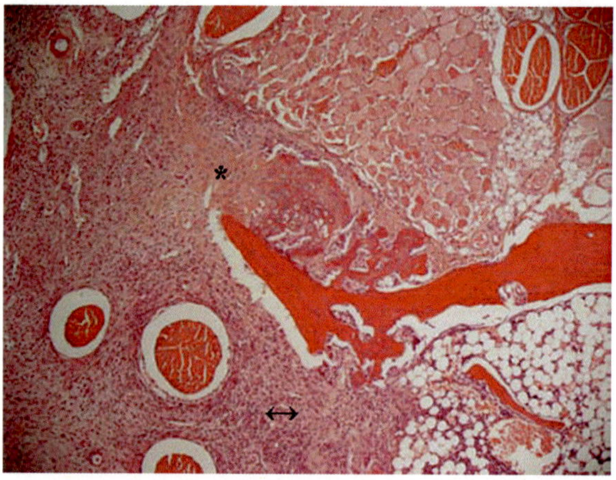

Figure 2.16 Catheter exteriorisation site (close to tail-cuff wire), Study Day 8 (H&E ×10 objective). *Osteomyelitis, slight. ↔ Fasciitis/cellulitis, moderately severe.

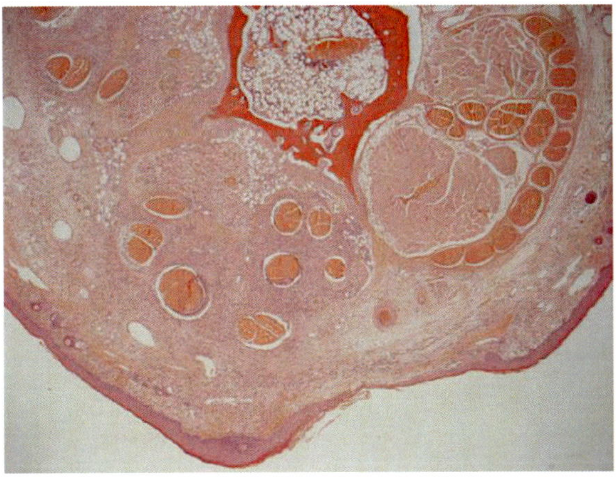

Figure 2.17 Tail-cuff wire, tail-cuff model, Study Day 8, (H&E ×4 objective).

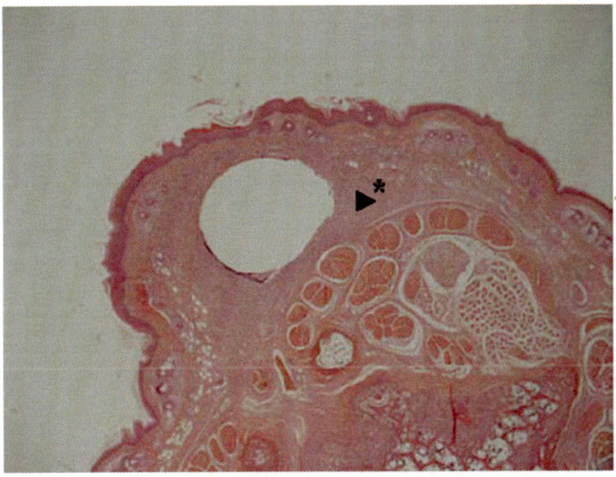

Figure 2.18 Catheter exteriorisation site, tail-cuff model, Study Day 15, (H&E ×4 objective). * Fasciitis/cellulitis, moderate. Dermatitis, minimal.

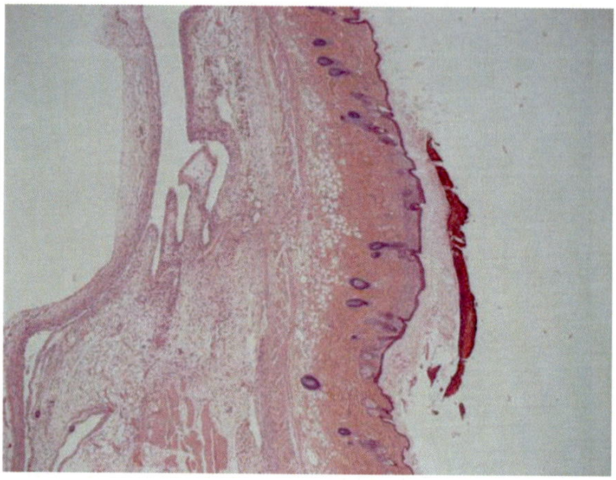

Figure 2.19 Catheter exteriorisation site, harness method, 8 days, (H&E ×4 objective).

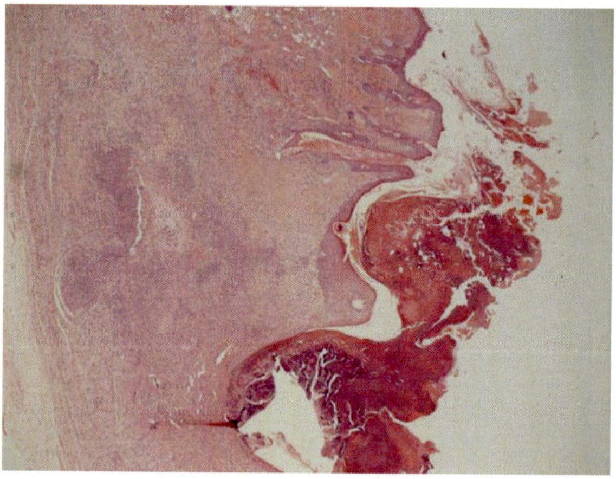

Figure 2.20 Catheter exteriorisation site, harness/skin-button model, Study Day 92, (H&E ×10 objective).

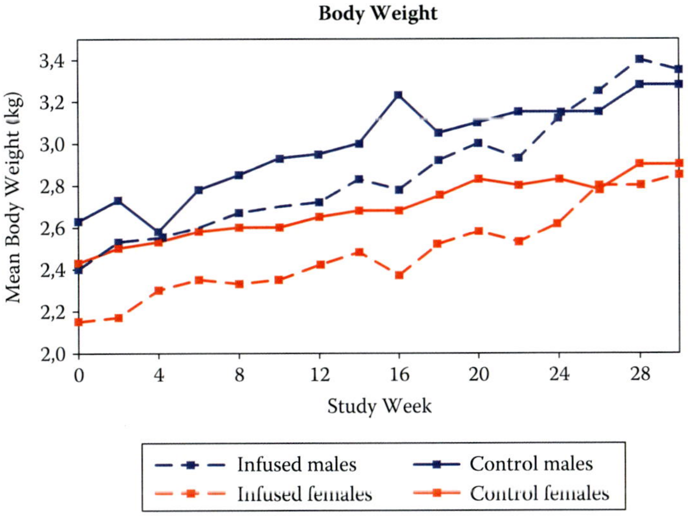

Figure 4.1 Mean body weights comparing cannulated animals on continuous infusion to non-cannulated animals over a period of more than 28 weeks.

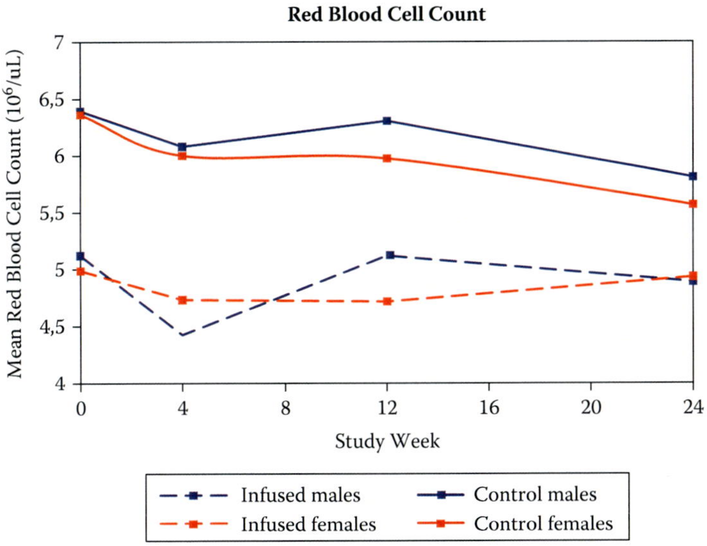

Figure 4.2 Group mean red blood cell counts.

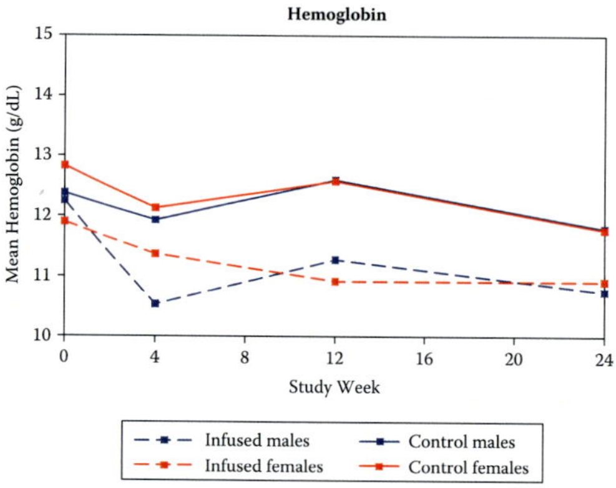

Figure 4. 3 Group mean haemoglobin.

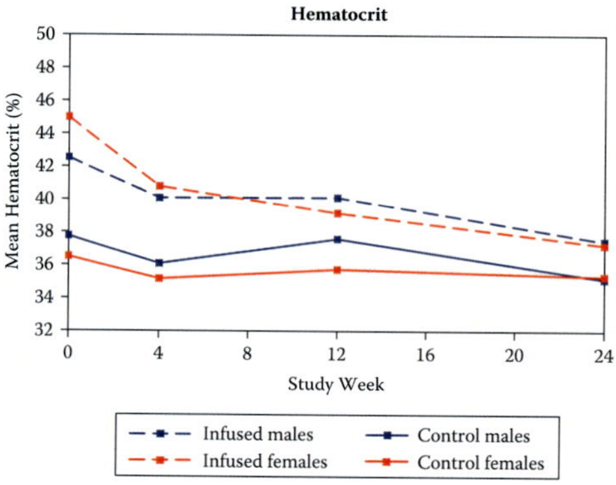

Figure 4.4 Group mean haematocrit.

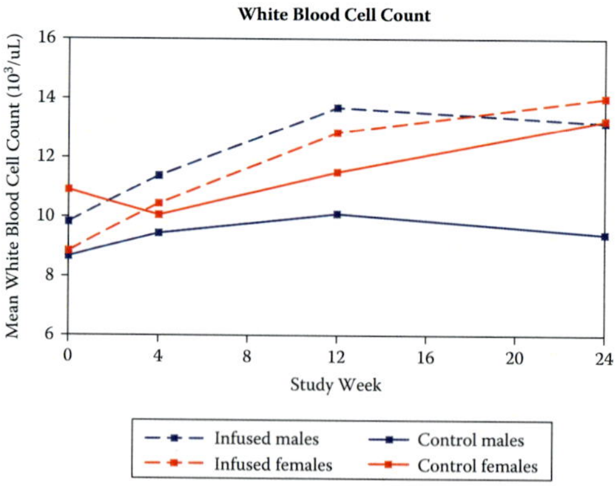

Figure 4.5 Group mean white blood cell count.

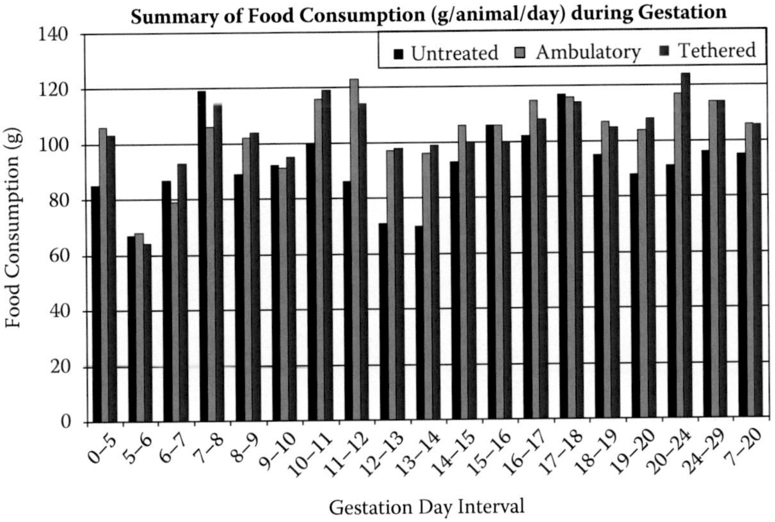

Figure 5.9 Food consumption summary on in-house comparison study of ambulatory vs. tethered continuous intravenous infusion in Dutch Belted rabbits.

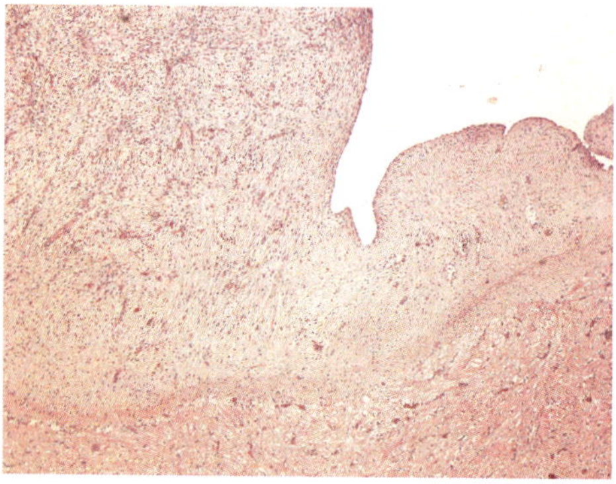

Figure 6.8 Well-organised thrombus in vena cava cranialis. (Treated with NaCl; H&E ×5).

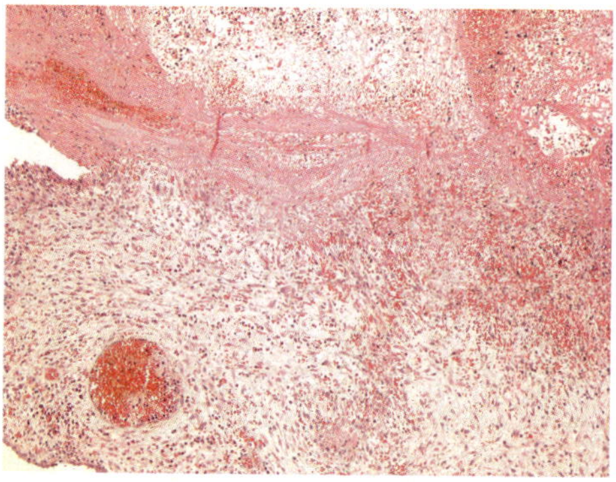

Figure 6.9 Active thrombus with inflammation of vessel wall. (Treated with NaCl; H&E ×10).

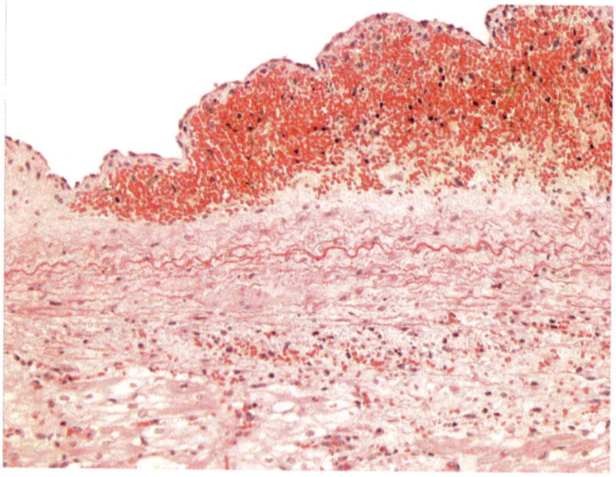

Figure 6.10 Subendothelial haemorrhage with inflammatory cells. (Treated with NaCl; H&E ×20).

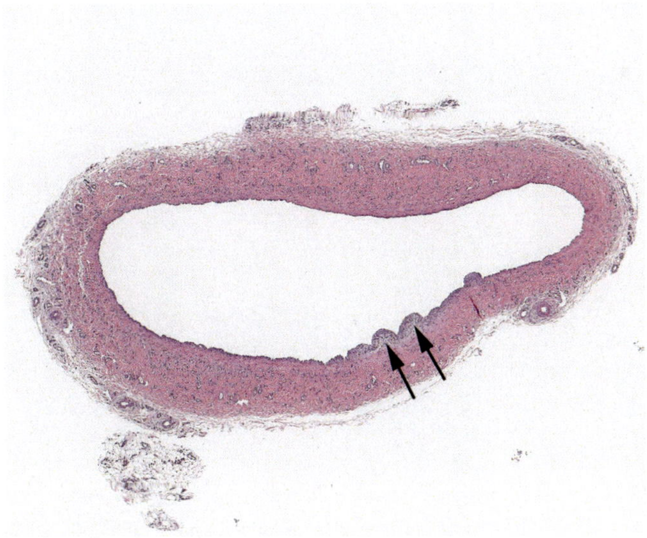

Figure 7.8 Jugular vein infusion site from beagle dog, Day 5. Note multifocal areas of intimal proliferation (arrows). (H&E stain, 25×).

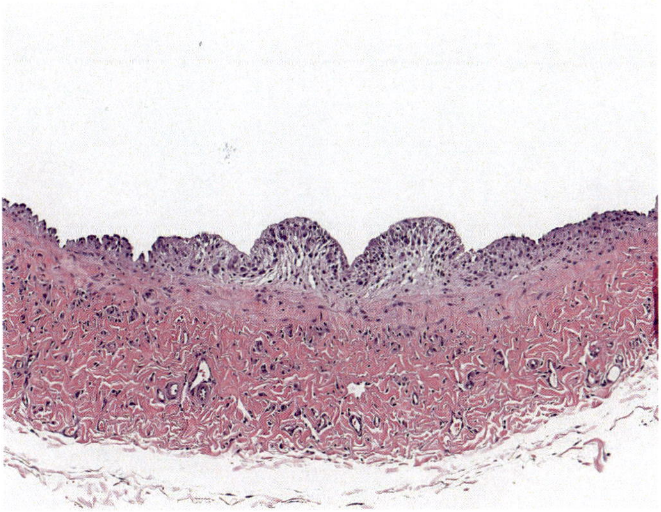

Figure 7.9 Jugular vein infusion site from beagle dog, Day 5. Note multifocal intimal proliferation. (H&E stain, 100×).

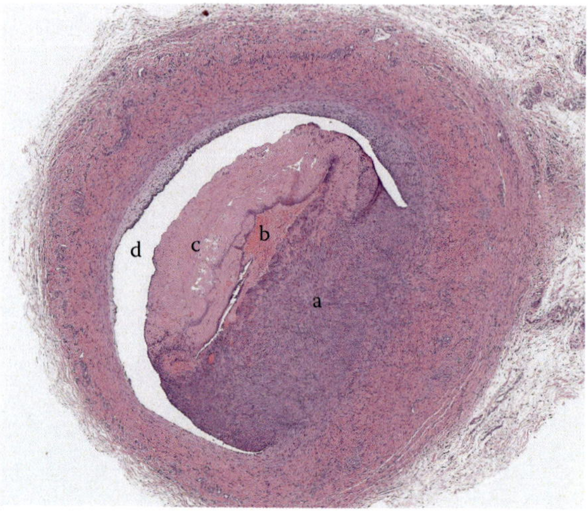

Figure 7.10 Jugular vein infusion site from beagle dog, Day 5. Note pronounced segmental intimal proliferation (a) with catheter tract (b), luminal accumulation of fibrinous material (c), and severely compromised lumen of jugular vein (d). (H&E stain, 25×.)

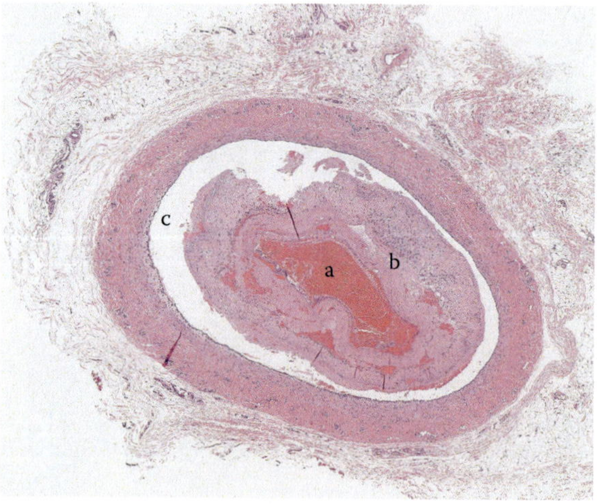

Figure 7.11 Jugular vein infusion site from beagle dog, Day 5. Note catheter tract (a) surrounded by zone of fibrin and leukocytes (b) and severely compromised femoral vein lumen (c). (H&E stain, 25×.)

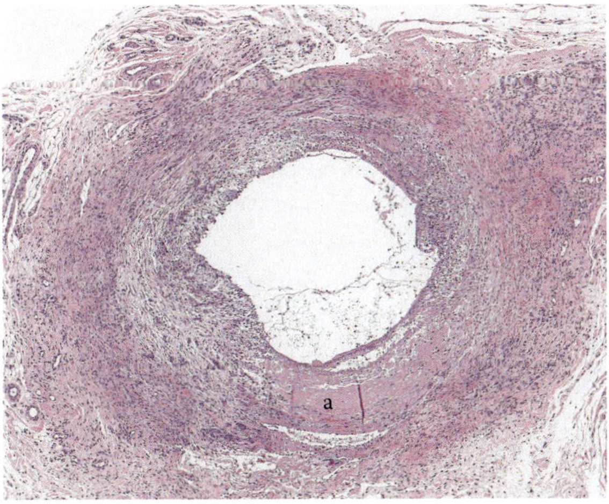

Figure 7.12 Jugular vein infusion site from beagle dog, Day 5. Note extensive disruption of vein wall with diffuse inflammatory cell infiltration and area of necrosis (a). H&E stain, 50×.

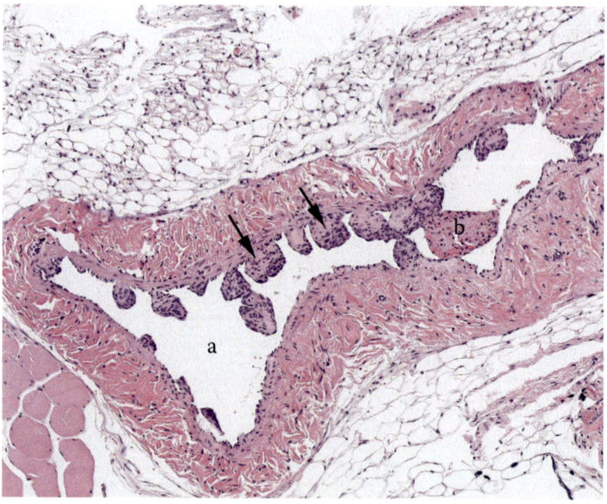

Figure 7.13 Femoral vein infusion site from Sprague-Dawley rat, Day 14. Note multifocal intimal proliferation forming villous projections (arrows) into femoral vein lumen (a) and a small thromboembolus (b) composed of fibrin and leukocytes. (H&E stain, 100×.)

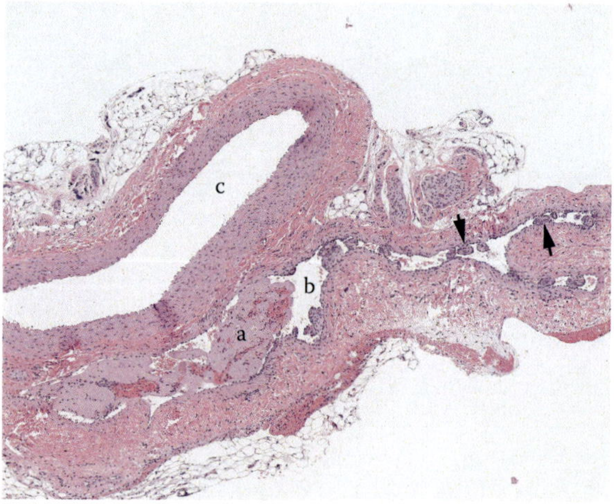

Figure 7.14 Femoral vein infusion site from Sprague Dawley rat, Day 14. Note intimal proliferation forming frond-like luminal projections (arrows) and aggregations of fibrin mixed with leukocytes (a) in femoral vein lumen (b). Adjacent femoral artery (c) is unaffected. (H&E stain, 50×.)

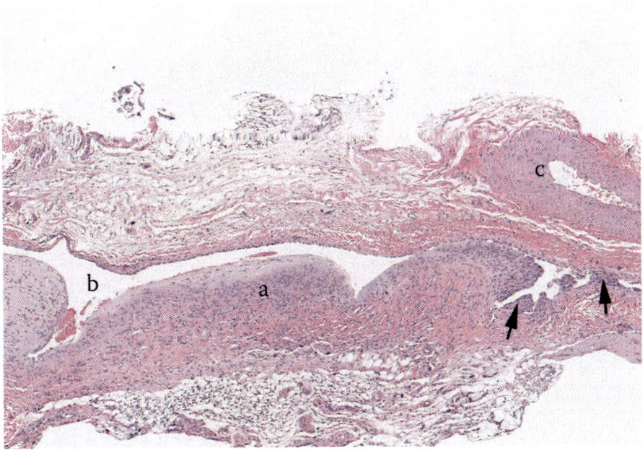

Figure 7.15 Femoral vein infusion site from Sprague-Dawley rat, Day 14. Note plaque-like area of intimal proliferation (a) with villous projections (arrows) into femoral vein lumen (b). The adjacent femoral artery (c) is unaffected. (H&E stain, 50×.)

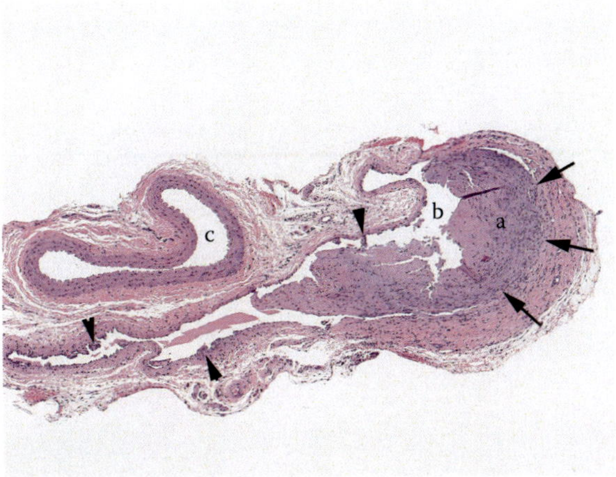

Figure 7.16 Femoral vein infusion site from Sprague Dawley rat, Day 14. Note segmental endothelial disruption (arrows) with attached mural thrombus (a). Villous intimal proliferation (arrowheads) is present. The lumen of the femoral vein (b) is moderately compromised. The femoral artery is unaffected (c). (H&E stain, 50×).

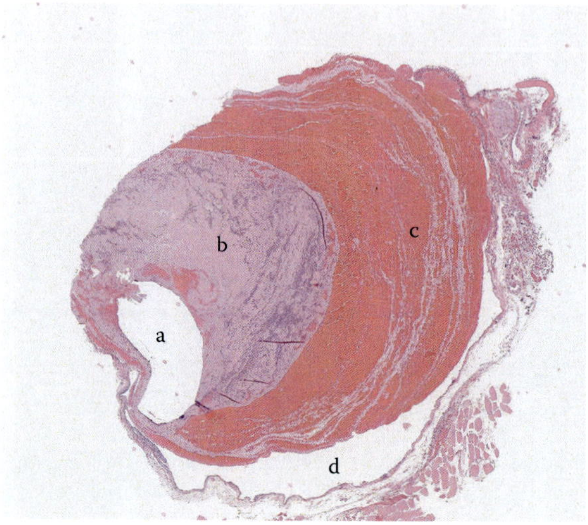

Figure 7.17 Specimen of femoral vein taken 2 cm anterior to infusion site of Sprague Dawley rat, Day 14. Note catheter tract (a) surrounded by zone of fibrin, necrotic debris, and leukocytes (b) with external coagulum (c) composed of erythrocytes and fibrin (c). Lumen of femoral vein (d) is severely compromised. (H&E stain, 25×).

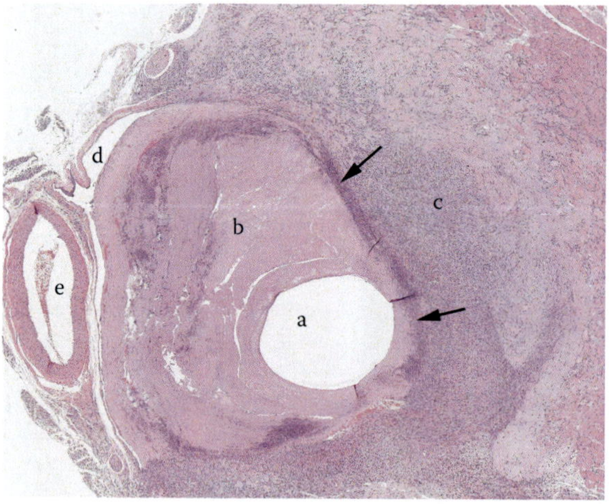

Figure 7.18 Femoral vein infusion site from Sprague Dawley rat, Day 14. Note catheter tract (a) surrounded by zone of necrotic debris and fibrin (b). Extensive discontinuity in vessel wall (arrows) is associated with pronounced inflammatory cell infiltration (c). Lumen of femoral vein (d) is severely compromised. Femoral artery (e) is unaffected. (H&E stain, 25×.)

chapter five

Reproductive models

Contents

The objective of this chapter is to discuss vascular infusion technology in the gestating animal. Developmental and reproductive toxicology (DART) studies require the use of both rodent and non-rodent species. The species most often used for DART studies are the rat or mouse as the rodent species and the rabbit as the non-rodent species. Other non-rodent species that may be used are the dog and the minipig. This chapter will give insights into the use of the rabbit, dog, rat, and mouse for this specialist area of toxicological assessment using this route of administration.

For pre-clinical pharmaceutical testing, three standard designs, as dictated by the International Conference on Harmonisation Tripartite Guideline are typically conducted: the ICH 4.1.1 study is aimed at evaluating effects on fertility and the reproductive process. Animals are exposed beginning prior to mating, through mating and implantation; the ICH 4.1.2 study is aimed at understanding the effects of exposure on the pregnant/lactating female and on the development of the conceptus and the offspring. The maternal animal is exposed from implantation through weaning of the offspring, which continue to be evaluated through sexual maturity; lastly, the ICH 4.1.3 study is designed to evaluate effects on the pregnant female and the developing foetus during the period of major organogenesis (implantation through closure of the hard palate)—the developmental toxicity study. While the infusion methods described in this chapter could be adapted for all three protocols, the focus of this chapter will be on developmental toxicity studies.

This chapter is divided into three parts (5.1: Rabbit and Dog; 5.2: Rat; 5.3: Mouse) reflecting the expertise of three different laboratories in vascular infusion animal models used in the assessment of reproductive toxicology endpoints. Whilst there may be some overlap with particular chapters of this book concerning these species (Chapters 1, 2 and 3), more specific details are provided here to support these specific study types.

5.1 *Rabbit and dog*

Teresa R Gleason BS, LVT, SRS, RLATG

WIL Research, USA

5.1.1 *Introduction*

5.1.1.1 *Choice and relevance of the species*

Mice, rats, and rabbits are the most commonly used species for DART studies. Due to their extensive use, historical control databases for these three species are extensive although the strength of individual databases does vary based on the choice of the strain. In the United States, the CD1 mouse (Crl:CD1), Sprague Dawley [Crl:CD(SD)] rat and the New Zealand White [Hra:(NZW)SPF] rabbit are most commonly used. Continuous vascular infusion in a gestating mouse is technically much more challenging as compared to the rat, and thus the rat is generally the more commonly used species. Non-rodent species such as the dog (Stump et al. 2002) may be used under special circumstances where the use of the rat and/or rabbit models is not recommended.

5.1.1.2 *Regulatory guidelines*

Guidelines for the assessment of toxic effects of pharmaceuticals, industrial and agricultural chemicals and food additives and ingredients on reproduction and development were developed in response to human health tragedies resulting from *in utero* exposures. The three main instances providing momentum for improvement of existing guidelines: the thalidomide epidemic, increasing epidemiological evidence for the link between *in utero* diethylstilbesterol exposure and vaginal clear-cell adenocarcinoma and identification of behavioural and functional alterations associated with exposure to methyl mercury (Christian et al. 2006). The history of the development of DART guidelines has been reviewed extensively in previous manuscripts (Hood 2006).

Regulatory guidelines for the conduct of DART studies are categorized into three broad groups: pharmaceuticals (US FDA 1994, 1996 and

ICH 2005; food and food additives (US FDA 1982, 2000; and industrial and agricultural chemicals (OECD 1983, 1995, 1996, 2001 and US EPA 1998). The Guidelines for Toxicity to Reproduction of Medicinal Products, put forth by the International Conference for Harmonisation (ICH) provide guidance for registration of pharmaceuticals intended for human use and is accepted by regulatory bodies of the European Union, Japan and the United States. International harmonisation efforts are also ongoing between the US EPA and the Organisation for Economic and Cooperative Development (OECD) (Christian et al. 2006).

5.1.1.3 *Choice of infusion model*

When conducting a developmental toxicity study, continuous intravascular infusion is one of the ways of ensuring that the foetus is exposed to a test article during all stages of organogenesis. This is especially true for labile materials with short half-lives where bolus dosing via oral gavage or intravenous injection may not be ideal. Depending upon the test species, each stage of embryonic and foetal development may only be a few hours in duration. Here continuous intravascular infusion provides one way of ensuring that the embryo/foetus is exposed to the test substance during each developmental stage.

Additionally, although the equipment necessary for continuous intravascular infusion (jackets, tethers, etc.) may cause some stress to the animal, with proper acclimation this should prove less stressful than the prolonged restraint or repeated needle sticks required in a bolus or intermittent dosing regimen.

5.1.1.4 *Limitations of available models*

One of the biggest limitations of available animal models for developmental toxicity studies, both rodent and non-rodent, is the timing of breeding relative to surgery. Our laboratory has had success with breeding of catheterized animals after allowing sufficient time for the females to recover from the surgical procedure. Rabbits are generally inseminated artificially. Rats however require natural mating and there is always an increased risk of damaging or dislodging the implanted catheter when a tethered female rat is placed with a male for breeding. With the longer gestation period in the canine model, female dogs are generally received from the supplier time-mated. However, this complicates the catheterisation as pregnancy limits the choice and use of available anaesthetics and analgesics.

Other limitations of continuous intravascular infusion include delivery of partial doses due to equipment failure or malfunction. This is especially true for labile materials with short half-lives and may result in individual animals being under-dosed during critical periods of organogenesis. However, given that developmental toxicity studies are typically conducted with a larger sample size (a minimum of 16–20 litter evaluated at term) as compared to general toxicity studies, such incidents are unlikely to jeopardise the outcome of the study.

When considering a non-surgical model, peripheral vessel size and restraint of the animal through gestation are the greatest limitations. Misplacement or dislodgement of a temporary catheter may complicate the dose delivery and potentially limit exposure of the foetus to the drug that could cause an adverse effect to be missed.

5.1.2 *Best practice*

5.1.2.1 *Surgical models*

5.1.2.1.1 Training. Training of the surgical staff and support staff is critically important to the success of a surgical model for vascular infusion. Personnel performing surgical procedures must demonstrate a proficiency in proper sterile technique, tissue handling, haemostasis, and closure techniques. In addition, these individuals are usually responsible for peri-operative care, including anaesthesia, analgesia, and post-operative care. There are few regulations that discuss in detail the manner in which surgical training or how a qualification process for surgical personnel is to be performed in the research setting. In the United States, the Animal Welfare Act provides some guidance as to specific topics related to training for surgery. These requirements are useful for the development of an institutional surgical training programme. A comprehensive training programme should include the following components and employ both didactic and practical training and evaluation: regulatory/ethical considerations; facility requirements; pre-surgical planning; analgesia; anaesthesia; aseptic technique; surgical technique; and post-operative care. Additionally, it is necessary to include post-training education and re-evaluation to ensure that procedural drift has not occurred. Certification programmes are available for non-doctoral degree personnel through the Academy of Surgical Research to ensure competency in anaesthesia (Surgical Research Anaesthetist), minor surgical procedures (Surgical Research Technician), and more complex surgical procedures (Surgical Research Specialist) (ASR 2009).

Training is also necessary for individuals responsible for dosing vascular infusion animals. Infusion studies are most successful when the maintenance and dosing of the animals is limited to a highly trained team of technicians. Ideally, this infusion team should be trained in aseptic technique, handling and maintenance of the surgically prepared animal, maintenance of all external equipment on the surgical animal models and non-surgical animal models, and use of the infusion pumps. These individuals must be additionally trained on the needs of gestating animals and be mindful of any restraint equipment (jackets, etc.) that may affect the dam. Post-training evaluation is critical for these individuals as well.

5.1.2.1.2 Applicability of the methodology: Surgical model of rabbit and dog. Surgical records must be maintained for each animal to accurately document the surgical procedure, including all complications that may have occurred. Keeping in mind that the surgery itself is not what is being evaluated, surgical complications that could affect the health of the animal, such as excessive bleeding or breaks in sterility, should be documented and may be reasons to exclude the animal from study.

Recommended infusion rates for vascular infusion are dependent upon the volume and duration of the dose. Guidelines state that for continuous intravenous infusion, the rate should not exceed 40–60 mL/kg/day (Hull 1995). An accurate way of calculating the dose administered is to weigh the dosing reservoir before and after dosing in grams, and then convert the weight to millilitres by dividing the difference by the density of the test article (g/mL) to derive the volume administered. The actual delivered volume is then divided by the expected volume and multiplied by 100 to get a percentage of dose delivered.

5.1.2.1.2.1 Rabbit femoral vein. The choice of vascular access in the rabbit for the developmental toxicity study in our laboratory is the femoral vein. After the catheters are placed, the animals are allowed to recover from surgery for a minimum of 7 days, at which time they are artificially inseminated. The animals are dosed from gestation day (GD) 7 to GD 20. As this is an ambulatory infusion model, standard rabbit caging with approved environmental enrichment will suffice.

Analgesics should be administered prior to surgery, and the animal should be observed closely post-operatively for signs of pain and distress. Post-operative analgesics are administered on an as-needed basis, as many can cause a decrease in appetite and it can be difficult to stimulate a rabbit's appetite once it has become anorexic. Antibiotics are not necessary if strict aseptic technique is observed and may be contraindicated in the rabbit, as it can disrupt the natural gut flora; however, we have found that enrofloxacin (5 mg/kg) may be given if necessary.

5.1.2.1.2.2 Dog jugular vein (percutaneous method). The choice of vascular access in the dog for a developmental toxicity study is the jugular vein. Pregnant animals arrive from the supplier between GD 4 and GD 10, and a percutaneous catheter is placed in the jugular vein before GD 18. The animals are dosed from GD 18 to GD 35. The percutaneous method is chosen, as it requires a much less invasive surgical procedure than the traditional vascular access port. The pregnant animal is exposed to less potential toxins, as the catheters are placed under the inhalant isoflurane only, and antibiotics and analgesics are generally not necessary. It also has the advantage that the entire system may be removed once dosing is completed.

5.1.2.1.3 Risks/advantages/disadvantages. The main risk of the described surgical procedures is the potential for introducing infectious pathogens during the surgical procedure or post-operatively. Replacement animals may be necessary in case of catheter dislodgement or patency issues.

5.1.2.2 Non-surgical models

5.1.2.2.1 Methods of restraint. The non-surgical methods of dosing the rabbit and the dog intravenously necessitate the use of restraint. For the rabbit, the marginal ear vein needs to be accessible, and there are commercially available restraints designed for this purpose (Figure 5.1). Because the animal must be securely restrained, it is not recommended to use this type of restrainer for longer than a 30-minute dosing period. If a longer dosing period is necessary, a cloth wrap-style restraint may prove useful with proper acclimation. Typically, developmental toxicity studies with an infusion period of greater than 30 minutes daily have not been conducted at our laboratory.

A sling style restraint (Figure 5.2) is generally used for short-term vascular infusion in the dog to allow access to the cephalic or saphenous veins. Our laboratory has not conducted intravenous DART studies on dogs which were not surgically implanted, but would not recommend a restraint period of longer than 30 minutes, because of the pressure on the abdomen. Alternate restraint designs may need to be considered for periods longer than 30 minutes.

5.1.2.2.2 Applicability of the methodology. Recommended infusion rates for vascular infusion are dependent upon the volume and duration of the dose. Guidelines state that for a period of infusion less than 6 hours,

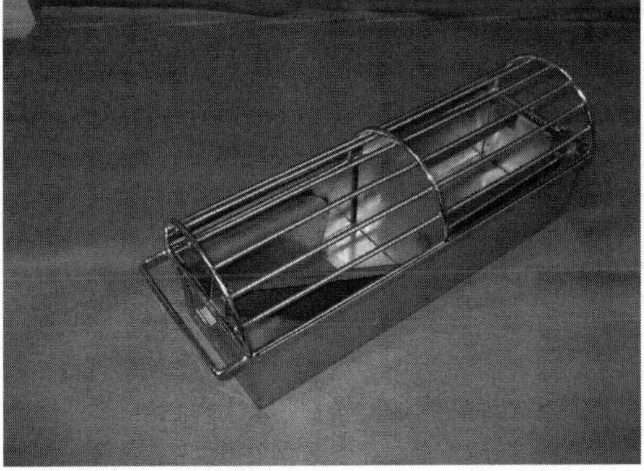

Figure 5.1 Rabbit restrainer for vascular dosing.

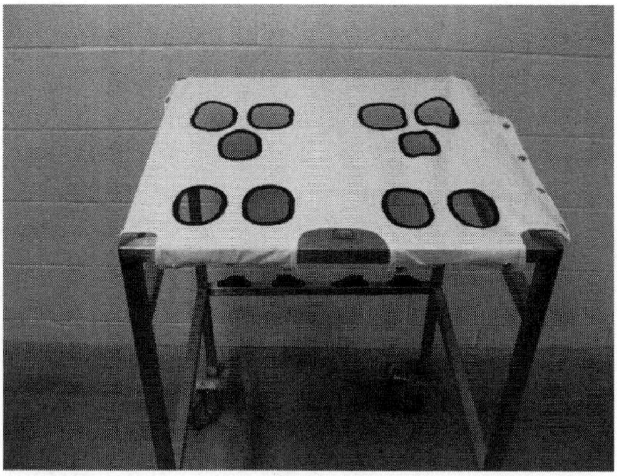

Figure 5.2 Dog sling-style restrainer (Lomir Biomedical.)

10 mL/kg/hr is an acceptable rate, not to exceed 60 mL/kg/day. If the infusion is over a period of less than one hour, a maximum rate of 0.25 mL/kg/min may be used (Hull 1995.)

5.1.2.2.3 Risks/advantages/disadvantages. The non-surgical model may be more appropriate for intermittent dosing of test articles with a longer half-life. The primary disadvantage of placing a catheter daily is that incorrect placement or dislodgement of the catheter will result in extravascular dose administration. This will result in the animal receiving an incomplete dose and may result in data interpretation errors. The prolonged restraint necessary for the non-surgical models may also cause unwanted stress on the gestating animal.

5.1.3 Practical techniques

5.1.3.1 Surgical model: Rabbit femoral vein catheterization

5.1.3.1.1 Preparation. Several days prior to the surgical procedure, the rabbits are incrementally acclimated to the infusion jacket and cervical collar. The day prior to the surgery, the dorsal and scapular areas could be clipped to cut down on anaesthesia time during the presurgical preparation period. Rabbits are not fasted overnight prior to surgery.

5.1.3.1.1.1 Premedication. Rabbits are premedicated with a sedative (acepromazine, 3 mg/kg), an analgesic (buprenorphine, 0.01 mg/kg), and an anticholinergic (glycopyrrolate, 0.01 mg/kg) delivered subcutaneously. If an antibiotic (enrofloxacin, 5 mg/kg) is to be administered, it should be given pre-operatively. Premedication and presurgical preparation should be conducted in a prep area separate from the surgical suite at least until all surgical sites have been clipped.

Anaesthesia is induced using 5% isoflurane inhalant anaesthesia in 100% oxygen delivered via a calibrated vaporized and face mask.

5.1.3.1.1.2 Presurgical preparation. Presurgical preparation begins once the animal reaches an adequate level of anaesthesia. Lubricating eye ointment is applied, and a temporary over-the-needle catheter is placed in an ear vein for supplemental fluid delivery and emergency use during the surgical procedure. The left inguinal surgical site is clipped and the scapular surgical site is clipped or touched up as necessary if clipping had been performed the previous day. The animal is moved to the surgical suite and placed on a warm-water-circulating blanket.

The animal is instrumented for intra-operative monitoring of electrocardiography, indirect blood pressure, pulse oximetry, body temperature, expired CO_2, heart rate, and respiratory rate. Warmed lactated Ringer's solution is delivered intravenously at a maintenance rate of approximately 10 mL/kg/hr.

The animal is positioned in left lateral recumbancy with the right hind limb retracted to allow access to the left inguinal and scapular areas. The surgical sites are prepared with a minimum of three alternating chlorhexadine and 70% isopropyl alcohol scrubs followed by a betadyne solution paint. All skin disinfectants are kept in a warming bath to avoid chilling the rabbit during the surgical scrub procedure.

Once the surgical scrub is completed, the animal is draped using a four-corner technique and sterile adhesive drape over the incision sites.

The surgeons prepare by donning mask and hair covering and performing a surgical scrub on forearms and hands. A leave-on surgical sterilant is applied, and once it is dry, the surgeons don sterile gowns and gloves.

5.1.3.1.1.3 Anaesthesia. Isoflurane anaesthesia (approximately 2%) is used throughout the procedure and delivered via face mask. The anaesthetist monitors vital signs and reflexes continuously. Emergency drugs such as lidocaine, epinephrine, atropine, and doxapram should be available with calculated doses for use in case of a cardiovascular or respiratory emergency.

5.1.3.1.2 Surgical procedure. The skin parallel vascular access port and a 3- to 6-French tapered rounded-tip polyurethane catheter are attached and filled with sterile saline. Using a #10 scalpel blade, a shallow incision approximately 2–3 cm in length is made over the left femoral region, and an incision approximately 4–5 cm long is made in the skin perpendicular to the spine, approximately 4 cm caudal to the shoulder blades.

A trocar is used to tunnel the catheter subcutaneously across the back from the dorsal incision to the ventral incision. The femoral vein is isolated from the surrounding muscle and interstitial tissue by blunt dissection with mosquito haemostats and forceps, and three 4-0 non-absorbable (braided polyester) ligatures are passed under the vein. The femoral vein is ligated with the distal suture, and the cranial suture is used to restrict blood flow. The adventitia of the vessel is grasped with micro-dissecting forceps, and a small venotomy is made using Castroviejo scissors. The catheter is advanced approximately 6 cm into the caudal vena cava via the femoral vein to the approximate level of the kidney. Suture beads are located at 6 and 6.5 cm from the tip of the catheter, and the catheter is secured in place by tying the cranial suture around the vein and catheter as well as securing the remaining two ligatures to the catheter between the suture beads. To further secure the catheter, an additional suture (4-0 non-absorbable with attached needle) is passed through the underlying musculature and secured to the catheter at the area of the suture beads. Patency should be checked after catheter placement and after ligatures are tied to ensure good blood flow.

A subcutaneous pocket is made cranial to the dorsal incision, and the port is placed into the pocket and secured to the underlying musculature using 2-0 non-absorbable suture (braided polyester.) This is to ensure the port is not directly beneath the incision, which could lead to dehiscence.

The subcutaneous and subcuticular layers of the incisions are closed with 4-0 absorbable suture (Vicryl or equivalent.) The inguinal skin incision is sealed with tissue adhesive, and the dorsal skin incision with surgical staples.

Anaesthesia is discontinued, and the animal is moved to a heated recovery cage until fully ambulatory. A cervical collar is placed on the animal prior to returning to its home cage.

The above listed procedure could also be followed to allow implantation of pumps that could be directly attached to a catheter. The pump (an Alzet® osmotic pump or iPRECIO™ refillable battery-operated pump) could be placed subcutaneously in the scapular area and attached to an appropriately sized catheter as described. This would allow for continuous infusion; however, there are limitations on volume and rates when using these types of pumps.

5.1.3.2 Surgical model: Dog jugular vein catheterization (percutaneous method)

5.1.3.2.1 Preparation. Animals are time mated at the supplier and shipped to the lab to arrive between gestation days (GD) 4 and 10. Several days prior to the surgical procedure, the dogs are incrementally acclimated to the infusion jacket and pillow collar. The dogs are fasted overnight prior to surgery.

5.1.3.2.1.1 Premedication. As the described percutaneous method is minimally invasive and the animals are gestating, no pre-surgical medications are administered. Anaesthesia is induced using 5% isoflurane inhalant anaesthesia in 100% oxygen delivered via a calibrated vaporizer and face mask.

5.1.3.2.1.2 Presurgical preparation. Presurgical preparation begins once the animal reaches an adequate level of anaesthesia. Lubricating eye ointment is applied, and a temporary over-the-needle catheter is placed in a cephalic vein for emergency use during the surgical procedure. The hair is clipped over the right jugular vein and extended to the midline of dorsal and ventral neck to ensure adequate exposure. The animal is then moved to the surgical suite and placed on a warm-water circulating blanket.

The animal is instrumented for intra-operative monitoring of electrocardiography, indirect blood pressure, pulse oximetry, body temperature, expired CO_2, heart rate, and respiratory rate.

The animal is positioned in left lateral recumbency to allow access to the right jugular vein. The animal is loosely positioned in a two-pouch backpack jacket that is instrumented with a validated infusion pump, saline reservoir, and appropriate connective tubing threaded through the mesh of the jacket from the right pocket to the area of the collar. The surgical site is prepared with a minimum of three alternating chlorhexadine and 70% isopropyl alcohol scrubs followed by a betadyne solution paint.

Once the surgical scrub is completed, the animal is draped using a four-corner technique and sterile adhesive drape over the incision site.

The surgeons prepare by donning mask and hair covering and performing a surgical scrub on forearms and hands. A leave-on surgical sterilant is applied, and once it is dry, the surgeons don sterile gowns and gloves.

5.1.3.2.1.3 Anaesthesia. Isoflurane anaesthesia (approximately 2–3%) is used throughout the procedure delivered via face mask. As the animal has no sedative or analgesic drugs on board, the maintenance rate of anaesthesia may be higher than typically used during surgery. The anaesthetist monitors vital signs and reflexes continuously. Emergency drugs such as lidocaine, epinephrine, atropine, doxapram, dexamethasone, and diphenhydramine should be available with calculated doses for use in case of a cardiovascular or respiratory emergency.

5.1.3.2.2 Surgical procedure. Using a #10 scalpel blade, a shallow incision approximately 2 cm in length is made over the right jugular vein approximately mid-neck. An 18-g introducer needle with a peel-away sheath is inserted into the jugular vein in the direction of blood flow. The needle stylet is removed from the peel-away sheath and the 20-g polyurethane catheter is advanced into the jugular vein. The placement sheath is then peeled away and the catheter is fully advanced (approximately 10 cm) to the suture wing. The catheter stylus is removed and a prefilled 7" extension set, with an attached BD Posiflow™ and syringe, are attached to the catheter. The catheter is checked for patency by drawing up a small amount of blood and then thoroughly flushing the catheter. The catheter is then secured to the neck using 2-0 non-absorbable suture (monofilament nylon preferred to prevent bacteria from wicking through a braided suture.) The incision is closed over the bottom end of the suture wing to prevent the catheter from being dislodged when the dog moves her neck.

The catheter insertion site is thoroughly cleaned and dried and then covered with a pad of sterile gauze treated with antiseptic powder (McKillips). Secure the bandage with roll gauze and elastoplast tape.

Anaesthesia is discontinued, and the animal is fitted into its infusion jacket. The extension tubing is connected to the catheter at the Posiflow connector, and the pump is turned on to deliver saline at a keep-vein-open (KVO) rate of 0.5 mL/hr. The animal is moved to a heated recovery cage until fully ambulatory. A pillow collar is placed on the animal and attached to the jacket prior to return to its home cage.

5.1.3.2.3 Maintenance: Rabbit and dog surgical models

5.1.3.2.3.1 Recovery procedures. Surgically prepared animals must be observed a minimum of twice daily for signs of pain and distress, and surgical sites must be closely monitored for signs of irritation and infection. Any questionable observation must be brought to the attention of the veterinary staff to determine a course of action. Surgical staples for the rabbits should be removed 10–14 days post-operatively.

5.1.3.2.3.2 Housing. Because of the external equipment, animals must be single-housed, but there is no need for specialized caging, as the rabbit and dog surgical models described in this chapter are ambulatory. Care must be taken at time of cage or kennel cleaning that all external equipment remains clean and dry.

Environmental enrichment in the form of toys should be available at all times and should be rotated at least weekly to stimulate the animals. Chew toys provide a distraction for the dogs, but they still may prefer chewing on their jacket sleeves. Dogs may be exercised in groups with supervision.

5.1.3.2.3.3 Checks and troubleshooting. During the dosing period, all external equipment should be checked a minimum of four times daily. Items to be considered at the routine checks include infusion pump alarms; level of the reservoir; kinks or leaking extension tubing; condition of the catheter insertion sites; fit and condition of the infusion jackets and collars; and condition of any bandaging material.

Any repairs to the systems require strict aseptic technique. Dosing catheters (Cath-in-Cath external catheter for the rabbit, and jugular catheter for the dog) should not need to be replaced unless they become dislodged.

5.1.3.2.4 Record keeping. Detailed records must be kept for all events that occur on a study. All health records, surgical records, details of dosing, and equipment repairs must be recorded so that the data may be accurately interpreted.

5.1.3.3 Non-surgical models: Rabbit and dog

5.1.3.3.1 Methods of restraint and acclimatisation. Animals must be restrained for short-term infusion studies, as described in Section 5.1.2.2.1 In order to decrease the stress involved with prolonged restraint, the animals must be acclimated a minimum of four times on an incremental basis to the restraint method. For an infusion period of 30 minutes, an acclimation regimen may consist of two 10-minute sessions, one 20-minute session, and one 30-minute session prior to initiation of dosing.

5.1.3.3.2 Methods of vascular access. Prior to placement of daily over-the-needle catheters, the dose site (marginal ear vein for the rabbit and cephalic and saphenous veins for the dog) needs to be clipped and all excess hair removed. The site should then be prepared with 70% isopropyl alcohol, and the catheter placed in an aseptic manner.

5.1.3.3.3 Housing. Standard housing with approved enrichment devices is acceptable for these models. Dogs may be exercised in groups with supervision and may be pair housed.

5.1.3.3.4 Checks and troubleshooting. During the infusion period, the animals are monitored continuously. The infusion sites are checked for signs of swelling during dosing, which if present would indicate extravascular dose administration and require placement of a catheter in an alternate location for the remainder of the dose administration. Animals must never be left unattended in a restraint apparatus, and if an animal begins to struggle, it must be calmed to avoid injury.

5.1.4 Equipment

5.1.4.1 Surgical models

5.1.4.1.1 Surgical facilities. Catheter implantation as described in this chapter is considered a minor surgical procedure, as a body cavity is not penetrated. This does not, however, diminish the need for dedicated surgical facilities to conduct these procedures. Implantation of a foreign body increases the chances for surgical complications. In human cardiovascular implantation surgery, infection is the most common complication and ultimately leads to device removal (Wilkoff 2008).

Dedicated surgical facilities should be located in an area that will minimize traffic and thus decrease the potential for contamination. The facilities must have limited fixed equipment and be constructed of materials that are easily cleaned and sanitized. The design should include separate areas for instrument and pack preparation, pre-surgical preparation of the animals, surgeon preparation, surgery, and recovery. Each area must be equipped appropriately to maximize surgical support for the animals (NAS 2011).

5.1.4.1.2 Catheters/vascular access ports. The Cath-in-Cath II Port™ (AVA Biomedical) (Figure 5.3) is the preferred vascular access port for rabbits in our laboratory. One advantage over traditional vascular access ports is a low profile, skin parallel port that minimizes the occurrence of pressure necrosis of the skin over the port access site. The port can be attached to a tapered 6-French–3-French catheter, which is ideal for placement in the rabbit femoral vein. Additionally, the port is accessed with a flexible infusion catheter that minimizes skin injury and lessens the likelihood of needle dislodgement, which is commonly seen with traditional Huber needles. This external catheter can also be replaced if it is damaged or dislodged.

The percutaneous jugular catheter (SAI Infusion Technologies) (Figure 5.4) works well for the gestating dog, as it requires a much less invasive surgical procedure than placing a traditional subcutaneous catheter and vascular access port. The catheter and all bandaging materials are removed once dosing is complete. This preparation should not be used for long-term vascular infusion (greater than two or three weeks), but is ideal for dosing over the major organogenesis period of the dog (GD 18–35.)

5.1.4.1.3 External equipment. The external equipment for the gestating rabbit surgical model begins with the external dosing catheter

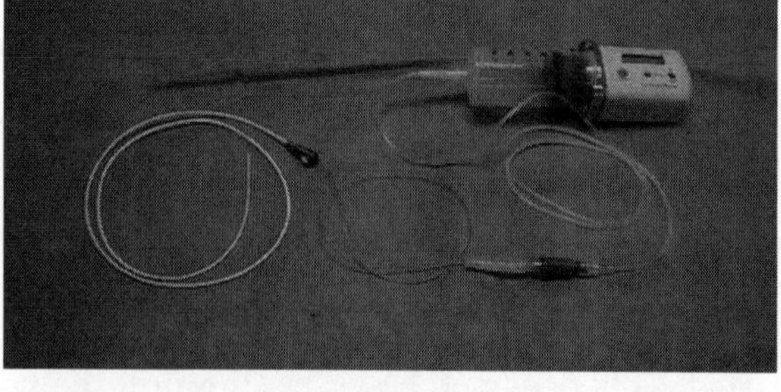

Figure 5.3 PCA 50 infusion pump attached to a Cath-in-Cath™ Infusion System (AVA Biomedical.)

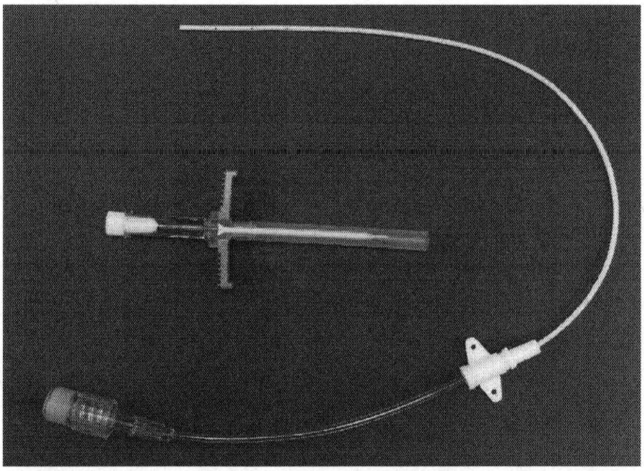

Figure 5.4 Percutaneous jugular catheter (SAI Infusion Technologies) with peel-away introducer needle.

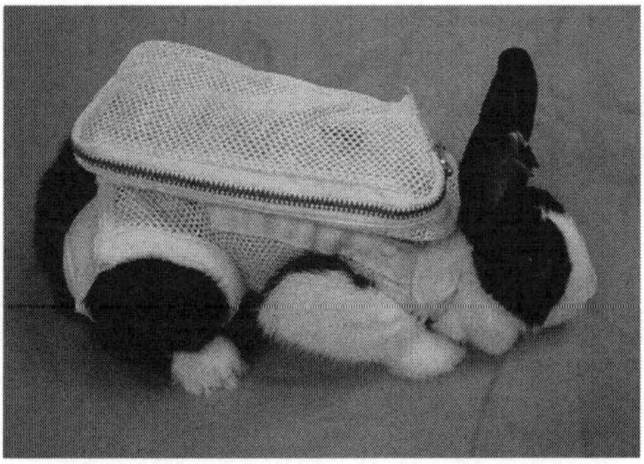

Figure 5.5 Dutch Belted rabbit with ambulatory infusion jacket (Lomir Biomedical). Pocket houses PCA-50 infusion pump and reservoir.

(AVA Biomedical). This is part of the Cath-in-Cath II system for accessing the subcutaneous vascular access port. An 18-g introducer is used to place the 24-g nylon dosing catheter through the skin and into the implanted catheter via the skin parallel port. A locking Luer adaptor is then attached to the catheter, and the system is closed using a Posiflow injection cap (Becton Dickinson). The external dosing catheter is then routed into the pocket of the infusion jacket (Lomir Biomedical) and the Posiflow is attached directly to the dosing syringe of the PCA 50 pump (Figures 5.3 and 5.5). Note that the jacket should be ordered with a spandex abdominal panel to allow growth of the pregnant rabbit. Additionally, the cervical collars used during the post-op recovery period come in various sizes from Lomir Biomedical.

For the dog, an extension set (appropriate for the type of pump used) is weaved through the mesh of the infusion jacket (Lomir Biomedical) and is connected to the Posiflow injection cap (Becton Dickinson) that was placed during the surgical procedure (Figure 5.6). A pillow collar that is attached to the jacket is then applied. As with the rabbit, the infusion jacket should be ordered with a spandex abdominal panel.

5.1.4.1.4 Infusion pumps. Traditionally the tethered infusion model (Figure 5.7) has been the only choice for the rabbit because of their inability to carry the weight of the pump and reservoir in an infusion

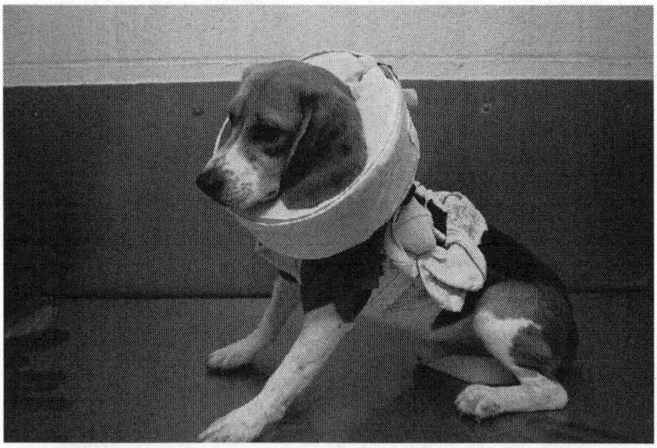

Figure 5.6 Dog in Lomir Biomedical pillow collar and ambulatory infusion jacket. Collar protects bandaged surgical site as well as keeping the animal from accessing the jacket.

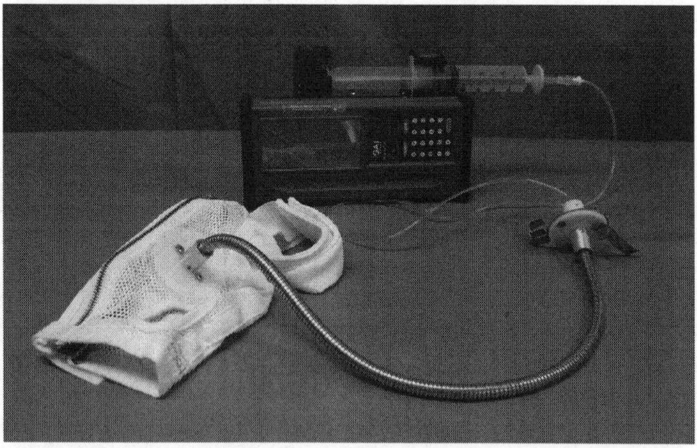

Figure 5.7 Traditional rabbit jacket with tether/swivel and SAI 3D syrine pump.

jacket. With the availability of smaller and lighter infusion pumps, our laboratory has gone exclusively to the ambulatory model. The PCA 50 pump is approximately 140 grams (with an empty reservoir) and has been used successfully on New Zealand white and Dutch Belted rabbits alike.

As our dog model is also ambulatory, any ambulatory infusion pump could be used depending upon the necessary rates or volumes needed for the study. Typically our laboratory uses the CADD Legacy Plus pump for dogs with either a cassette reservoir or a bag reservoir for larger volumes.

Infusion pumps should be calibrated before use on each study at rates that represent the range of settings and reservoir type that will be utilized. Battery life is often dependent upon rate of infusion, and should be considered for each study. Yearly manufacturer maintenance routines are also recommended.

5.1.4.2 *Non-surgical models*

5.1.4.2.1 Restraining devices. The restraint device used for the rabbit for intermittent infusion is made by Draper Metal Fabrication (Figure 5.1). This provides good access to the marginal ear veins or central ear arteries. Soft, wrap style restrainers are commercially available as well (Bunny Snuggle from Lomir Biomedical, or cat restraint bags from veterinary suppliers) and may be more appropriate for longer infusion periods.

A sling style restraint device is used for dog intermittent infusion. Figure 5.2 illustrates a dog sling designed to house two dogs at a time; however, Lomir Biomedical can customize sling frames and canvases to accommodate single animals as well. As stated earlier, our laboratory has not had experience with dosing gestating dogs that were not surgically implanted, so while this sling style restraint is used on general toxicology studies, alternative designs may need to be considered to alleviate pressure on the pregnant animal's abdomen.

5.1.4.2.2 Vascular access. For the rabbit non-surgical model, the marginal ear vein is most often used for infusion. This would be accessed with a Terumo® Surflash™ over-the-needle catheter (24G × ¾"). A prefilled extension set would then be connected to the catheter, and a splint would be placed in the ear to provide support for taping the catheter in place. The splint also keeps the ear from bending so that the catheter is not dislodged. Splints are commercially available through veterinary supply companies, or they can be created by rolling disposable towels and wrapping with tape.

The cephalic vein is the easiest peripheral vessel to access in the dog; however, the saphenous and jugular veins may be used as well. Some protocols call for a rotation of infusion sites, and while this seems appropriate, it can place unnecessary stress on the technical staff in that they must hit the left cephalic vein on all animals on a particular dosing day.

Our experience has shown that it is best to allow the technician to evaluate the condition of the dosing sites and decide where best to place the catheter on a given day. Individual dosing sites must be recorded on a daily basis for each animal. For the cephalic or saphenous veins, a Terumo Surflash over-the-needle catheter (22G × 1") would be used, and for the jugular vein, a 2–3" catheter is recommended. These are attached to an appropriate length pre-filled extension set and are easily secured with tape. It is advisable to use a long enough extension set to allow the addition of a shock loop to prevent the animal from dislodging the catheter.

5.1.4.2.3 Infusion pumps. Various syringe pumps could be used for this model depending upon the necessary delivery rate. Syringe pumps should be calibrated before use on each study at rates that represent the range of settings and syringe sizes that will be utilized. Yearly manufacturer maintenance routines are also recommended.

5.1.5 Background data: Resultant pathologies

5.1.5.1 Surgical models

5.1.5.1.1 Complications associated with the surgical process. The surgical process is a stressful event for the non-pregnant rabbit and for the pregnant dog. Rabbits tend to over-groom the surgical sites and may chew at the inguinal incision. Using tissue adhesive to close this wound in combination with a cervical collar (or an Elizabethan collar) to limit access has alleviated this problem. Trimming the rabbits' hind toenails prevents excessive damage to the port incision.

Because of the location of the catheter on the lateral neck of the dog, if the insertion site is not covered adequately by the bandage and protective collar, the dog may access the catheter and dislodge it while scratching. Frequent nail trimming prevents excessive damage.

5.1.5.1.2 Issues associated with study design. The main issue with the study design for DART models of continuous infusion is that any disruption of the dosing process could negatively affect the study. If a dosing catheter is pulled out, or a pump malfunctions and a full dose is not delivered during any period of organogenesis, an adverse affect may be missed.

An additional complication with the rabbit model is that positioning the animal for artificial insemination requires stretching the hind limbs back, and if the catheter is not securely anchored in the femoral vein, it may become dislodged. This should be evident when the external dosing catheter of the Cath-in-Cath system is put in place, as a patency check is performed at that time. If the animal is not patent, it needs to be removed from study.

5.1.5.1.3 In-life changes associated with the model. Decreases in body weight gains are expected with all jacketed animals. Both the pregnant rabbit and dog experience decrease in body weight gains compared to unrestrained animals. Expected clinical signs include swelling and reddening of the surgical sites as part of the normal healing process. Some reddening of the skin at jacket pressure points may be seen, and it is imperative to observe the growth of the animals and change jackets as the animals grow, paying particular attention to the abdominal area of the dams.

Rabbits are prone to anorexia following surgery, and in our laboratory we routinely offer kale as a dietary supplement for approximately five days following surgery.

Tables 5.1–5.3 and Figures 5.8–5.9 demonstrate the results of an in-house validation study that was run at our laboratory. Briefly, two groups of 25 female Dutch Belted rabbits underwent femoral vein catheterization, as described in Section 5.1.3.1, and were artificially inseminated after an approximately 2-week recovery period. The control group consisted of 10 inseminated females that were not surgically manipulated or treated.

On gestation day (GD) 5, the surgically catheterized rabbits were fitted with either an ambulatory jacket (Figure 5.5) or tether jacket (Figure 5.7), and the ports were aseptically accessed with the external dosing catheter. Rabbits received 0.9% sodium chloride for injection at a target rate of 0.4 mL/kg/hr continuously from GD 7 to GD 20.

Jacket systems were removed following the infusion period, and a necropsy and laparohysterectomy was conducted on all animals on GD 29. Pregnancy rates in control, ambulatory, and tethered groups were 70%, 96%, and 88%, respectively, compared to 90 ±5% in our historical control database for naturally mated Dutch Belted rabbits. Mean numbers of viable foetuses were 6.0, 5.3, and 5.0 in each respective group, with mean post-implantation losses of 10.7%, 19.0%, and 26.3%. Numbers of corpora lutea per doe were similar across groups. Foetal weights were also similar across

Table 5.1 Maternal Observations and Survival on In-House Comparison Study of Ambulatory vs. Tethered Continuous Intravenous Infusion in Dutch Belted Rabbits

	Control*	Ambulatory	Tethered
Number of gravid	7/10 (70%)	24/25 (96%)	22/25 (88%)
Number of aborted	1 on GD 25 (only 2 implants)	None	1 on GD 22 (only 1 implant)
Survived to GD 29 necropsy	9/10	25/25	24/25

*Control animals were untreated.

Table 5.2 Laparohysterectomy Data, GD 29, on In-House Comparison Study of Ambulatory vs. Tethered Continuous Intravenous Infusion in Dutch Belted Rabbits

	Untreated	Ambulatory	Tethered
Corpora lutea	8.5	8.5	7.9
Implantation sites	6.7	6.5	6.0
Viable foetuses (%)	89.3	81.0	73.7
Dead foetuses (%)	0.0	0.0	0.0
Early resorptions (%)	10.7	18.5	26.3
Females with 100% resorptions	0	2*	4*
Mean number of early resorptions	0.7	1.1	1.0
Late resorptions (%)	0.0	0.5	0.0
Total resorptions (%)	10.7	19.0	26.3
Pre-implantation loss (%)	16.8	23.8	26.8
Post-implantation loss (%)	10.7	19.0	26.3
Mean foetal weight (g)	42.9	41.3	42.3

*= 1 to 5 implantation sites in each of these females.

Table 5.3 External Foetal Malformations on In-House Comparison Study of Ambulatory vs. Tethered Continuous Intravenous Infusion in Dutch Belted Rabbits

	Untreated control	Ambulatory	Tethered
Foetuses (litters) examined	36 (6)	128 (22)	104 (17)
Acephaly*		1#	
Syndactyly*		1#	
Omphalocele*		1#	1
Exencephaly*		1	
Maxillary micrognathia*			1#
Single naris*			1#
Total foetuses affected	0	2	2

*= No occurrences in the current WIL HC database for time-mated DB rabbits (5 studies).

#= Same foetus.

groups. External malformations were present in 0/36, 2/128, and 2/104 foetuses in the control, ambulatory, and tethered groups, respectively. The ambulatory continuous infusion model was found to be comparable to the tethered system with no significant effects on pregnancy or embryo-foetal development relative to the control group (Gleason et al. 2008).

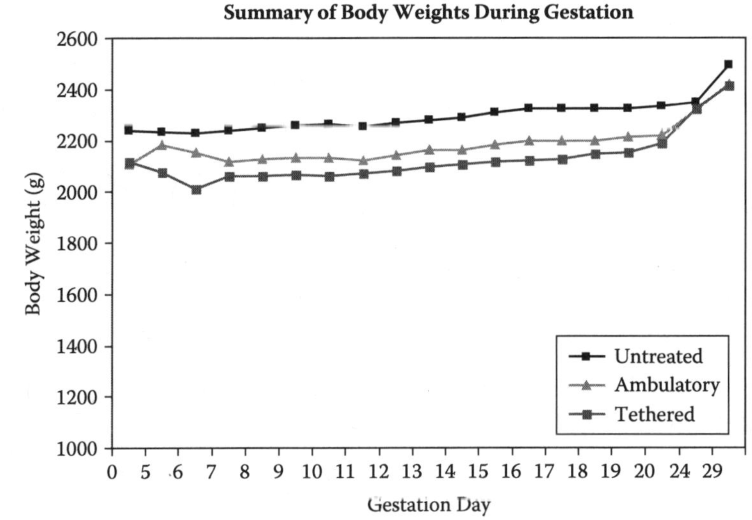

Figure 5.8 Summary of body weights during gestation.

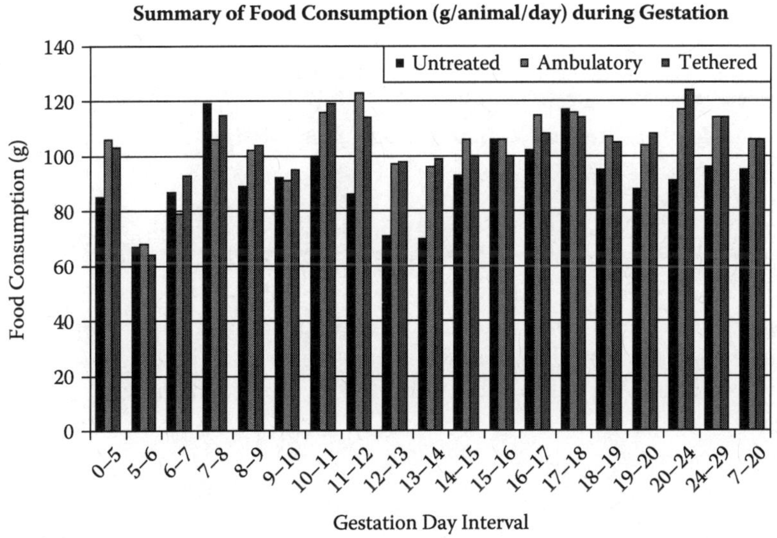

Figure 5.9 (See colour insert.) Food consumption summary on in-house comparison study of ambulatory vs. tethered continuous intravenous infusion in Dutch Belted rabbits.

5.1.5.1.4 Pathology. Pathology considerations for vascular infusion are discussed in detail in Chapter 7, Juvenile Animals; however, particular concerns for the gestating animal revolve primarily around infection. Infection of the dam during gestation may impact foetal development and could result in abortion. When multiple foetuses are present *in utero*, infectious pathogens may result in different outcomes in different foetuses (Givens and Marley 2008.) This could result in incorrect interpretation of the study data.

5.1.5.1.4.1 Clinical pathology. Clinical pathology (particularly white blood cell parameters) is affected by the healing process and the presence of a foreign body, and potentially by infection if bacterial contamination occurs during the surgical procedure or upon return to the home cage.

5.1.5.2 Non-surgical models: Rabbit and dog

5.1.5.2.1 Issues associated with study design. Repeated vascular access may be difficult in the marginal ear vein of the rabbit, particularly if the test article is irritating. As the dog has more peripheral sites to choose from, it is less problematic. Misplacement or dislodgement of the temporary catheter may further complicate subsequent doses and ultimately data interpretation.

5.1.5.2.2 In-life changes associated with the model. Restraint for short-term vascular infusion should not affect the growth curve of the gestating animal nor inhibit food consumption. Clinical signs may include irritation and swelling at the injection sites.

5.1.5.2.3 Pathology. As referred to in Section 5.1.5.1.4, infection could cause issues with the developing foetuses and could ultimately result in abortion or incorrect study interpretation. Although the non-surgical models for the rabbit and dog involve a less invasive procedure than the surgical models, it is still a parenteral technique and strict asepsis should be followed to avoid introducing pathogens during peripheral vascular access.

5.1.6 Conclusion

Regulatory guidelines require the testing of all chemicals that a human foetus may be exposed to. Rodent and non-rodent species must be tested to satisfy these guidelines. Continuous intravascular infusion provides exposure during all stages of foetal development, and the surgical preparation is less stressful than prolonged repeated restraint required of non-surgical models. Improvements in catheter designs and infusion pumps have made continuous infusion studies in gestating animals a viable option. The ambulatory method of infusion is considered less stressful for the animals than the traditional tethered and is possible in virtually all species used on DART studies.

References

ASR. 2009. Guidelines for training in surgical research with animals. Academy of Surgical Research. *Journal of Investigative Surgery* 22: 218–225.

Christian MS, Hoberman AM and Lewis EM. 2006. Nonclinical juvenile toxicity testing. In *Developmental and Reproductive Toxicology: A Practical Approach*, 2nd edition, Hood RD (ed.). Boca Raton, FL: CRC Press/Taylor & Francis Group.

Givens D and Marley MSD. 2008. Infectious causes of embryonic and fetal mortality. *Theriogenology* 70(3): 270–285. (Doi:10.1016/J.Theriogenology.2008.04.018).

Gleason T, Edwards T, Miller W, Sloter E, Stump D and Chengelis C. 2008 Ambulatory continuous intravenous infusion model for pregnant Dutch Belted rabbits. 24th Annual Academy of Surgical Research National Meeting, New Orleans, LA.

Hood, RD. 2006. *Developmental and Reproductive Toxicology: A Practical Approach*, Boca Raton, FL: CRC Press/Taylor & Francis.

Hull, RM 1995. Guideline limit volumes for dosing animals in the preclinical stage of safety evaluation. Toxicology Subcommittee of the Association of the British Pharmaceutical Industry. *Human and Experimental Toxicology* 14: 305–307.

NAS. 2011. *Guide for the Care and Use of Laboratory Animals*. Washington, DC: The National Academies Press.

Stump DG, Nemec MD, Wally CS, Chengelis CP. 2002. Embryo/fetal developmental toxicity studies in dogs via intravenous jugular vein infusion. *Journal of Toxicological Sciences* 27: 390.

Wilkoff BL. 2008. Materials and design challenges for cardiovascular implantable electronic devices. Materials and Processes for Medical Devices Conference, Cleveland, Ohio.

5.2 Rat

Melissa Chapman, BSc (Hons)

Huntingdon Life Sciences, UK

David P Myers, BSc (Hons), PhD, IDT

Huntingdon Life Sciences, UK

Karen A Critchell, HNC

Huntingdon Life Sciences, UK

5.2.1 Introduction

The details given in this section on the application of vascular infusion technology in the rat will provide a useful comparison of some contemporary reproductive endpoint data from work carried out using the model.

There will be overlap with some sections of the practical procedures, and in these circumstances appropriate reference is made to Chapter 1, Rat.

5.2.1.1 *Choice and relevance of the species*

Fertility and pre- and post-natal studies are required in at least one species, preferably the rat (ICH 2005). Studies of embryo-foetal development are usually required in two species: a rodent, preferably rats, and a non-rodent, preferably rabbits (ICH 2005).

As stated in the ICH guidelines, the rat is the preferred rodent species for the testing of medicinal products (biopharmaceuticals will be considered later), because of practicality, the ability to compare the results of reproductive studies with the results of other studies performed with this species, and the availability of a large amount of background knowledge.

The rat is viewed in the ICH guidelines as being unsuitable for testing dopamine agonists owing to its dependence on prolactin as the major hormone involved in the establishment and maintenance of early pregnancy. The only practical rodent alternative to the rat is the mouse. However, most if not all laboratories have only limited background control data in this species, and, as noted in the ICH guidelines, mice are very sensitive to stress: malformation clusters within a litter are particularly evident in this species, and this could complicate interpretation of results from embryo-foetal studies.

For the testing of biopharmaceuticals, the rat may not be a suitable species if the test material is not pharmacologically active in rats (Bussiere et al. 2009). Testing should be performed, if possible, in a species in which the test substance is pharmacologically active, and if this is not possible then other approaches such as the use of transgenic/knockout mice may be used (Bussiere et al. 2009).

5.2.1.2 *Regulatory guidelines*

Regulatory studies are performed in accordance with the ICH Harmonised Guideline S5(R2), *Detection of Toxicity to Reproduction for Medicinal Products and Toxicity to Male Fertility* (ICH 2005). The guideline is flexible with the testing strategy for each medicinal product, determined by anticipated use, the intended route of clinical administration, and previous toxicity and pharmacokinetic data.

The guideline divides one complete life cycle into 6 stages:

A. Premating to conception
B. Conception to implantation
C. Implantation to closure of the hard palate
D. Closure of the hard palate to the end of pregnancy
E. Birth to weaning
F. Weaning to sexual maturity

It is expected that studies will cover this complete life cycle, unless this is not necessary based on the intended use of the drug; for example, the testing of a drug intended to be administered only in the third trimester of pregnancy to prevent premature delivery need not cover stages A–C of the life cycle.

The Guideline details what it calls the most probable option to cover one complete life cycle. This is a set of three studies:

Study 4.1.1 Study of fertility and early embryonic development to implantation (covers Stages A and B)—sometimes called a Segment I study

Study 4.1.2 Study for effects on pre- and postnatal development, including maternal function (covers Stages C–F)—sometimes called a Segment III study

Study 4.1.3 Study for effects on embryo-foetal development (covers Stages C–D)—sometimes called a Segment II study

It is quite common to combine studies 4.1.1 and 4.1.3 into a single study of fertility and embryo-foetal development. This reduces animal and test product use and is a particularly useful approach if dose escalation is required to allow the animals to develop tolerance to the pharmacological effects of the test product; for the continuous infusion model this ensures that any surgical procedure is performed in advance of mating. Some pharmaceutical companies routinely include the assessment of male fertility on a 13- or 26-week rat toxicity study, pairing treated males with untreated females at a convenient point in the study where the 2-week pairing period will not interfere with other study procedures, and then evaluating the mating performance of the animals and the pregnancy status and litter data in the untreated females. However, if the intended route of administration is continuous infusion, it may be more appropriate to perform this as a stand-alone study because of the potential limitations of the longevity of the model to 26 weeks.

It is common to include blood sampling for toxicokinetic evaluation in the embryo-foetal development study, when samples are routinely taken after dosing on the first and last days of treatment. It is less common to include toxicokinetics in a study for effects on pre- and post-natal development; but when included, blood samples are often taken on the first day of dosing and then at peak milk production around Day 12 of lactation, before the offspring start to eat solid food. Blood sampling for toxicokinetic evaluation is rarely included in the fertility study, since data for the males is available from repeat dose toxicity studies, and treatment of the females ends during early pregnancy at implantation, before the period of embryonic/foetal organ development.

5.2.1.2.1 Study designs

5.2.1.2.1.1 Female fertility studies and combined female fertility/embryo-foetal development studies. Female rats are treated for at least two weeks before pairing, during pairing and mating, and up to the morning of Day 8 after mating to cover the completion of implantation. Group size is at least 22 animals. Sprague Dawley females are approximately 8 weeks old at the start of treatment, while Han Wistar females are 1–2 weeks older. Oestrus cycles are evaluated from the start of dosing by dry vaginal smearing using cotton buds before pairing with untreated stock males. Only females found to be in pro-oestrus are paired in order to minimise the length of time the male is with the female and thus any potential interference with the infusion system by the male partner. After pairing, wet vaginal smears are taken using pipette lavage until positive indication of mating is recorded. After completion of the final infusion on the morning of Day 8 after mating, the animal is detached from the infusion system, leaving the implanted catheter *in situ*.

The females are then retained untreated until Day 14 after mating, when they are killed, subject to caesarean section, and examined macroscopically. Corpora lutea, implantations, embryonic resorptions, and live and dead embryos are counted to allow assessment of any potential effects upon fertility or early embryonic development.

It is quite common to perform a combined study of female fertility and embryo-foetal development. In this case the final infusion would be completed on the morning of Day 17 or 18 after mating. The females would be killed on Day 20 after mating, litter parameters recorded, and foetuses weighed and examined for abnormalities, as described in the section on embryo-foetal studies.

5.2.1.2.1.2 Male fertility studies. Male rats are routinely treated for 4 weeks before pairing (the minimum guideline requirement is for 2 weeks of treatment before pairing), during pairing and mating, and up to the time when the majority of their untreated female partners have been killed, a total of at least 7 full weeks of treatment. Group size is at least 22 animals. Sprague Dawley males are approximately 13 weeks old at pairing, while Han Wistar males are 1–2 weeks older. Oestrus cycles in the untreated females are evaluated for four days before pairing with treated males. Only females found to be pro-oestrus are paired in order to minimise the length of time the female is with the male and thus any potential interference with the infusion system by the female partner. After pairing, wet vaginal smears are taken using pipette lavage until positive indication of mating is recorded.

The females are then retained until Day 14 after mating when they are killed, subject to caesarean section and examined macroscopically. Corpora lutea, implantations, embryonic resorptions and live and dead embryos are counted to allow assessment of any potential influence of the male on fertility or early embryonic development. The treated males are killed and examined macroscopically, and the male reproductive organs are weighed and retained in appropriate fixative.

5.2.1.2.1.3 Embryo-foetal development studies. At this laboratory, surgery is performed before pairing with stock males in house. At some other laboratories, time-mated females are bought from a supplier, surgically prepared following delivery, and then treated from Day 6 after mating. Group size is at least 22 animals. Sprague Dawley females are approximately 10 weeks old at pairing, while Han Wistar females are 1–2 weeks older. Female rats are treated from Day 6 after mating until the morning of Day 17 or 18 after mating, dependent upon the treatment regimen, which equates to treatment from implantation to closure of the hard palate. After completion of the final infusion the animal is detached from the infusion system, leaving the implanted catheter *in situ*.

The females are then retained untreated until Day 20 or 21 after mating (that is, 1–2 days before the expected time of parturition), when they are killed, subject to caesarean section, and examined macroscopically. Corpora lutea, implantations, embryo-foetal resorptions, and live and dead foetuses are counted. All foetuses are individually weighed, sexed, and examined for external abnormalities. At least half of the foetuses in each litter are examined for visceral abnormalities, either after preservation in Bouin's solution followed by free-hand serial sectioning, or by fresh microdissection on Day 21 of gestation. The other half of the foetuses in each litter are subject to detailed skeletal examination following staining of the skeleton with Alizarin red.

5.2.1.2.1.4 Pre- and post-natal studies. At this laboratory, surgery is performed before pairing with stock males in house. At some other laboratories, time mated females are bought from a supplier, surgically prepared following delivery, and then treated from Day 6 after mating. Group size is at least 22 animals. Sprague Dawley females are approximately 10 weeks old at pairing, while Han Wistar females are 1–2 weeks older. Female rats are treated from Day 6 after mating until the morning of Day 21 of lactation, which equates to treatment from implantation to the end of lactation. After completion of the final infusion, the animal is detached from the infusion system, leaving the implanted catheter *in situ*.

From Day 20 after mating, females are checked three times per day for evidence of parturition. If difficulties are observed, the progress of parturition will be monitored at more frequent intervals. Numbers of live and dead offspring are recorded during and at completion of parturition. Litters are examined daily for evidence of ill health or reaction to maternal treatment. Pups are individually sexed, and weighed on Days 1, 4, 7,

11, 14, 18, 21, and 25 of age. Each pup is assessed for the ability to show surface righting from Day 2 of age, air righting from Day 14–19 of age, and for the presence of the startle response and pupil closure reflex on Day 20 of age. The females are then killed on Day 21 of lactation, when they are examined macroscopically and the numbers of uterine implantation sites recorded. Females that are not observed to give birth to a litter by Day 25 after mating are killed and examined. One male and one female pup per litter are allocated to form the untreated F1 generation and are reared to maturity, during which time their age at sexual maturation, motor activity, learning/memory, growth, and reproductive capacity are evaluated. Unselected pups are killed and examined macroscopically on Day 21 of age.

5.2.1.3 *Choice of infusion model*

In addition to the general considerations of animal welfare, altered physiology, and local tolerance at the route of administration (as discussed in Chapter 1), consideration should also be given to the plasma drug half-life when considering the model for reproductive studies. Because of its relatively small size, the rat has a faster metabolism than the human, resulting in faster elimination of potential toxicants and therefore an associated reduction in exposure. Where this is considerably lower in rats than in humans, it is of particular concern for the embryo-foetal toxicity study, as the primary objective of this study type is to expose the processes of major organogenesis in the developing embryo. Embryonic development is a continuous process, with the various organ systems being formed according to a fixed timetable within that process. This gives rise to the so-called 'critical periods' for embryotoxicity, which are themselves shorter in rats than they are in humans. If a 'drug holiday' occurs between doses, the conceptuses may not be adequately exposed to the test article. It is therefore apparent that whilst continuous infusion may not necessarily be the intended clinical route of administration, this model provides an opportunity to continually expose the developing embryo/foetus and therefore a possible enhancement for the pharmaceutical testing strategy. In addition, this model allows not only the continued exposure during organogenesis but also the withdrawal of treatment at a very particular stage of organogenesis, thus enabling the targeted treatment of developing organ systems at certain phases.

For fertility studies, the continuous infusion model ensures that the animals are exposed at the time of copulation. As rats are nocturnal and copulate at night, administration for shorter periods via a peripheral vein may result in a delay between dosing and copulation when the plasma levels will have decreased or may have been totally eliminated. Thus using other administration routes could result in possible effects on mating behaviour being undetected.

For pre- and post-natal studies, the continuous exposure of the dam during lactation provides an opportunity for the pups to receive constant exposure to the test article, and exposure is not influenced by the time of test article administration, as in other routes of administration.

In addition to these points, with all reproductive study designs, the use of the surgical model provides benefits in terms of animal welfare over a non-surgical restrained infusion model. After the initial surgical preparation, there are no subsequent stresses on the dam and the developing embryo, as may be apparent after a period of daily restraint.

5.2.1.4 *Limitations of available models*

For pre-clinical studies, where the drug is intended for clinical administration by vascular infusion, a surgical model can be used to overcome the technical limitations associated with bolus administration via small peripheral veins. In these studies, where high doses are needed to assess toxicity, solubility restrictions may affect dose volumes, and high drug concentrations may result in dose irritancy, haemolysis at the interface of the infusion site and circulating blood, or effects on blood-buffering capacity. The surgical model allows delivery at a slower rate, and the resulting lower localised concentrations of the test article can often overcome these problems.

For rat reproductive studies, a surgical model has few limitations, predominantly owing to the relatively short durations of the reproductive study designs in comparison to general toxicity studies, so long-term catheter patency is rarely an issue. For laboratories that purchase time-mated animals and then subsequently surgically prepare the animals, additional consideration must be given to the day of mating and subsequent arrival and surgical capabilities so as to ensure the correct allocation of time on study.

5.2.2 *Best practice*

5.2.2.1 *Surgical models*

5.2.2.1.1 Training. Refer to Chapter 1.

5.2.2.1.2 Applicability of the methodology. The choice of the site of vascular access is, as with the standard toxicity model, dependent on the preference of the surgeon or investigator. The standard vessels used (jugular or femoral vein) offer no limitations to the reproductive model.

Surgical records should be maintained so that the investigator can link the surgical occasion and any subsequent surgical intervention to the particular day of gestation or lactation, thus allowing any impact of the antibiotic, analgesic, or anaesthetic regimen to be assessed.

A KVO maintenance rate of 0.5–1.0 mL/hour is acceptable and is sufficient to maintain catheter patency either post-surgery or during periods of intermittent dosing if a catheter lock is not used. The duration of daily infusion will vary with each project: it can range from a short duration of one hour up to a continuous 24-hour model. The rates and volumes that are considered 'good practice' for reproductive rodent studies are comparable to those recommended for dosing in general rat toxicity studies by continuous infusion, as discussed in a number of publications (Diehl 2000; Smith 1999 and Morton 2001). At this laboratory, recommended infusion rates for all reproductive studies are typically in the range of 2–3.3 mL/kg hour, equivalent to a total daily volume of 48–80 mL/kg/day; it is recognised that the lower rate of 2 mL/kg/hour provides a flow rate for a continued period of time that ensures catheter patency, and the higher rate of 3.3 mL/kg/hour is considered acceptable in relation to the mean circulating blood volume. However, all studies and their dosing regimens and flow rates and volumes should be considered on a case by case basis, and if values outside of these ranges can be justified based on the test article properties and the length of infusion, their use should not be discounted. Flow rates of 0.625 mL/kg/hour and up to 5 mL/kg/hour have been successfully employed on embryo-foetal studies at this laboratory; however, their suitability for use on longer-term toxicity studies may not be appropriate, and consideration to the entire programme of work should be given before selecting flow rates outside of the recommended ranges.

As with all routes of administration, dosing accuracy is paramount to the successful running of any study or project. As previously discussed, accurate record keeping during the treatment phase is crucial especially if there have been any interruptions to infusion during this time. Any subsequent discrepancies in embryo-foetal development or in maternal performance can then be correlated with interruptions.

5.2.2.1.3 Risks/advantages/disadvantages. Refer to Chapter 1.

5.2.2.2 Non-surgical models

5.2.2.2.1 Methods of restraint. The method of restraint for reproductive rodent studies is identical to that described for standard toxicity studies in this species (refer to Chapter 1). Whilst it is stated that most review committees would accept restraint up to 4 hours, for reproductive studies the restraint time would be considerably less. In addition to the consideration of the regular provision of food and water, the impact of prolonged daily restraint on the developing foetuses is unknown, and as such, restraint of only up to one hour is recommended; for longer restraint times, a surgical model must be considered. For pre- and post-natal studies, prolonged restraint is not appropriate, owing to the requirement of the pups to suckle during the lactation phase.

5.2.2.2.2 Applicability of the methodology. Refer to Chapter 1.

5.2.2.2.3 Risks/advantages/disadvantages. When choosing to dose into a peripheral vein, there is always the potential for these vessels to become damaged through haemorrhage or local tolerance issues. This is of particular concern for long-term toxicity studies but often may not be of such concern for shorter-term reproductive studies.

An advantage of this model is that animals do not have to undergo a surgical procedure, which may be of particular advantage for laboratories that do not have their own surgical or stud-male colony, as some study designs could not be conducted without these *in situ*.

The disadvantages, as discussed in Section 5.1.2.2.1, are the physiological impact on the dam and developing foetuses by being restrained in a tube for long periods of time on a daily basis and the effects on pre- and post-natal studies of the mother being away from the litter for protracted periods of time during lactation. Therefore, whilst it may be appropriate to restrain a non-pregnant animal in a tube for peripheral vein dosing for periods up to four hours, there are greater considerations for pregnant animals, and for both the toxicity and reproductive study route of administration ensuring consistent drug exposure levels is a consideration.

5.2.3 Practical techniques

5.2.3.1 Surgical models

5.2.3.1.1 Preparation

5.2.3.1.1.1 Premedication. The antibiotic and analgesic regimen employed in the surgical preparation of rodents for reproductive studies at this laboratory is identical to that employed for general toxicity studies (see Chapter 1). There are, however, a number of points to consider for the reproductive model. Many laboratories will use opioids that have a relatively short duration of action, which is compounded by the disproportionately faster rates of drug metabolism in small rodents (Morris 1995). Therefore, buprenorphine, which has a prolonged duration of action, is often the widely used analgesia of choice in small animal surgery, with dosing at 8–12 hour intervals being effective in controlling most post-surgical pain. It is also reported that the administration of buprenorphine to rodents prior to anaesthesia with isoflurane may result in up to a 0.50% reduction in the concentration of anaesthetic required for maintenance during the surgical procedure (Flecknell and Waterman-Pearson 2000). Whilst the use of the opioid buprenorphine clearly has advantages in rodent surgery, at this laboratory we have also experienced an increased incidence of post-surgery wound interference that was most prevalent amongst female animals and therefore of most detriment to reproductive studies. It is reported that some of the clinical side effects of buprenorphine include excessive licking or biting of the limbs and cage, which may also be directed to the surgical incision (Lee-Parritz 2007). The removal of buprenorphine from the analgesia regimen at this laboratory reduced the incidence of wound interference in adult female rats compared to animals whose regimens previously had included buprenorphine. Effective pain relief is provided by the administration of the NSAID carprofen (Rimadyl) and bupivacaine, with animals showing no signs of discomfort and an improvement in post-operative body weight profiles, thus reducing post-surgery recovery times.

5.2.3.1.1.2 Presurgical preparation. Refer to Chapter 1.

5.2.3.1.1.3 Anaesthesia. Whilst it is widely accepted that both injectable and gaseous anaesthetics can be employed during the surgical preparation of rodents, it is acknowledged that injectable anaesthetics are less controllable and the period of post-operative recovery is protracted compared with inhaled anaesthetics. The implications of this to a pregnant animal must be considered if the surgical procedure or any subsequent reparative surgery is performed during the period of organogenesis. In addition, if the dam requires reparative surgery during the period of lactation, prolonged recovery periods from anaesthesia have the potential to impact maternal performance and milk production and hence directly impact the offspring.

5.2.3.1.2 Surgical procedure. Refer to Chapter 1.

5.2.3.1.3 Maintenance

5.2.3.1.3.1 Recovery procedures. The preferred anaesthetic regimen employed for rat reproductive studies is inhaled gaseous anaesthesia, as this allows the animal to recover rapidly and return to normal behaviour within 10–15 minutes of surgery. This is of particular benefit for laboratories that allow reparative surgery during the period of treatment to coincide with gestation or lactation in order to minimise the potential impact of maternal anaesthesia on the developing embryo/foetus or the pups and to minimise maternal body weight loss.

Owing to the protracted recovery time observed by the use of injectable anaesthetics, their use on reproductive studies should be considered with caution, in particular if surgical preparation occurs after mating.

At this laboratory, as surgical procedures are carried out in advance of mating, animals have a period of recovery prior to body weight being recorded on the Day 0 of gestation. This ensures that the surgical procedure has no impact on the maternal body weight gain and food consumption measurements. For laboratories performing the surgical procedure on time-mated animals, the expectation would be for there to be an immediate reduction in body weight and food consumption, with a continued impact for up to 7 days during the subsequent surgical recovery period.

As the critical time for embryo-foetal studies is from Day 6 of gestation, the investigator must be aware of the potential effect on maternal weight gain. This approach is also not practical should the study design require dose escalation from a low starting dose to ensure that the target dose is achieved by Day 6 of gestation.

If the surgical procedure is performed in advance of mating, in order to reduce the risk of interference with the infusion system by the stock males whilst the females are in pairing for mating, females should be paired only when vaginal smears indicate that they are likely to mate that night, that is, when they are in pro-oestrus. To aid in the assessment of the stages of oestrus, wet smears are taken from all females for at least four days before the first scheduled day of pairing. Smearing continues throughout the pairing period to determine those suitable for mating and to confirm the presence of spermatozoa post-mating, prior to allocation. Females in pro-oestrus are paired on a 1:1 basis with stock males from the same source. After pairing, checks are made for evidence of mating, including ejected copulation plugs in cage trays and the presence of sperm in a vaginal smear. Animals are allocated to study on Day 0 of gestation, when positive evidence of mating is detected. Only females with a sperm-positive vaginal smear or at least two copulation plugs are selected. Females are allocated to group and cage position in sequence of mating, ensuring that animals mated on any one day are evenly distributed amongst the groups. The sequence of allocation is controlled to prevent stock males from providing more than one mated female to each treatment group.

There is the potential for this approach to impact the oestrous cycles. At this laboratory, it has been demonstrated that most oestrous cycles before surgery are of four days' duration. Immediately following surgery, there is a tendency for the oestrous cycle to be slightly longer; however, for the majority of females the second cycle post-surgery returns to a typical four-day cycle (Table 5.4). This is attributed to the effects of surgery, either as an effect of the medication given to the animals, or the stress of the procedure, or a combination of these two factors. It is therefore recommended that females be allowed approximately 10 days' recovery (two oestrous cycles) following surgery prior to pairing for mating to allow both healing of the surgical sites and recovery from any transient disturbance of the oestrous cycle.

Oestrous cycles were classified and reported as 'Before surgery' or 'After surgery' for individual cycles that both started and finished within either of these two phases. Calculation of cycle length was started from the first recorded late oestrus (E2) smear, that is, the first observed cycle; cycles incomplete and in progress at the start of smearing and therefore of unknown duration were excluded from calculations.

Table 5.4 Oestrous Cycles – Incidence of Cycle Lengths

Flow rate (mL/kg/hour)	Cycle length		Cycles before surgery		Cycles after surgery			
	(days)		1st	2nd	1st	2nd	3rd	4th
2	Total Number of Cycles		16	0	16	16	14	10
	4	n	16	0	5	13	14	9
		(%)	(100)		(31)	(81)	(100)	(90)
	5	n	0	0	7	2	0	0
		(%)			(44)	(13)		
	3 or 6–10	n	0	0	3	1	0	1
		(%)			(19)	(6)		(10)
	11–20	n	0	0	1	0	0	0
		(%)			(6)			
5	Total Number of Cycles		17	7	17	17	16	6
	4	n	17	6	7	15	15	6
		(%)	(100)	(86)	(41)	(88)	(94)	(100)
	5	n	0	1	7	1	0	0
		(%)		(14)	(41)	(6)		
	3 or 6–10	n	0	0	3	0	1	0
		(%)			(18)		(6)	
	11–20	n	0	0	0	1	0	0
		(%)				(6)		

5.2.3.1.3.2 Housing. Identical housing to that used for standard rodent toxicity studies is considered appropriate for fertility and embryo-foetal studies where the cage design is able to accommodate the tether and the swivel. If animals are mated after the surgical preparation, then during the period of pairing with the stock male the animals need to be housed in a cage with a grid floor. Thus a modified version of this cage should be used, so that after pairing, checks can be made for evidence of mating by looking for ejected copulation plugs in cage trays. For reasons of animal welfare, the time on gridded bottom cages should be kept to a minimum wherever possible.

As with general toxicity studies, shelters are rarely suitable for tethered animals. Other enrichment products are readily available, such as nylon chew items, wooden chew blocks, and nestlets; these have been provided for tethered rats without any reported issues. However, additional consideration should be given to the type of environmental enrichment available during the period of pairing with a stock male on an embryofoetal or fertility study, and during the period of lactation on a pre- and post-natal study; enrichment such as chew blocks are not considered appropriate during these phases.

It is frequently reported that there is a slight increased incidence of wound interference post-surgically amongst females. The introduction of an increased number or array of environmental enrichment products has at our laboratory served to combat this issue.

5.2.3.1.3.3 Checks and troubleshooting. The frequency of the infusion system checks should be tailored to the dosing regimen employed in the study. As discussed with the rat model, regular checks throughout the study day are required to ensure the continuing function of all the infusion system components and to maintain catheter patency. Particular importance must also be placed on the satisfactory fitting of the jacket/harness because of the rapid growth phase during the period of gestation. In pre- and post-natal studies, additional checks during the lactation phase must be employed to ensure that the pups have continued access to the nipples. When jackets are used, there is the potential for partial occlusion of the nipples, which may result in the reduction in the quantity of milk available, with a direct effect on the rate of post-natal weight gain in the rat.

5.2.3.1.4 Record keeping. Record keeping should be consistent with those on general toxicity studies and should include details of pre-, peri-, and post-operative treatments. Post-surgical pain/distress assessments should also be documented, as well as any associated supplementary care. Careful monitoring and recording of these events will help a laboratory to establish what they consider the appropriate frequency of adverse effects, such as post-operative infections, haemorrhage, and wound healing and breakdown. Where a laboratory allows reparative surgery under general anaesthesia, for example as a result of a breakdown in the infusion system or to allow sutures to be replaced, additional consideration should be given to the potential effect of the additional period of anaesthesia on the stage of gestation or lactation.

As standard for all reproductive studies, records of mating performance and the sequence of mating are required to ensure that animals mated on any one day are evenly distributed amongst the groups. The sequence of allocation should be controlled to prevent stock males from providing more than one mated female to each treatment group.

The importance of the maintenance of detailed and quality record keeping is of paramount importance in aiding the identification of reasons for problems that may manifest later.

5.2.3.2 *Non-surgical models*

Refer to Chapter 1.

5.2.4 *Equipment*

Refer to Chapter 1.

5.2.5 *Background data*

The data presented in this section has been collated from the Control data from both Han Wistar and Sprague Dawley rats infused with physiological saline. In addition, data comparing a standard infusion rate of 2 mL/kg/hour with that of a high infusion rate of 5 mL/kg/hour are included for comparison to demonstrate the potential effect of an increased flow rate.

5.2.5.1 *Maternal responses*

5.2.5.1.1 Mortality. Losses on infusion studies at this laboratory are typically in the range of 3–4% over the duration of the in-life phase. A slight increase may be observed that is associated with the potential interference from the male during the mating procedure, but this is minimal and of no consequence to the overall integrity of the studies.

5.2.5.1.2 Clinical signs. At the time of placing an animal on study, providing there has been acceptable recovery from the surgical procedure, no associated clinical signs would be anticipated.

5.2.5.1.3 Body weight performance. A marginal reduction in body weight gain is observed for both strains of rat for infusion models. These differences are slight and are considered of no consequence, as is shown in Figures 5.10 and 5.11.

When comparing the standard infusion rate of 2 mL/kg/hour with the higher rate of 5 mL/kg/hour, there is no discernible difference in maternal weight gain. However, the absolute weights and body weight gain during gestation period for the infused females remain lower compared with the Control data, as seen in Figure 5.12.

5.2.5.1.4 Food consumption. Food consumption values were similar for the two infusion groups and showed no adverse effect of continuous

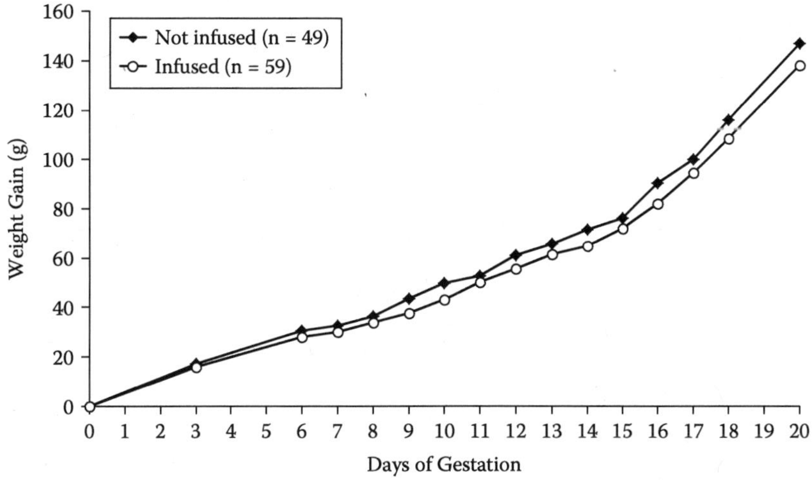

Figure 5.10 Body weight gain during gestation in the Sprague Dawley rat.

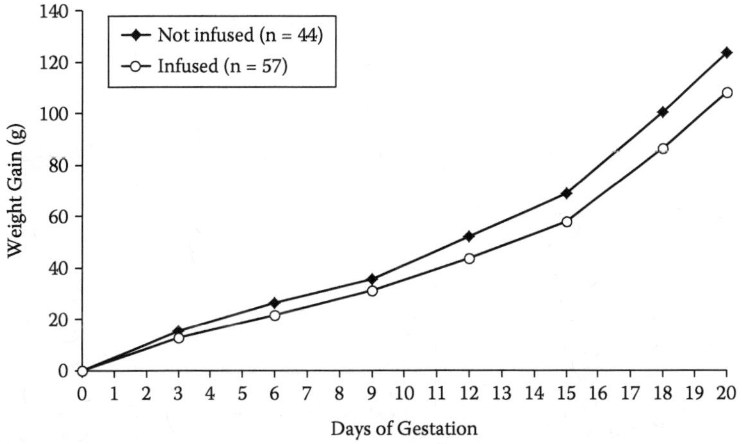

Figure 5.11 Body weight gain during gestation in the Han Wistar rat.

infusion compared with the Control data, as is shown in Figures 5.13 and 5.14.

5.2.5.1.5 Mating performance. On fertility studies, continuous infusion has no effect compared to a non-infused model on the percentage of mating, conception rate, and fertility index, with data showing mating performance of 97% on infusion studies (326 pregnant females from 336

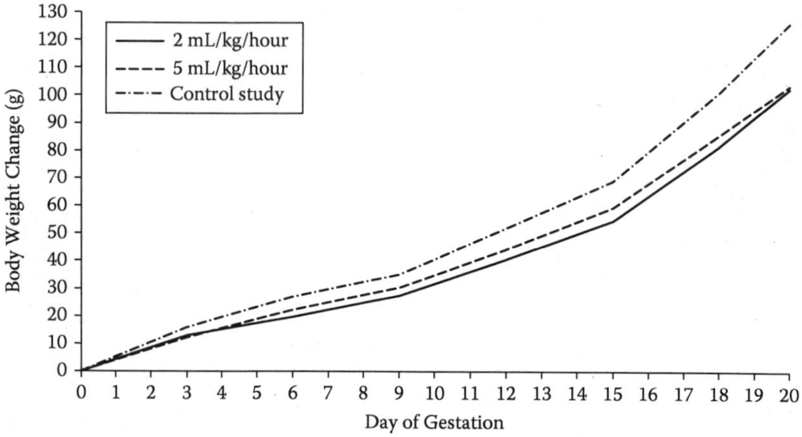

Figure 5.12 Maternal weight gain at standard infusion rate of 2 mL/kg/hour with the higher rate of 5 mL/kg/hour.

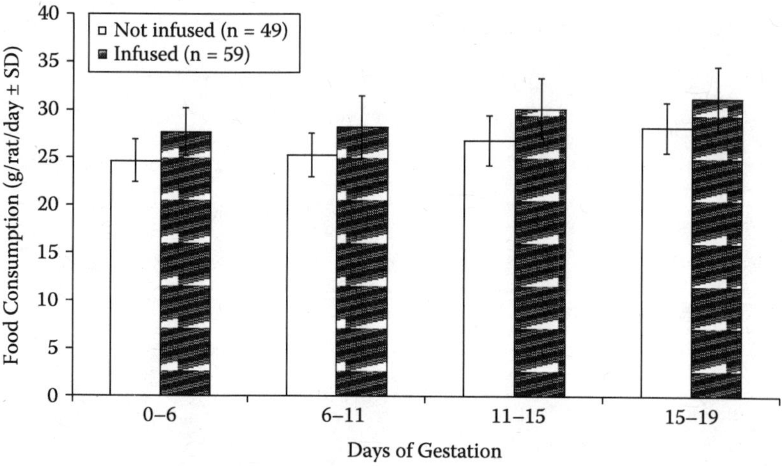

Figure 5.13 Food consumption during gestation in the Sprague Dawley rat.

paired) and 99.3% on comparable non-infusion studies (1769 pregnant females from 1781 paired). Thus, typically all females are pregnant at scheduled termination, and therefore an increase in group size for infusion studies is not needed.

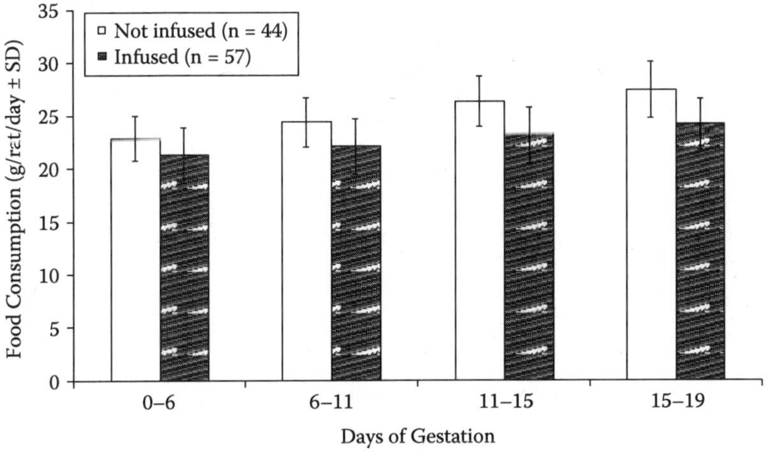

Figure 5.14 Food consumption during gestation in the Han Wistar rat

5.2.5.1.6 Macropathology and associated histopathological changes. The purpose of the developmental and reproductive studies is not to gain an in-depth understanding of the macro- and histopathological changes, as these will be evaluated as part of the associated general toxicity studies. However, the dam is subject to macroscopic examination, and if abnormalities are detected, these tissues can be retained for histological processing as required.

Changes observed are comparable to those reported in general rat toxicity studies up to 13 weeks in duration. Macroscopically these are scabs and thickening at the interscapular exteriorisation site of the catheter due to granulation tissue formation, and slight thickening and white material at the catheter tip. Providing there has been no occlusion of the catheter tip, this finding is considered of no consequence to the accurate delivery of the test article.

Typically, histopathological findings are comparable with those in general rat toxicity studies. These are slight to moderate levels of granulation tissue with associated low levels of inflammation at the interscapular exteriorisation site and along the catheter track, suture granuloma at the femoral vein access site, minimal to mild endothelial proliferation, intimal thickening and small thrombi at the catheter tip, and the presence of thin fibrin sleeves. All these findings are considered in low incidences to be acceptable procedure-related pathologies associated with this route of administration and of no consequence to the developing foetuses.

5.2.5.2 *Litter responses*

5.2.5.2.1 Litter data. Litter data as assessed by the mean number of corpora lutea, implantations, resorptions, and live young and the extent of pre-and post-implantation loss are comparable for infused and non-infused animals (Tables 5.5–5.7). The number of corpora lutea, implantations, and live litter size are slightly lower in the infusion groups; however, pre-and post-implantation survival does not show any adverse effect of continuous infusion or the flow rate employed.

5.2.5.2.2 Placental, litter, and foetal weights. Foetal and placental weights are broadly similar for both infusion rates. The placental weight from the infusion groups was similar to the Control data; however, the foetal weight in the infusion groups was slightly superior, and overall the litter weight was slightly lower than the Control data, which is consistent with the slightly smaller litter size observed for the infusion groups (Tables 5.8–5.10).

5.2.5.2.3 Foetal observations. External examination of foetuses revealed no significant changes from the expected background changes. In addition, data was correlated from 132 foetuses (24 litters) for the

Table 5.5 Maternal Performance and Litter Data on Day 20 from Sprague Dawley Embryo Foetal and Combined Fertility and Embryo Foetal Studies

	Infused	Non-infused
Number of females mated	61	52
Number of females pregnant	60 (98%)	50 (96%)
Females with viable foetuses	59 (98%)	49 (98%)
Corpora lutea – mean (SD)	16.1 (2.46)	16.3 (1.72)
Number of implantations – mean (SD)	15.0 (2.24)	15.9 (1.76)
Early resorptions – mean	0.9	0.8
Late resorptions – mean	0.0	0.0
Total resorptions – mean	1.0	0.8
Number of male live young – mean (SD)	7.3 (2.37)	7.2 (2.03)
Number of female live young – mean (SD)	6.7 (2.08)	7.9 (2.52)
Total live young – mean (SD)	14.0 (2.75)	15.1 (1.85)
Percentage male sex ratio	51.6	48.2
Pre-implantation loss – mean (%)	6.8	3.9
Post-implantation loss – mean (%)	7.1	5.1

Table 5.6 Maternal Performance and Litter Data on Day 20 from Han Wistar Embryo Foetal Studies

	Infused	Non-infused
Number of females mated	57	44
Number of females pregnant	57 (100%)	44 (100%)
Females with viable foetuses	57 (100%)	44 (100%)
Corpora lutea – mean (SD)	12.5 (1.59)	14.1 (2.07)
Number of implantations – mean (SD)	11.1 (2.51)	12.0 (3.46)
Early resorptions – mean	0.5	0.7
Late resorptions – mean	0.0	0.0
Total resorptions – mean	0.6	0.7
Number of male live young – mean (SD)	5.4 (2.09)	5.5 (2.15)
Number of female live young – mean (SD)	5.1 (2.17)	5.8 (2.62)
Total live young – mean (SD)	10.5 (2.56)	11.3 (3.59)
Percentage male sex ratio	51.9	49.9
Pre-implantation loss – mean	12.5	15.8
Post-implantation loss – mean	5.2	8.0

Table 5.7 Maternal Performance and Litter Data on Day 20 from Han Wistar Embryo Foetal Studies Comparing Differing Rates of Infusion

	2 mL/kg/hour	5 mL/kg/hour	Non-infused
Number of females mated	15	14	44
Number of females pregnant	15 (100%)	14 (100%)	44
Females with viable foetuses	15 (100%)	14 (100%)	44 (100%)
Corpora lutea – mean (SD)	12.1 (1.4)	12.4 (1.8)	14.4 (1.5)
Number of implantations – mean (SD)	10.9 (1.4)	10.7 (2.6)	12.7 (2.9)
Early resorptions – mean	0.4	0.5	0.8
Late resorptions – mean	0.1	0.0	0.0
Total resorptions – mean	0.5	0.5	0.8
Number of male live young – mean (SD)	5.9 (1.8)	5.0 (2.0)	5.5 (2.3)
Number of female live young – mean (SD)	4.5 (1.4)	5.2 (1.9)	6.5 (2.4)
Total live young – mean (SD)	10.4 (1.3)	10.2 (2.7)	12.0 (3.3)
Percentage male sex ratio	55.8	48.2	44.7
Pre-implantation loss – mean	10.2	14.8	12.1
Post-implantation loss – mean	4.1	4.7	7.7

Table 5.8 Placental Litter and Foetal Weights (g) on Day 20 from Sprague Dawley Embryo Foetal and Combined Fertility and Embryo Foetal Studies

	Infused	Non-infused
Placental weight – mean (SD)	0.59 (0.061)	0.55 (0.062)
Litter weight – mean (SD)	54.49 (11.056)	56.90 (7.083)
Male foetal weight – mean (SD)	4.00 (0.266)	3.89 (0.239)
Female foetal weight – mean (SD)	3.79 (0.252)	3.69 (0.224)
Overall foetal weight – mean (SD)	3.90 (0.255)	3.79 (0.229)

Table 5.9 Placental Litter and Foetal Weights (g) on Day 20 from Han Wistar Embryo Foetal Studies

	Infused	Non-infused
Placental weight – mean (SD)	0.50 (0.061)	0.49 (0.084)
Litter weight – mean (SD)	38.01 (9.534)	39.95 (12.601)
Male foetal weight mean (SD)	3.72 (0.245)	3.64 (0.230)
Female foetal weight – mean (SD)	3.53 (0.198)	3.44 (0.202)
Overall foetal weight – mean (SD)	3.63 (0.197)	3.55 (0.199)

Table 5.10 Placental Litter and Foetal Weights (g) on Day 20 from Han Wistar Embryo Foetal Studies Comparing Differing Rates of Infusion

	2 mL/kg/hour	5 mL/kg/hour	Non-infused
Placental weight – mean (SD)	0.48 (0.005)	0.48 (0.02)	0.49 (0.05)
Litter weight – mean (SD)	37.09 (4.82)	36.66 (9.70)	40.75 (11.10)
Litter size – mean (SD)	10.4 (1.3)	10.2 (2.7)	12.0 (3.3)
Male foetal weight – mean (SD)	3.65 (0.20)	3.72 (0.24)	3.49 (0.23)
Female foetal weight – mean (SD)	3.47 (0.20)	3.52 (0.27)	3.34 (0.21)
Overall foetal weight – mean (SD)	3.57 (0.18)	3.62 (0.24)	3.41 (0.20)

Han Wistar rat and 221 foetuses (31 litters) for the Sprague Dawley rat and compared with contemporaneous non-infused studies for major, skeletal, and visceral abnormalities. There were no noteworthy differences in major or visceral findings in either strain of rat. There were some marginal differences in some minor skeletal findings in the Han Wistar strain of rat, consisting of partially fused/bridge of ossification maxilla to jugal, offset alignment of pelvic girdle, and incompletely ossified cranial centres and sacro-caudal vertebral arches (Table 5.11).

Table 5.11 Foetal Skeletal Abnormalities Observed at a Marginal Increased Incidence in the Han Wistar Rat

	Infusion		Non-infusion	
	Number of foetuses	Number of litters	Number of foetuses	Number of litters
Number of foetuses/ litters examined	132	24	245	44
	Minor skeletal abnormalities			
Partially fused/bridge of ossification maxilla to jugal	8 (6.06%)	7 (29.17%)	11 (4.49%)	8 (18.18%)
	Rib and vertebral configuration			
Pelvic girdle – offset alignment	8 (6.06%)	5 (20.83%)	5 (2.04%)	5 (11.36%)
	Incomplete ossification/unossified			
Head/neck – cranial centres	34 (25.76%)	14 (58.33%)	35 (14.29%)	17 (38.64%)
Vertebrae – sacrocaudal	7 (5.30%)	3(12.50%)	2 (0.82%)	2 (4.55%)

5.2.6 Conclusion

The continuous infusion model in rodent reproductive studies has now been in use for a number of years in a variety of laboratory settings and as such is well characterised. Whilst the use of this model may result in some minor deviations from what are considered normal background ranges, providing the laboratory has a good understanding of what is deemed acceptable and continually collates and monitors these data in relation to concurrent control data, then interpretation of data from continuous infusion studies is not adversely impacted and therefore provides a robust model to support the reproductive testing strategy.

References

Bussiere JL et al. 2009. Alternative strategies for toxicity testing of species-specific biopharmaceuticals. *International Journal of Toxicology* 28(3): 230–253.

ICH. 2005. Harmonised guideline S5(R2): *Detection of Toxicity to Reproduction for Medicinal Products and Toxicity to Male Fertility.* International Conference on Harmonization. November 2005.

Diehl K-H et al. 2001. A good practice guide to the administration of substances and removal of blood, including routes and volumes. *Journal of Applied Toxicology* 21: 15–23.
Flecknell P and Waterman-Pearson A. 2000. Management of postoperative pain and other acute pain. In *Pain Management in Animals*, Chapter 5. Flecknell P and Waterman-Pearson A (eds.), pp. 132–135. London: WB Saunders.
Lee-Parritz D. 2007. Analgesia for rodent experimental surgery. *Israel Journal of Veterinary Medicine* 62: 3–4.
Morris TH. 1995. Antibiotic therapeutics in laboratory animals. *Laboratory Animals* 29: 16–36.
Morton DB et al. 2001. Refining procedures for the administration of substances. *Laboratory Animals* 35: 1–41.
Patten D. 2011. Reducing post-operative wound interference in adult female Sprague Dawley rats. Infusion Technology Organisation Conference, Barcelona.
Smith D. 1999. Dosing limit volumes: A European view. Refinement Workshop, New Orleans, LA, March 1999.

5.3 *Mouse*

Hans van Wijk, FIAT, RAnTech, Biotechnicus
Covance, UK

Richard S Bartlett, BSc, CBiol, MSB
Covance, UK

5.3.1 *Introduction*

For all medicinal products, pre-clinical studies for the assessment of developmental and reproductive toxicity (DART) are performed in accordance with the ICH Harmonised Tripartite Guideline S5(R2), of 24 June 1993. The combination of studies performed is required to cover all stages of reproduction from conception to sexual maturity, with a three-study design being most common, as follows:

1. Study of fertility and early embryonic development to implantation (one rodent species)
2. Study for effects on embryo-foetal development (rodent and non-rodent)
3. Study for effects on pre- and postnatal development, including maternal function (one rodent species)

Alternative study designs may be adopted, such as the combination of the female fertility and embryo-foetal development study (1 and 2).

The rat (rodent) and rabbit (non-rodent) are routinely used owing to their reproductive characteristics and the quantity of published background data on them. There are, however, instances where they are not suitable models; rats are sensitive to sexual hormones and are highly susceptible to non-steroidal anti-inflammatory drugs, and rabbits are susceptible to some antibiotics and disturbances to the alimentary tract (S5(R2), 24 June 1993). The mouse is therefore used as the alternative species. For studies requiring intravenous administration in mice, the test article can be administered as a slow bolus injection via the tail vein. This approach, however, is recommended only for short-term studies (up to seven consecutive days) and for administration durations of less than five minutes, because of the adverse effects on tail quality with longer dosing (personal observation). The mouse vascular catheterised continuous intravenous infusion technique, using a surgically implanted femoral vein catheter with tail-cuff exteriorisation, has shown its efficacy on general toxicology studies (in this chapter). However, the reliability of this method decreases over time as a result of tail-cuff constraints to animal growth (refer to Chapter 2, Mouse).

5.3.2 *Study design/aim*

The feasibility and reliability of using the mouse for continuous intravenous infusion in reproductive toxicology studies were assessed by performing a combined female fertility and embryo-foetal development study using a larger-diameter tail-cuff (6 mm) with the aim of extending the duration of continuous infusion. The tail-cuff exteriorisation method was favoured over the jacket/harness exteriorisation method for this study, as it provides a greater range of movement for mating (Figure 5.15) and does not restrict abdominal growth during pregnancy (Figures 5.16 and 5.17). Sufficient female mice of the Crl:CD-1(ICR) strain aged approximately

Figure 5.15 Female tethered mouse housed together with non-infused male mouse.

Figure 5.16 Pregnant female mouse ready to climb its tether.

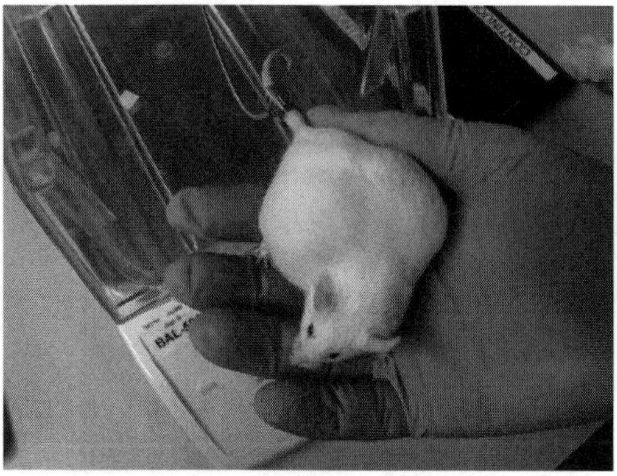

Figure 5.17 Heavily pregnant female mouse.

7–9 weeks old were surgically implanted with a femoral vein catheter exteriorised with a 6-mm tail-cuff (see Chapter 2, Mouse), to provide 33 animals for study Day 1. These mice were allowed to recover for at least 6 days post-surgery. During this period the animals were infused with saline at an infusion rate of 0.1 mL/hour to keep the catheter patent. On Day 1 of the study, females were intravenously administered 0.9% sodium chloride by continuous infusion (4 mL/kg/hr) for at least 2 weeks prior to pairing, during pairing (with uncatheterised males), and through to gestation day (GD) 18 (a total of 42 days). On GD 18, females were killed and the uterine contents examined.

5.3.3 Results

Throughout the study several animals were removed because of catheter failures. The failures from Day 1 to the end of the study (Stage 3) are categorised in Table 5.12.

General failure rates over Stages 1 (during surgery) and 2 (during the post-surgical recovery period until Day 1 of study) are described in Section 5.1.5, Pathology, of this chapter.

The majority of failures were attributed to the larger diameter of the tail-cuff, which slipped down the tail and allowed the catheter to become exposed and chewed by the mouse. During the study the majority of females also had sores and lesions around the tail-cuff and swollen tails; however, there were no adverse effects on body weight gains or food consumption. In addition, there were no effects of catheterisation and infusion on mating performance, fertility, or fecundity (Table 5.13); all females mated within six days, with a pregnancy rate of 95.7%.

In females surviving to necropsy, uterine and foetal data were generally within background ranges (Table 5.14), indicating no effects of continuous infusion of saline at an infusion rate of 4 mL/kg/hr on any reproductive or developmental parameters. The increase in post-implantation loss was attributed to the higher number of corpora lutea, compared with

Table 5.12 Failure Rate and Reasons for Failure of the 6 mm Tail-cuff Infusion Method from Day 1 until the End of Study (42 Days)

Reason for loss of animal	Females (total on day 1 = 33)
Catheter-related problems due to tail-cuff size	12 (36.4%)
Detached tail-cuff	1 (3.0%)
Died during patency check	1 (3.0%)
Other (non-infusion related)	1 (3.0%)
Total loss Stage 3	15 (45.5%)

Table 5.13 Mating Performance

Number of animals paired	25
Mated	23
Not-mated	1#
Mated with undetermined pregnancy	1
Number pregnant	22 (95.7%)
Number with live embryos	17

#Animal taken off study prior to implantation.

***Table* 5.14** Caesarean and Foetal Data

	Infused Females	Historical control data (non-infused)
Number of litters evaluated#	15	162
Mean number of corpora lutea	19.9	15.4 (14.4–16.3)
Mean number of implantation sites	15.5	13.6 (12.7–14.6)
Mean pre-implantation loss (%)	20.2	10.9 (7.3–15.6)
Mean number of early intrauterine deaths	0.5	0.7 (0.2–1.1)
Mean number of late intrauterine deaths	0.7	0.2 (0.1–0.4)
Mean number of dead foetuses	0.1	0.04 (0.0–0.2)
Mean post-implantation loss (%)	8.4	6.9 (3.1–11.0)
Mean number of foetuses per female	14.2	12.7 (11.5–14.0)
Mean % male foetuses	55.0	51.0 (45.4–55.7)
Mean foetal weight	1.35	1.43 (1.39–1.46)

#Excludes two females with live embryos killed on the wrong GD (GD 16 and 17) in error.

historical ranges; this was therefore considered not to be an adverse effect of the infusion technique.

5.3.4 *Conclusion*

It is concluded that continuous intravenous infusion of saline (4 mL/kg/hr) via surgical vascular catheterisation of the femoral vein and exiting at the tail using a tail-cuff is a viable method of intravenous dosing for mice in DART studies. Reproductive parameters were not adversely affected by the infusion technique. However, the larger tail-cuff did not increase reliability of the infusion technique, as it slipped down the tail and allowed the catheter to become exposed and chewed by the mouse. Therefore, separate fertility and embryo-foetal development studies would be preferable for mice DART studies. Performing these studies separately would result in study durations of approximately 22–28 days (post-surgery recovery) in a female fertility study and approximately 17–22 days (post-surgery recovery) in an embryo-foetal development study, based on current mating performance. A 5-mm tail-cuff is also recommended to avoid the cuff slipping down the tail and exposing the catheter; this would therefore increase the success rate to levels described in this chapter.

Reference

International Conference on Harmonisation (ICH). 1993. Harmonised Tripartite Guideline S5(R2): *Detection of Toxicity to Reproduction for Medicinal Products and Toxicity to Male Fertility* (24 June 1993).

chapter six

Minipig

Peter Glerup, DVM, MSc
CiToxLAB Scantox, Denmark

Mikala Skydsgaard, DVM
Novo Nordisk, Denmark

Gitte Jeppesen, DVM
CiToxLAB Scantox, Denmark

Contents

6.1 Introduction

6.1.1 Choice and relevance of model

The choice of animal species in relation to non-clinical research should always be carefully considered and justified. Pigs, including minipigs, are considered good models for humans because the two species share some important characteristics. Anatomically, the cardiovascular system, the skin, the major part of the gastrointestinal tract, the nasal cavity, and the urogenital system all show a high degree of resemblance to humans. Also, similarities are often found in the metabolism of drugs and in physiological parameters.

Minipigs offer an advantage over other pigs with their smaller size and thereby a markedly reduced test substance requirement. Today more than 13 strains of minipigs exist globally. The Göttingen minipig (Figure 6.1) is bred to the highest SPF standards, reducing background findings to a minimum. This makes this strain of minipigs ideal for research purposes. The information given in this chapter will focus on the use of the Göttingen minipig.

For regulatory toxicity studies, the minipig should therefore always be considered a relevant test species. The minipig is fully accepted globally by regulatory authorities, and considerable historical control data for the Göttingen minipig has been generated now for more than two decades.

Figure 6.1 Göttingen minipigs.

In drug development, the intended clinical route of dosing should always be applied in preclinical studies, wherever possible. All general routes of dosing (oral, dermal, subcutaneous, intramuscular, and intravenous) are feasible in the minipig, including continuous intravenous infusion. The access to peripheral veins of minipigs is limited; the auricular veins can be used for single intravenous bolus injections but are not recommended for repeated intravenous injections or continuous intravenous infusion. This necessitates surgical implantation of catheters to be used for these purposes. The surgical procedures and other practical aspects of continuous intravenous infusion studies are described in the following.

6.1.2 *Regulatory guidelines*

For preclinical safety studies, relevant guideline requirements should be consulted prior to study initiation. The ICH guideline M3(R2), *Guidance on Nonclinical Safety Studies for the Conduct of Human Clinical Trials and Marketing Authorization for Pharmaceuticals,* is of major importance in relation to development of small molecules. For biotechnology-derived substances, the ICH guideline S6(R1), *Preclinical Safety Evaluation of Biotechnology-Derived Pharmaceuticals,* should be studied. Inputs for the specific study designs can be gained from relevant OECD guidelines.

While proof-of-concept studies do not need to be conducted in compliance with Good Laboratory Practice (GLP), preclinical studies required for marketing of intravenous drugs and studies aimed at supporting product safety and efficacy for clinical trial initiation must be in compliance.

6.2 *Animals and equipment*

6.2.1 *Animals*

Prior to surgery, the animals should be acclimatised at the facility for at least a couple of weeks, during which time the animals are observed clinically in order to evaluate their general health status and suitability for surgery. Blood samples for clinical pathological analyses may also be taken for evaluation of organ functionality, as well as electrocardiograms for evaluation of heart function.

After ingestion of diet, it will usually take 12 hours or more before the stomach is completely emptied. In order to reduce the risk for aspiration during anaesthesia, it is recommended that the animals fast for at least 12 hours prior to surgery. As some bedding material is often ingested too, it should be removed in the period prior to surgery. The animals should have *ad libitum* access to drinking water until surgery.

Owing to the weight and size of the equipment (ambulatory infusion pump plus counterbalance), animals for continuous infusion studies should be of a certain size. Depending on the specific equipment used, minimum body weight should be approximately 10 kg (corresponding to an age of 4–5 months) or higher. However, intravenous bolus injection can be performed at much younger ages (at our laboratory from 7 days of age) using a surgically implanted vascular access port.

For continuous infusion studies, the animals must be housed individually, as there is a potential risk of pen-mates damaging the equipment. The infusion pump and the exteriorised part of the catheter are placed on the dorsum of each animal, protecting the equipment from self-inflicted damage. Animals with implanted vascular access ports can be group-housed again after recovery and as soon as the bandages are removed over the surgical areas.

6.2.2 Ethics and animal welfare

Pigs and minipigs are highly intelligent animals and should be housed in an enriched environment. An adequate floor area should be provided, taking the size of the animals included in continuous infusion into consideration. National legislative housing requirements should always be consulted. Because the animals are bandaged following surgery, it is not possible to group-house them. However, visual, auditory, and olfactory contact can still be maintained and should be prioritised. Upon recovery and removal of bandages and sutures, it will again be possible to group-house animals with their implanted vascular access ports, whereas this is not possible for animals with catheters exteriorised for continuous infusion.

Bedding material (sawdust, for example) must be offered in order to provide a clean and dry environment. This is of major importance in keeping the dressings clean and thereby reducing the risk of post-surgical infection. In addition, hay can be provided, not only as a dietary supplement, but also as an excellent enrichment tool, as the animals enjoy chewing the hay, moving it around within their pens, and using it for nest building. Furthermore, various plastics toys may be provided, but in most cases the animals will lose interest in these toys within a short period of time. Therefore, it is important to switch to new toys frequently. Finally, for cognitive enrichment, the animals may be trained (by use of positive enrichment training) in various study-related procedures, such as weighing and placement in slings during infusion cassette changes. Pigs and minipigs are easy learners and will be stimulated when challenged to learn new things. The contact to humans will also be strengthened during the training sessions.

An anaesthetic protocol providing good and safe anaesthesia for the necessary duration and a quick and smooth recovery should be used (see Section 6.3.1). Following surgery, analgesics must always be used (see Section 6.3.6). A pain-treatment programme should be individually based, which makes frequent and detailed observations of each animal very important. Clinical signs, such as reduced food consumption, aggressiveness, vocalisation, and reluctance to move, may be indicators of inadequate pain treatment, and corrective actions must be taken immediately. As surgery in relation to experimental procedures is a planned process, treatment with analgesics should always be started prior to surgery (pre-emptively), resulting in less activation of the pain sensory system. The body temperature is in most cases reduced during anaesthesia, and therefore it is important to use heating pads during surgery and to provide a warm recovery area after surgery. More details of anaesthesia and analgesia can be found later in this chapter (Sections 6.3.1 and 6.3.6).

6.2.3 *Infusion equipment*

In order to allow free movement of the animals within their pen, the preference is to use ambulatory infusion pumps, which can be carried by the animals in specially designed pig jackets. These pig jackets, containing pockets on each side of the back for infusion pump and counterbalance, are commercially available in various sizes. Several types of infusion pumps for use in this type of study are also commercially available; the infusion pumps routinely used at our laboratory are CADD-Legacy® 1, to which 50- or 100-mL single-use reservoirs can be connected. For these pumps the infusion rate can be varied from 1 to 125 mL/hour, which will be sufficient for use in minipigs. Ambulatory infusion pumps (iPRECIO® Dual infusion pump) with two reservoirs (one for the test item solution and one for a second test item solution or saline) are now commercially available in a size and weight also suitable for minipig studies. These pumps can be controlled via Bluetooth® or manually. Regardless of the type and brand of infusion equipment, all infusion systems should always have attached alarm systems to provide an alert in case of high pressure within the system.

A tethered system connected to a pump not carried by the animal can be considered an alternative, but such a system will markedly reduce the ability of the animals to move around freely. Also, the length of the catheter will increase catheter dead space, and thereby also the test item requirements. On the other hand, the infusion pumps, being outside the pen area, can easily be handled without interfering with the animals.

6.2.4 *Catheters and vascular access ports*

Silicone or polyurethane catheters can both be used in these studies. The advantage of silicone catheters is the high degree of flexibility and the softness that renders them less traumatic to the vessel wall. To some extent this is not the case for polyurethane catheters. In addition, polyurethane catheters are not autoclavable, a procedure that is possible for silicone catheters. However, the polyurethane catheters are stronger and with their larger internal diameter are less prone to catheter blockage. Also, anti-coagulant agents can be applied to the internal surface of polyurethane catheters, helping maintain patency. This is more difficult in silicone catheters.

The size and length of the catheters must be based on the age and size of the animals. For example, a 7-Fr catheter with an approximate length of 30 cm is suitable for a minipig weighing about 10 kg. To immobilise the catheter once it is inserted to the correct position in the vein, retention beads should be attached to the proximal end of the catheter. For a position in the cranial caval vein in a 10-kg minipig, the catheter will be introduced approximately 10 cm, and the retention beads should therefore be placed at this distance from the catheter tip. Also, towards the point of catheter exit from the animal, it is advisable to attach retention material such as Dacron cuffs, which allow tissue ingrowth, or suture disks. In order to minimise damage of the venous wall, the tip of the proximal end of the catheter should be rounded.

Vascular access ports (VAPs) are also available in various materials. Plastic ports are generally not recommended in minipigs owing to their rather fragile structure, and should be considered for short-term usage only. While titanium and polysulfone ports are comparable with respect to biocompatibility, the titanium allows for use of a larger and denser septum, which grips the Huber needles used for dosing procedures better. Accordingly, titanium ports may be more appropriate for protracted or continuous access. However, the titanium ports are more expensive than polysulfone ports.

The VAP should fit tightly to the catheter implanted into the vein, avoiding leakage and potential separation. Sets of VAPs and catheters, with catheters produced exactly to fit the VAPs, are commercially available. For the correct positioning of the VAP within the subcutaneous tissue and to keep the VAP from turning and the catheter from kinking, the VAP should be constructed to allow for suture fixation to the surrounding tissue, for instance by having exhibiting holes for the placement of ligatures.

The size of the VAP should be based on the size of the animal in which it is to be implanted. For a minipig weighing 10 kg, a VAP with a diameter of about 2 cm can easily be implanted. The height of the VAP should be large enough to allow for external identification during dosing procedures. However, it should not induce pressure necrosis of the overlying skin. For juvenile minipigs, rat-size VAPs should be used.

For dosing procedures, VAPs require special Huber-point needles, which will increase the lifespan of the silicone membrane of the VAP.

Huber needles prevent silicone particles from being 'cut' from the membrane upon introduction. If normal needles are used, there is a higher risk of such damage to the membrane, and thus a risk of introducing small silicone particles into the bloodstream.

6.2.5 *The infusion system and the infusate*

Components of the infusion system coming in direct contact with the infusate (e.g. catheters, tubing, syringes, and infusion cassettes) should ideally be hospital-grade materials, since these are typically compatible with most formulations. However, testing the formulations for compatibility with, and absorption by, components of the infusion system, as well as for leaching of chemicals out of the infusion system into the infusate should be considered.

Special attention should be given to selection of a formulation vehicle that has minimal potential for vascular irritancy. Formulations with a pH in the range of 5.5 and 7.0 and an approximate osmolality of 300 mOsm/L are ideal. Outside of the preferred pH range the risk of vascular irritation increases, which potentially can also lead to catheter occlusion. Low osmolality will increase the risk of haemolysis and associated adverse effects. However, if infused over a shorter time and in low volumes, formulations with a low pH and/or low osmolality may still be tolerated.

6.2.6 *Surgical facilities*

Adequate surgical facilities, including a pre-operative room, a surgical room, and a room for recovery, should be available. These facilities should comply with normal standards for surgical facilities, including availability of ventilators and equipment for inhalation anaesthesia. Also, adequate facilities with a high level of hygiene should be available for changing infusion cassettes.

6.3 *Best practice*

6.3.1 *Anaesthesia*

Anaesthesia can be induced using injection or inhalation anaesthesia. Regardless of what method is being used, it is always advisable to intubate the animals, enabling assisted respiration should that be necessary. For intubation, a long laryngoscope is placed over the tongue and the epiglottis is exposed after lifting the soft palate. Introduction of the endotracheal tube can be facilitated by spray application of local anaesthetics (e.g. xylocaine), reducing the risk of laryngo-spasms. Slightly rotating the tube will ease the procedure.

Prior to anaesthesia, sedatives should be given. For pigs and minipigs, azaperone is a very efficient sedative. Azaperone is administered as a single intramuscular injection about 20 minutes prior to induction of anaesthesia. In order to avoid detrimental effects of the parasympathetic nervous system, atropine should be co-administered with the sedative. Sedation of the animals should be performed in the animals' usual surroundings, as this will facilitate the following anaesthesia.

A good, efficient and safe anaesthetic protocol for injection anaesthesia is a mixture of zolazepam (125 mg), tiletamine (125 mg), ketamine (100 mg/mL, 1.5 mL), xylazine (20 mg/mL, 6.5 mL), and methadone (10 mg/ml, 1.5 mL). All substances are mixed and administered as a single intramuscular injection in a dose volume of 0.1 mL/kg. Anaesthesia will be present within about 5 minutes and will last for about 45 to 60 minutes, which should be more than sufficient for the implantation of vascular catheters. Supplements of 0.033 mL/kg can be administered to prolong anaesthesia. Alternatives are intramuscular combinations of ketamine (20 mg/kg) and xylazine (2 mg/kg), ketamine (33 mg/kg) and midazolam (0.5 mg/kg), and ketamine (10 mg/kg) and medetomidine (0.2 mg/kg). Propofol (12–20 mg/kg i.v./hour) can also be used to provide stable general anaesthesia.

Isoflurane, desflurane, and sevoflurane can all be used as efficient and relatively safe inhalant anaesthetics. The use of halothane is not recommended owing to the sensitising effect of the myocardium to catecholamine-induced arrhythmias and because of the more severe depressant effect on the myocardium compared with other inhalant anaesthetics. Isoflurane, desflurane, and sevoflurane exhibit similar physiological effects, with only minor differences in efficiency and safety. However, isoflurane is considerably less expensive compared with desflurane and sevoflurane and is therefore often the inhalation agent of choice. By co-administrating nitrous oxide, the percentage of the inhalant agent can be reduced while still maintaining the same level of anaesthesia. Moreover, a reduced myocardial depressant effect is seen.

6.3.2 *Preparation of the animals for surgery*

Following anaesthesia, the animals are clipped and shaved in the neck and on the anterior part of the back (or other relevant areas if veins other than the external or internal jugular veins are used for catheterisation). It is recommended to prepare an area larger than the minimum space required for surgery, should it become necessary to extend the surgical incision and in order to ensure sufficient space for attachment of bandages post-surgery. After washing has been completed, the surgical area is swabbed with 70% ethanol, which is rinsed off with sterile water. Finally, an iodine solution is applied for disinfection.

Surgery (and subsequent handling of catheters and injections into a vascular access port) should be performed using aseptic procedures to avoid catheter-related local or systemic infections. All bandaging material coming into direct contact with the catheter and the wound tissue, as well as all equipment used for dosing (such as infusion cassettes for continuous infusion and Huber needles for dosing into VAPs), should be sterilised.

Prophylactic antibiotic treatment may be worth considering; however, it should not be used as a replacement for aseptic techniques.

6.3.3 *Surgical procedure: Jugular vein*

The superior vein for catheterisation is the external jugular vein, with placement of the catheter tip in the cranial caval vein in front of the entrance into the right atrium. Placement within one of the largest veins in the body ensures immediate dilution of the dosed test compound owing to the high blood velocity and the immediate distribution from the heart into the entire blood stream. The surgical procedure for catheterisation of the external jugular vein is described in the following section. Catheterisation of other veins will be described later in this chapter.

The animal is placed in lateral recumbency, with the neck extended and slightly turned, exposing the jugular furrow. The upper foreleg should be pulled backwards. For a right-handed surgeon it is most convenient to place the animal in left recumbency. This position allows the surgical procedure to be conducted without any re-positioning of the animal during surgery.

An incision (length approximately 5–6 cm) is made over the jugular furrow, and the right external jugular vein is located and exposed by blunt dissection (Figure 6.2). A length of about 1–2 cm of the vein is isolated and dissected free of adventitious tissue. The vessel should be handled gently in order to avoid vasospasm and/or rupture. Two ligatures are placed cranially under the vein, cutting off the blood flow through the vein. Sacrificing one of the external jugular veins does not cause any problems concerning venous drainage from the head region, as there is massive reserve capacity in all the remaining veins. Another ligature, not yet tightened, is placed under the caudal end of the vein, assisting in manipulation of the vein during catheter insertion.

Another incision (about 1–2 cm), extending into the subcutaneous tissue, is made caudal to the right scapula towards the dorsum. At this site, the catheter exit is best protected from self-inflicted damage and it is furthermore the shortest distance to the infusion pump placed in a pocket of the minipig jacket applied immediately after surgery. The catheter is tunnelled subcutaneously from the dorsal incision to the incision in the jugular furrow using a trocar. At the jugular furrow incision, a small cut is

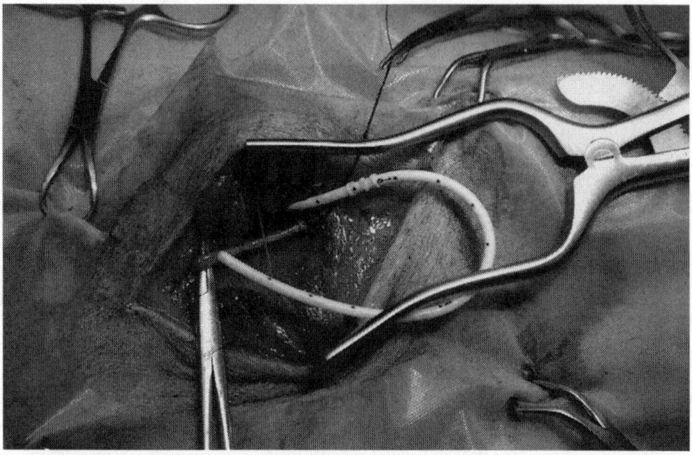

Figure 6.2 Isolation of the right external jugular vein. The catheter has been tunnelled subcutaneously from the incision in the back.

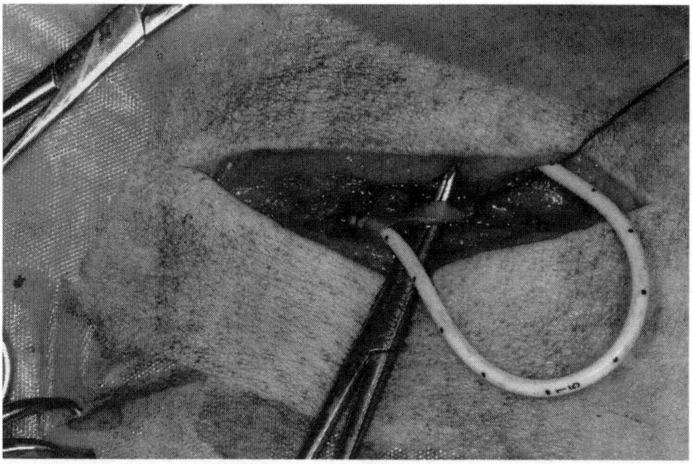

Figure 6.3 Catheter inserted and anchored by holding sutures into the external jugular vein.

made by sharp scissors into the isolated part of the vein, and the catheter is introduced into the vein to a position in the cranial caval vein (Figure 6.3). A sterile plastic catheter vessel introducer might be a good helping tool for the introduction. For a 10-kg minipig, approximately 10-cm of the catheter should be introduced into the vein. This ensures correct position within the cranial caval vein in front of the entrance to the right atrium. Correct catheter placement can be checked radiographically. A position within the

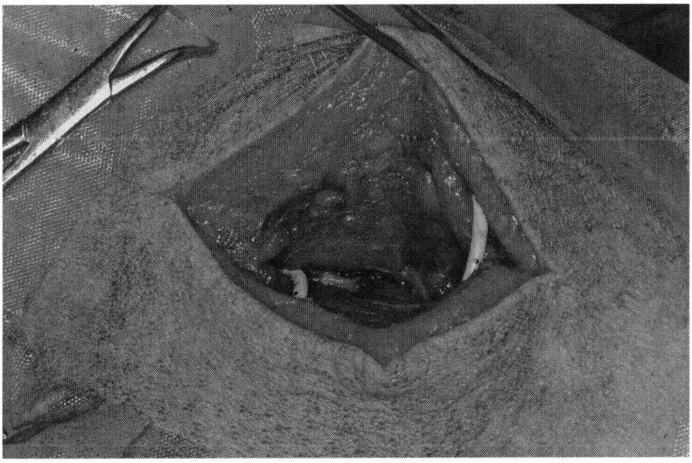

Figure 6.4 Loop of catheter, allowing for growth of the animal, has been placed subcutaneously before wound closure.

atrium is undesirable because of the risk of inducing cardiac arrhythmias and valvular damage. The catheter is secured in this position by application of 2 or 3 ligatures around the catheter and the vessel wall. Synthetic non-absorbable suture material is recommended; silk material is usually not recommended owing to its pro-inflammatory effect in swine.

To allow for growth of the animal and to prevent the catheter from being pulled backwards and out its position within the vein by growth of the animal, a catheter loop can be made at the point of catheter insertion (Figure 6.3). It is important to allow for sufficient space for this loop, as catheter kinking might otherwise be a problem. The loop is placed subcutaneously around the jugular furrow incision (Figure 6.4). Immediately after catheter introduction, the catheter is flushed with heparinised saline and the distal end of the catheter is attached to the infusion pump, infusing saline at a low infusion rate (2–3 mL/hour). The two incisions are closed by routine suture techniques. Skin sutures are placed closely around the free exteriorised end of the catheter, taking care not to puncture the catheter, obliterate the catheter lumen, or otherwise damage the catheter.

For implantation of a vascular access port (VAP), the surgical procedure in the area of the external jugular vein will be identical to what is described above (Figure 6.5). The VAP is placed over the scapula or in the neck region, where the VAP can easily be identified. It is important to select a localisation for the VAP ideal for dosing purposes in the conscious animal. A subcutaneous 'pocket' is prepared for the VAP. The skin incision should not be directly over the intended placement of the VAP, but to one side of it; this will reduce skin tension over the skin incision, which will promote skin

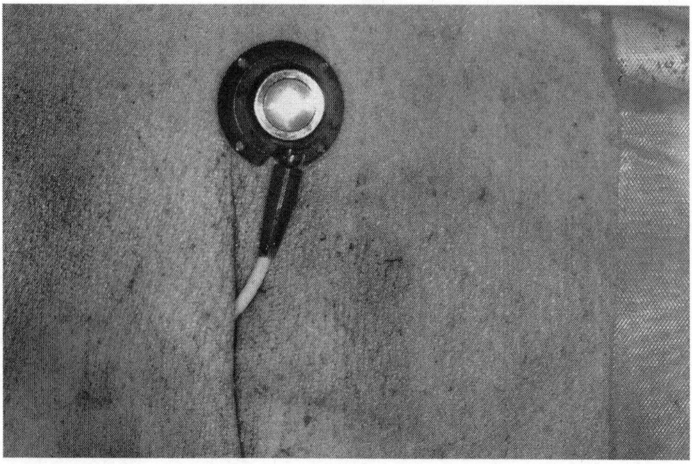

Figure 6.5 Connection of catheter to a VAP (back incision).

healing. The subcutaneous 'pocket' should be made just large enough for the VAP. If made too large, there will be an increased risk of VAP dislocation, including twisting; if too small, the skin tension will increase considerably, with the risk of wound rupture following surgery. If possible, the VAP should be secured by non-absorbable sutures to the underlying structures prior to skin closure in order to reduce the risk of VAP twisting. The VAP is connected to the catheter, which is tunnelled subcutaneously from the skin incision in the ventral part of the neck. The length of the catheter should be individually adjusted to make the best 'fit' for each animal. In order to allow for growth of each animal, one or more catheter loops can be prepared during the course of the catheter. Again, sufficient space for the loop should be provided, thus reducing the risk of catheter kinking.

6.3.4 *Surgical procedure: Other veins*

The external jugular vein is frequently used for blood sampling. For repeated blood sampling procedures, it may be an advantage to be able to sample from both external jugular veins, in which case an alternative placement of the infusion catheter becomes relevant.

The internal jugular vein is located deeper than the external jugular vein and can be accessed using the same incision over the jugular furrow, as described above. The internal jugular vein is situated slightly deeper and medial to the external jugular vein and will appear in close proximity to the carotid artery and the vagus nerve. The procedure for vein catheterisation is as described above, and catheter exteriorsation is also preferably performed in the caudo-dorsal part of the neck.

The cephalic vein of the pig runs subcutaneously from the flexor aspect of the elbow joint to the lateral thoracic furrow into the external jugular vein. The cephalic vein can be accessed at the junction of the cranial aspect of the front limb and the ventral surface of the neck. As described for the external jugular vein, a part of the vein is dissected free of adventitious tissue, and following ligation of the distal end of the vein the catheter is catheterised. From the point of catheter insertion, the catheter can be tunnelled subcutaneously, as previously described.

The femoral vein is located within the femoral canal at the medial aspect of the hind leg. For catheterisation of this vein, the animal is placed in dorsal recumbency with the hind leg for catheterisation retracted in a caudolateral position and the other hind leg retracted caudally. For a right-handed surgeon, it will be easier to catheterise the right femoral vein. Following a skin incision over the femoral vein, the muscle fascia of the sartorius and gracilis muscles are separated, and, by retracting the gracilis muscle, the femoral canal is exposed. The vein is found caudal to the femoral artery and medial to the saphenous nerve. Because of the branching of the vein, which becomes more pronounced at the distal part of the vein, only a very short part of the femoral vein (the proximal end running towards the external iliac vein) is suitable for catheterisation. Along with the deep position of the vein, this makes the catheterisation procedure difficult. The longest possible part of the vein should be isolated and dissected free of surrounding tissue. The distal part of the vein is ligated and the catheter is introduced proximal to the position, using the same principles as described for external jugular vein catheterisation. At least 10–15 cm of catheter length should be introduced depending on the size of the animal, ensuring a position within the caudal caval vein. From the point of insertion, the catheter should be tunnelled subcutaneously to the point of exteriorisation, preferable on the back of the animal. Alternatively, it should be tunnelled to the position of the vascular access port, which is optimally also placed in the back region of the animal. Depending on the specific site used for catheter exteriorsation, it should be noted that the catheter dead space using the femoral vein for catheterisation will be considerably larger compared with jugular vein catheterisation, owing to the length of the catheter implanted.

6.3.5 *Transcutaneous catheterisation*

Using the Seldinger technique, it is also possible to catheterise the external jugular vein by percutaneous puncture of the vessel. Following this procedure, surgery can be avoided, which will markedly reduce procedure recovery. However, the catheter is exteriorised at the ventral aspect of the neck, a position that is less desirable and has a higher risk of contamination and self-inflicted damage, even though the site of exteriorsation will be protected by bandages. Transcutane catheterisation is therefore recommended only for short-term infusions; for chronic use, surgical implantation is necessary.

The procedure is performed with the animal in full anaesthesia. The animal should be positioned in dorsal recumbency, with the forelegs restrained backwards and the neck slightly bent downwards. The area for catheterisation is prepared as for surgical procedures, and aseptic techniques should be applied. The external jugular vein is found within the jugular furrow; the needle is introduced in this area at a 45° angle. Upon venous puncture, which will be evidenced by appearance of blood from the needle, a guidewire is introduced into the lumen of the needle. It is important to introduce a sufficient length of the guidewire before the needle is withdrawn, ensuring correct positioning within the vessel lumen. The procedure can beneficially be performed using fluoroscopy as guidance.

Upon withdrawal of the needle, the catheter is introduced over the guidewire. A small skin incision will ease catheter penetration into the skin. Rotating movements of the catheter upon introduction over the guidewire will make the insertion procedure easier. After the desired length of the catheter has been correctly positioned within the vein, the guidewire is withdrawn and the catheter is filled with lock solution or connected to the infusion pump. By use of skin sutures, the catheter can be further secured to the skin. A vascular access port can also be fixed onto the skin, using a similar principle. The catheter should be protected by bandages upon insertion.

Other veins, such as the femoral vein, can be accessed using this procedure. It is important to study the anatomy in detail prior to catheter insertion in order to have a successful introduction.

6.3.6 *Post-surgical analgesia*

An intensive analgesic treatment programme must be followed subsequent to the surgical procedure. The optimal choice of analgesics is opioids, and owing to its half-life duration, buprenorphine is the drug of choice. Buprenorphine is administered intramuscularly at a dose of 0.01 mg/kg with intervals of administration of about 8 hours. The animals should be observed frequently to evaluate pain treatment, and especially the diet consumption, alertness, and general level of activity should be noted. If changes are noticed in the general behaviour of the animal, it might become necessary to prolong analgesic treatment.

Analgesic treatment can beneficially be instituted pre-emptively, with reduced activation of the pain sensory system as a result. With good effect, opioids can be combined with non-steroid anti-inflammatory drugs (NSAIDs), for example meloxicam, administered on a daily basis for 3–4 days in a dose of 0.4 mg/kg.

6.4 Practical techniques following surgery

6.4.1 Bandaging procedures

Following surgery, the wounds are bandaged and, if surgery with catheter exteriorsation has been performed, a pig jacket is attached to each animal. This jacket has two pockets: one pocket is for the ambulatory infusion pump to which the catheter is attached, while the other pocket can be used for placement of a counterbalance. The counterbalance will counteract the weight of the infusion pump and will prevent the pig jacket from sliding to the side.

Ideal dressings for placement over the surgical wounds are foam bandages with attached rims of adhesives. The central foam will absorb any leakages from the wound. In order to maintain placement of the dressings at the required site, it is recommended to apply a secondary stretch bandage over the primary dressing and the neck and cranial part of the thorax. Using this procedure, it will not be possible for the animal to interfere with the dressings and the wounds. The bandages applied should preferably be changed on a daily basis. Especially on the day following surgery, it is very important to frequently check and to change the dressings, as post-surgical haemorrhage may have occurred, possibly exceeding the maximum absorbent capacity of the primary dressing.

The animals are usually allowed at least a 7-day recovery period before start of treatment. During this period the animals are infused with sterile saline at low infusion rate (2–3 mL/hour), in order to maintain patency of the catheter. Vascular access ports are flushed with sterile heparinised saline at least twice weekly.

6.4.2 Maintenance of catheters

The pig is slightly more prone to thrombus formation compared with other animal species, and therefore correct maintenance of the catheters is of major importance. The infusion cassettes should always be filled under aseptic conditions. Catheters must be handled using sterile techniques, and correct training of staff is therefore essential. At all changes of infusion cassettes, it is recommended to flush the catheter with a sterile heparinised (at least 100 IU/mL) saline solution. Flushing should be performed at least twice weekly. Other catheter lock solutions can be considered, such as hypertonic glucose, dextrose solutions, or solutions containing taurolidine citrate. Taurolidine has antimicrobial effects against a wide range of infectious organisms, whereas the citrate has some anticoagulant properties.

When not used for dosing procedures, exteriorised catheters used for continuous infusion should be continuously infused with saline at a low infusion rate in order to maintain patency. An infusion rate of 2–3 mL/hour is usually adequate.

Using the correct techniques, patency of catheters can be maintained for several weeks without any problems. Continuous infusion studies of 4 weeks' duration can therefore easily be performed in the minipig.

6.4.3 Changing infusion cassettes

Sterilised infusion cassettes containing various volumes of liquids are commercially available. These cassettes are easily attached to the infusion pump, and the rate of infusion is set on the pump. When handling the catheters and the infusion equipment, it is extremely important to restrict movements of the animals, as the risk for sudden movements and thereby damage of the catheters by retraction (especially a risk in the first days after surgery, when the catheters are not yet firmly anchored within the surrounding tissue) is significantly increased. The animals are best controlled during these procedures by placement in a sling. After short periods of training, the animals usually fully accept placement within the sling, and handling the infusion equipment, including changing infusion cassettes, can easily be performed.

The infusion cassettes should be filled with test formulation of saline under aseptic conditions. Care should be taken to avoid occurrence of air bubbles within the liquid, which potentially could have fatal effects if dosed intravenously. Prior to the changing of cassettes, all external parts of the equipment, including the distal end of the catheter, the used cassette still connected to the catheter, and the new replacement cassette, should be sterilised by application of 70% ethanol solution. Sterile gloves should be worn during the changing procedure. It should always be kept in mind that the catheter gives direct access to the blood circulation of the animal.

6.5 Histopathological examination

At the end of the treatment period, the animals should be subjected to a thorough macroscopic and microscopic examination. At necropsy, selected organs and tissues are sampled, weighed, and fixed in 4% neutral buffered formaldehyde. The list should include the intravenous infusion site (the tissue/vein surrounding the tip of the catheter), the heart, the liver, the kidneys, and the lungs.

A detailed macroscopic description of all changes should be made, including size and location, as well as other relevant descriptors. When the lesions are extensive, the macroscopic details often offer an advantage over a microscopic examination restricted to a minor sample of the lesion. Photographs of the macroscopic changes could be considered (examples given in Figures 6.6 and 6.7).

The tissues intended for microscopic examination are trimmed, and representative specimens are taken for histological processing. The

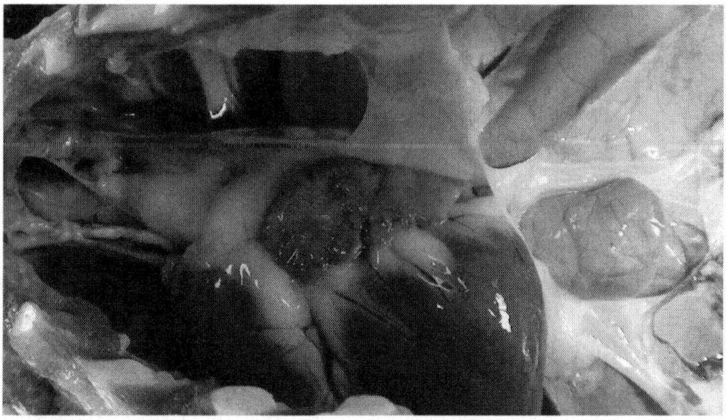

Figure 6.6 Tip of catheter in the vena cava cranialis at autopsy.

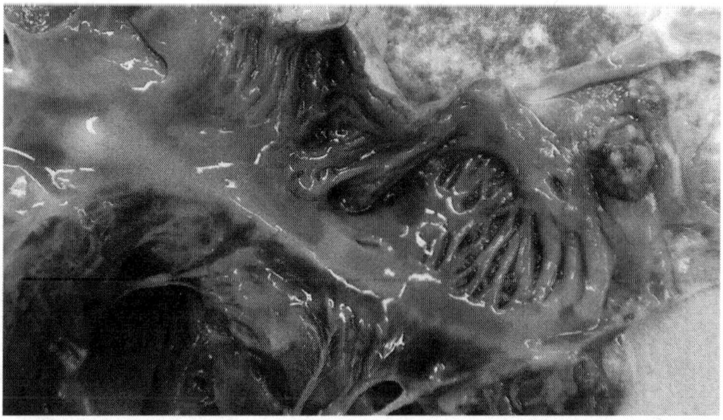

Figure 6.7 Endocardial surface of the heart and proximal part of the caval vein.

specimens are embedded in paraffin and cut at a nominal thickness of approximately 5 µm, stained with haematoxylin and eosin (H&E), and examined under a light microscope. If indicated, special stains such as Masson's trichrome (for fibrosis) and Mallory's PTAH (for fibrin) could be performed. In case of extensive lesions at the infusion site, several sections should be considered to obtain a better insight into the nature of the lesion; if possible to prepare, longitudinal sections, including the vessel wall, might offer an advantage to transverse sections.

The most common procedure-related change in continuous intravenous infusion studies in Göttingen minipigs is by far thrombus formation at the intravenous infusion site in the cranial caval vein (or occasionally in the right atrium if the tip of the catheter has been advanced too far and reached the heart). The morphology of the thrombus is most often a chronic well-organised part, associated with thickening/hypertrophy of the vessel wall, and a part with less-advanced organisation characterised by fibrin, leukocytes, and haemorrhage (Figures 6.8, 6.9, and 6.10). Abscessation might be present. In cases of major thromboses, microthrombi might be found elsewhere, mainly in the lungs but also occasionally in the kidneys (glomeruli) and the liver.

The cause of the thrombi might be a foreign-body reaction to the presence of the catheter (the type and material of the catheter is of importance), direct trauma to the vessel wall, changed haemodynamics, chemical and physical properties of the infusate (irritative potential), the rate and volume of the infusion, and/or microbial contamination/infection. It is important to bear in mind that when employing intravenous infusion, the common organs constituting the first line of defence of the body to invading microorganisms are bypassed, highlighting the need for strict aseptical procedures during surgery and treatment.

6.6 Conclusion

Continuous intravenous infusion is a well-established technique in minipigs. The pathological changes in the minipigs secondary to the continuous intravenous infusion are slight and acceptable. The most common findings in these studies are foreign body granulomas, mainly in the lungs.

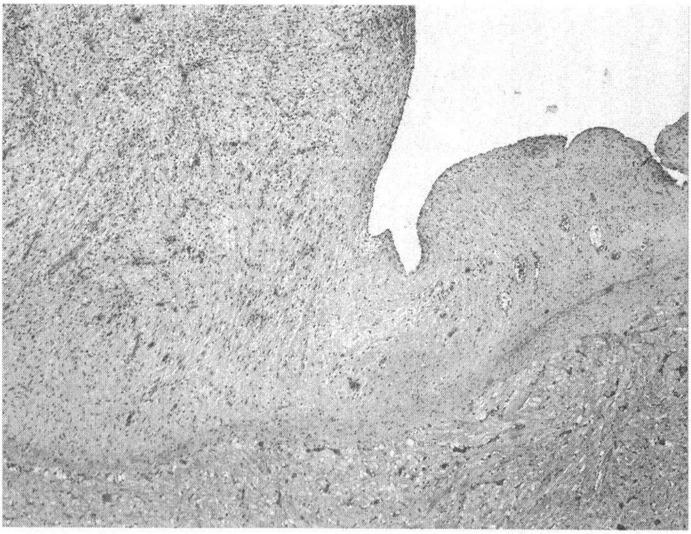

Figure 6.8 (See colour insert.) Well-organised thrombus in vena cava cranialis. (Treated with NaCl; H&E ×5.)

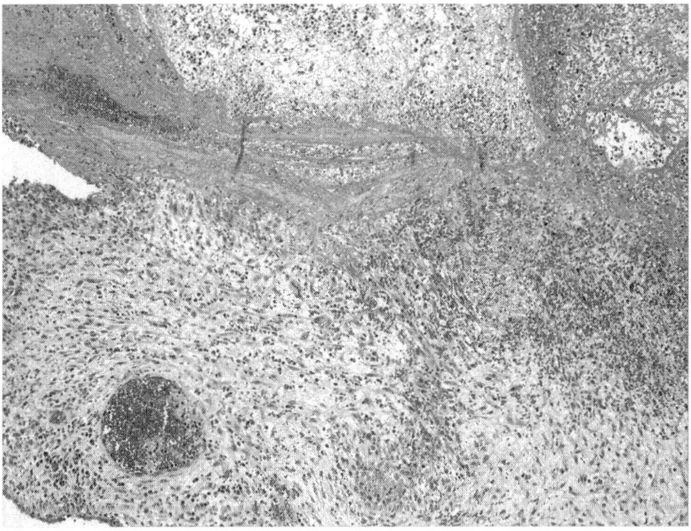

Figure 6.9 (See colour insert.) Active thrombus with inflammation of vessel wall. (Treated with NaCl; H&E ×10.)

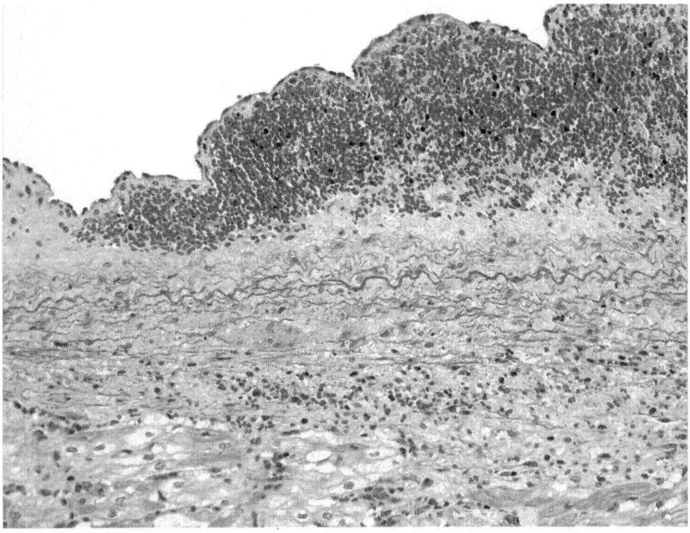

Figure 6.10 (See colour insert.) Subendothelial haemorrhage with inflammatory cells. (Treated with NaCl; H&E ×20.)

The use of minipigs in continuous intravenous infusion studies should always be considered as a valid alternative to other laboratory animal species.

References

Healing G and Smith D (eds). 2000. *Handbook of Pre-clinical Continuous Intravenous Infusion*. London and New York: Taylor & Francis.

Nickel R, Schummer A and Seiferle E. 1981. *The Anatomy of the Domestic Animals*, Volume 3. Berlin: Verlag Paul Parey.

Swindle M. 2007. *Swine in the Laboratory*. Boca Raton, FL: CRC Press.

Swindle M, Nolan T, Jacobsen A, Wolf P, Dalton MJ and Smith AC. 2005. Vascular access port (VAP) usage in large animal species. *Contemporary Topics in Laboratory Animal Science* 44(3): 7–17.

chapter seven

Juvenile animals

Teresa R Gleason, BS, LVT, SRS, RLATG
WIL Research Laboratories Inc., USA

George A Parker, DVM, PhD, DACVP, DABT
WIL Research Laboratories Inc., USA

Contents

7.1 Introduction

The purpose of this chapter is to discuss vascular infusion technology for the juvenile animal model. While juvenile toxicology studies are often conducted in the rat pup, this chapter will focus instead on two larger models: the piglet and the puppy. Repeated venous access is more feasible in these species than in the rat pup, with more sites to choose from (including the jugular, cephalic, saphenous, marginal ear, and femoral vein).

Physicians have traditionally determined the dosage of therapeutics for use in the paediatric population based on professional experience and the patient's body weight. The problem with this approach is that children are not just small versions of adults. Major functional differences exist between human neonates, infants, children, and adults. Developing organs, organ systems, and metabolic pathways in children can react very differently to chemicals from those in fully developed adults (Beck et al. 2006).

The development of the major systems in the human is age dependant (EMA 2008):

- Nervous system: development up to adulthood
- Reproductive system: development up to adulthood
- Pulmonary system: development up to two years of age

- Immune system: development up to twelve years of age
- Renal system: development up to one year of age
- Skeletal system: development up to adulthood
- Organs and systems involved in absorption and metabolism of substances: development of biotransformation enzymes up to adolescence

7.1.1 Choice and relevance of the species

When determining an appropriate species for a juvenile toxicology study, one is encouraged to consider the pharmacology, pharmacokinetics, and toxicology of the test article; the comparative developmental status of the major organs of concern between juvenile animals and juvenile humans; and the sensitivity of the selected species to a particular toxin (FDA 2006). The age of the intended human population relative to the comparative age categories based on development is also a factor when choosing a species (Figure 7.1) Traditionally, rats (≤ 21 days) and dogs (≤ 2 months) have been the rodent and non-rodent species of choice for evaluating toxicity in juvenile studies. Other species, such as pigs or non-human primates, may be more appropriate in cases where drug metabolism in rats and dogs may differ significantly from that in humans; however, litter size may contraindicate use of the non-human primate.

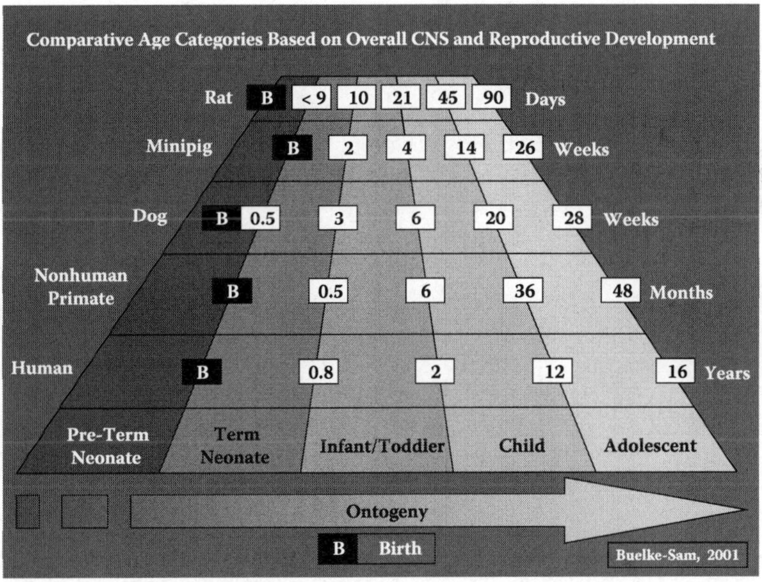

Figure 7.1 Comparative age categories based on overall CNS and reproductive development.

The route of administration may also play a significant role in species selection, particularly in the case of intravenous infusion in a juvenile animal. Ideally, the intended human route of administration should be used for juvenile toxicology studies. Intravascular infusion as the route of administration limits the choice of species, as does the length of infusion. Continuous intravascular infusion further complicates choices by being virtually impossible in the juvenile animal because of the activities of its siblings, the behaviour of the mother, and the animal's need to nurse. Limitations are also seen in the human paediatric patient, where in the instance of intravenous therapy multiple short-term infusions or bolus doses are often used to deliver therapeutic drugs rather than continuous intravenous infusion. The models discussed in this chapter (the juvenile minipig and dog) will cover both techniques: daily short-term infusion as well as multiple daily bolus injections. Table 7.1 describes the approximate juvenile age-range equivalents of these two species to the human.

Alternative techniques are used if the rat pup needs to be the rodent species, such as multiple daily intraperitoneal or subcutaneous injections. While these routes of administration do not mimic the intended human route of administration, they may provide adequate toxicological levels that will affect developing organ systems for determining whether the therapeutic agent is safe to advance to clinical trials in the human paediatric population. However, since recently, it has been considered possible to perform repeat intravenous dosing in rats from 21 days of age.

7.1.2 Regulatory guidelines for juvenile toxicology studies

Juvenile toxicology studies are mandated by both European and U.S. regulatory agencies prior to the initiation of paediatric clinical trials. The regulatory guidance documents are a recent development; most non-clinical

Table 7.1 Approximate Juvenile Age-Range Equivalents

	Human	Minipig	Dog
Preterm newborn infant	< 38 weeks gestation	Preterm	< 0.5 weeks old
Neonates	0–27 days old	Birth–2 weeks old	0.5–3 weeks old
Infant/toddler	28 days–23 months old	2–4 weeks old	3–6 weeks old
Child	2–11 years old	4–14 weeks old	6–20 weeks old
Adolescent	12–16 years old	14–26 weeks old	20–28 weeks old

Source: Modified from Beck MJ, et al. 2006. Nonclinical juvenile toxicity testing. In *Developmental and Reproductive Toxicology: A Practical Approach*, 2nd edition, Hood RD (ed). Boca Raton, FL: CRC Press/Taylor & Francis Group, and Stiltzlein EB, et al. 2011. Intravenous infusion and restraint in juvenile beagle dogs. 62nd Annual AALAS National Meeting, San Diego, CA.

and clinical testing for drug development had been previously conducted only in adult populations. There was also little difference in the way drugs were prescribed between the adult and paediatric populations despite the fact that science has long recognised developmental differences between the populations and their potential physiological responses to toxins. Adjustments for drug doses in children were based on calculations that used age or relative body size (e.g., Young's Rule, Cowling's Rule, or Clark's Rule) as well as clinical and nonclinical adult studies (Kearns et al. 2003; Blumer 1999). It is further presumed that there may have been apprehension on the part of physicians to conduct clinical trials on children, as many in the health community considered it unethical to put children at risk, especially since children could not give legal consent (AAP 1995; Halpern 1988).

In 1995, a survey conducted by the American Academy of Pediatrics Committee on Drugs found that the majority of drugs listed in the *Physician's Desk Reference* lack information on the safety or efficacy for paediatric use. The drug industry had little incentive to develop new therapeutics for the paediatric population or to evaluate the safety of existing drugs because physicians have the ability to legally prescribe drugs off-label to children. Many drug companies added a disclaimer on the label indicating that safety and effectiveness had not been established in the paediatric population.

Paediatric legislation (Best Pharmaceuticals for Children Act of 2002, and the Pediatric Research Equity Act of 2003) provided a mechanism to obtain the necessary paediatric safety and efficacy information on drug product labels. In 2003, the U.S. Food and Drug Administration (FDA) issued the draft guidance document 'Nonclinical Safety Evaluation of Pediatric Drug Products', which was finalised in 2006. The European Medicines Agency's Committee for Human Medicinal Products drafted the 'Guideline on the Need for Non-Clinical Testing in Juvenile Animals of Pharmaceuticals for Paediatric Indications' in 2005, which subsequently became effective in 2008 (EMA 2008).

7.1.3 Choice of infusion model

In human medicine, continuous intravenous infusion in the infant is not always an option—traditional therapy often involves administering medications by slow intravenous infusion every 4–6 hours for several days. This may be accomplished using temporary catheters placed in a superficial vein or via surgically implanted catheters with attached vascular access ports. The latter preparation is commonly utilised when the infant must continue intravenous therapy at home; further surgery is required to remove or replace the catheters and ports as the infant grows.

The juvenile minipig is a good choice for surgical catheter implantation for vascular infusion. Shortly after birth, these animals are able to tolerate anaesthesia and surgical procedures, and they are able to be fitted with a jacket to house external equipment. They are of sufficient size to allow implantation of a relatively large-bore catheter (3–4 French), and dosing can span the human age-range equivalents of neonate, infant/toddler, child, and adolescent (Beck et al. 2009). Additionally, experience has shown that the sow shows little interest in the external equipment on the piglets and does little to disrupt it.

A good choice for placement of a temporary catheter for daily short-term administration is the juvenile canine. A multitude of restraint methods are available commercially to accommodate the growth of the puppies through each developmental stage, and by post-natal day (PND) 14 the animals are large enough to accommodate catheter placement in the cephalic, saphenous, and jugular veins (Stiltzlein et al. 2011).

7.1.4 Limitations of available models

There are many limitations of available juvenile animal models of vascular infusion. As stated previously, the rat is a commonly used species for preclinical juvenile toxicology studies. However, there are many issues precluding the widespread use of the rat for juvenile safety assessment compared with the larger laboratory animal species. The size of the peripheral vessels limits accessibility, and rapid growth of the animal prohibits surgical placement of a catheter.

Growth of the animal is the primary limitation of the surgical model of vascular infusion, as it is difficult to span the entire juvenile age range from neonate to adolescent without repeated surgeries to replace equipment. Housing, interaction with siblings, the animal's ability to nurse, and maternal behaviours may also be problematic.

When considering a non-surgical model, peripheral vessel size and restraint of the animal through growth stages are the greatest limitations. Misplacement or dislodgement of a temporary catheter may complicate the dose delivery and potentially impact the resultant pathology results. The length of time the animal is restrained and away from the dam should be considered as well.

7.2 Best practice

Generally for vascular infusion, surgical placement of a catheter rather than repeated temporary catheter placement for test article delivery is considered best practice. With appropriate aseptic preparation, there is less chance of introducing foreign material to the animal, and the histopathology is less complex as there is only one infusion site. Additionally, surgical placement is considered more humane: the catheter is placed under general anaesthesia, and the animal does not have to undergo prolonged restraint and multiple intravenous sticks during the dosing procedure. Limitations of preferred species for juvenile toxicology studies (the rat and the dog) as described make the application of the acknowledged best practice procedures difficult to impossible, and thus alternatives were developed to achieve the goals of the studies. For example, surgical catheterisation of a blood vessel in a juvenile rat presents a significant surgical skills challenge, and the activity of juvenile dogs presents the challenge of making surgical catheterisations secure. There are also the considerable problems presented by the rapid growth rates of these two species.

7.2.1 Surgical model: Juvenile Göttingen minipig

Although all the details given in this section are applicable to the minipig model, the majority of the process, with minor adjustments, could also be applicable to the juvenile canine. It will become apparent why the minipig is the preferred model. The model described was used in our laboratory for multiple daily intravenous infusions. This model could also be utilised for daily short-term infusion depending upon the study design.

Surgical implantation of the jugular vein in the offspring of time-bred Göttingen minipig sows is usually performed on post-natal day 3 with dosing beginning on post-natal day 4. Via the implanted catheter, animals can be dosed intermittently from PND 4 to PND 13, which covers the human equivalent of neonate to toddler age.

7.2.1.1 Training

Training of the surgical staff and support staff is critically important to the success of a surgical model for vascular infusion. Personnel performing surgical procedures must demonstrate a proficiency in proper sterile technique, tissue handling, haemostasis, and closure techniques. In addition, these individuals are usually responsible for peri-operative care, including anaesthesia, analgesia, and post-operative care. There are few regulations that discuss in detail the manner in which surgical training or how a qualification process for surgical personnel is to be performed in the research setting. In the United States, the Animal Welfare Act provides some guidance to specific topics related to training for surgery. These requirements are useful for the development of an institutional surgical training programme. A comprehensive training programme should include the following components and employ both didactic and practical training and evaluation: regulatory/ethical considerations; facility requirements; pre-surgical planning; analgesia; anaesthesia; aseptic technique; surgical technique; and post-operative care. Additionally, it is necessary to include post-training education and re-evaluation to ensure that procedural drift has not occurred. Certification programmes are available for non-doctoral degree personnel through the Academy of Surgical Research to ensure competency in anaesthesia (Surgical Research Anaesthetist), minor surgical procedures (Surgical Research Technician), and more complex surgical procedures (Surgical Research Specialist) (ASR 2009).

Training is also necessary for individuals responsible for dosing the vascular infusion animals. Infusion studies are most successful when the maintenance and dosing of the animals are limited to a highly trained team of technicians. Ideally, this infusion team should be trained in aseptic technique, handling and maintenance of the surgically prepared animal, maintenance of all external equipment on the surgical animal models and non-surgical animal models, and use of the infusion pumps. Post-training evaluation is critical for these individuals as well.

7.2.1.2 *Applicability of the methodology*

Peripheral vessels are difficult to access repeatedly in adult minipigs (the ear vein is most readily accessible) and the problem is compounded in the juvenile. The jugular vein has been chosen as the site of surgical catheter placement for our studies, as it is easily accessible in the juvenile minipig and a relatively large-diameter catheter may be surgically implanted (3–4 Fr.) Additionally, the port is placed in the scapular area so there is less catheter dead space and less surgical trauma during the subcutaneous tunnelling procedure from the catheter insertion site to the vascular access port. A limitation to this placement is that advancing the catheter further than 3 cm could result in the catheter tip entering the heart.

The femoral vein is difficult to access owing to the anatomy of the inguinal area. The vein is relatively deep, and the hip has limited flexibility to allow surgical access. The size of the vessel limits the catheter choice to a 2-Fr. diameter, which is more prone to clot formation and particularly in the inguinal location is more prone to kinking.

Systemic antibiotics and analgesics are not used on the juvenile minipig in our laboratory. The goal is to conduct the surgical procedures in a sterile manner quickly and to return the piglet to the sow as soon as possible. Additionally, if the animal were to be sedated by analgesics, it could rapidly lose body heat and potentially be trampled by the sow. Animal room temperature is elevated and heat lamps are positioned over the creep area of the farrowing cage to allow the piglet supplemental heat as well as the ability to get away from the sow post-operatively (Figure 7.2).

Surgical records must be maintained for each animal to accurately document the surgical procedure, including all complications that may have occurred. Keeping in mind that the surgery itself is not what is being evaluated, surgical complications that could affect the health of

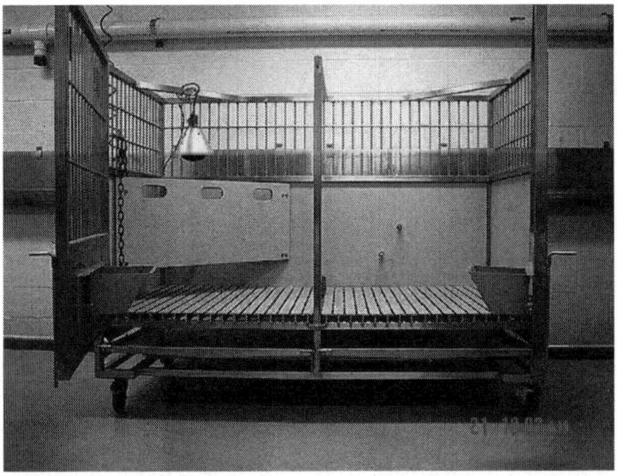

Figure 7.2 Caging for sow and litter with heated creep area for piglets.

the animal, such as excessive bleeding or breaks in sterility, should be documented and may be reasons to exclude the animal from study.

Recommended infusion rates for vascular infusion are dependent upon the volume and duration of the dose. Guidelines state that for a period of infusion less than 6 hours, 10 mL/kg/hr is an acceptable rate, not to exceed 60 mL/kg/day. For periods less than 1 hour, the rate should not exceed 0.25 mL/kg/min (Hull 1995). A study design for this model requires a relatively high infusion volume, 7 mL/kg/dose, repeated every 4 hours from post-natal days 4 to 13, resulting in a total daily volume of 42 mL/kg. Each infusion was delivered over an approximate 45- to 60-second period. As the catheter was being accessed so frequently, the use of an anticoagulant was not necessary during the dosing procedure, and the catheter was flushed with 0.9% saline following each dose.

7.2.1.3 Risks/advantages/disadvantages

The main risk of the described surgical model is the potential for infection in an animal with an implant. The animal is being returned to a farrowing cage with its dam and siblings, and although the cage is cleaned daily, the potential for bacterial contamination is high.

7.2.2 Non-surgical models: Juvenile canine (beagle)

The non-surgical juvenile model of vascular infusion discussed in this chapter is the canine. The juvenile dog is the most applicable laboratory animal species since the minipig (or non-human primate) cannot, ethically, be restrained for prolonged periods. The juvenile canine is restrained for daily intravenous infusion dosing from post-natal days 14 to 91, which allows dosing over the full juvenile toxicology study range from neonate to adolescent.

The juvenile beagles are whelped at the supplier and arrive with their dams at approximately 7–10 days of age. Animals are infusion-dosed daily for approximately 30 minutes using a temporarily placed 'over-the-needle' intravenous catheter.

7.2.2.1 *Methods of restraint*

Restraint devices vary and are dependent upon the size of the animal and the vessel chosen for infusion. A combination of veterinary-style examination restrain bags and limb splints can be used until the animal is large enough to be restrained in a sling restraint. As dosing normally begins on post-natal day 14 and restrain methods vary as the animal grows, it is not feasible to acclimate the animal to the restraint methods.

7.2.2.2 *Applicability of the methodology*

The maternal behaviour of the dam limits choices of infusion methods in the juvenile canine. The dam is extremely attentive to the pups and repeatedly grooms her young to not only stimulate and clean them, but also to promote bonding and more rapid maturation of their nervous systems. This grooming behaviour is also necessary to promote elimination of urine and faeces. This behaviour of the dam, as well as the natural instincts of a pup's littermates to play-fight as they age, limits the placement of external infusion equipment such as surgically placed catheters and protective jackets.

Surgically placed vascular access ports and implantable infusion pumps are a consideration, but are not commonly used owing to the rapid growth of the animal and the need for further surgical intervention to continually replace the implanted equipment.

Peripheral vessels are preferred for studies conducted at our laboratory on puppies from post-natal day 14 to post-natal day 91. This range enables dosing over the full juvenile toxicology study range from neonate to adolescent. Additionally, the peripheral veins (cephalic or saphenous) are easily accessed in neonate and adolescent animals, and the dosing equipment is easy to temporarily secure during the infusion period. The jugular vein may be used occasionally if the cephalic or saphenous cannot be accessed, but the temporary catheter is difficult to place because of the size of the neonate, and is difficult to secure for the infusion period.

7.2.2.3 *Risks/advantages/disadvantages*

The primary disadvantage of placing a catheter daily is that incorrect placement or dislodgement of the catheter will result in extravascular dose administration. This will result in the animal receiving an incorrect dose, and could complicate histopathology evaluation, particularly if the test article is irritating. It is difficult to determine exact dose locations for the histopathology sectioning as well. Despite the fact that the anatomical vein location is recorded daily (e.g., right cephalic vein), it is not possible to determine the exact location of infusion in the vessel.

7.3 Practical techniques

7.3.1 Surgical model: Juvenile Göttingen minipig

7.3.1.1 Preparation

On post-natal day 2, the piglets are fitted with an infusion jacket for acclimation prior to surgery on post-natal day 3.

7.3.1.1.1 Premedication. Since developing animals have difficulty maintaining core body temperature early in life and time away from the sow can have a negative impact on development, premedication with analgesics and sedatives is contraindicated. The prolonged recovery time required means more time away from the sow, and preliminary validation work demonstrated much better recovery results by returning the piglets to the sows as quickly as possible. At our laboratory, anaesthesia is induced using 5% isoflurane in 100% oxygen delivered via a calibrated vaporiser and rodent-sized nose cone. Also, a lubricating eye ointment is applied prior to surgical preparation.

7.3.1.1.2 Presurgical preparation. Presurgical preparation commences once the animal has reached an adequate level of anaesthesia. This includes clipping the hair over the surgical sites and scrubbing the incision sites with a minimum of three alternating chlorhexadine and 70% isopropyl alcohol scrubs followed by betadyne solution paint. All skin disinfectants are kept in a warming bath to avoid chilling the piglets during the surgical scrub procedure.

Once the surgical scrub is completed, the animal is instrumented for intra-operative monitoring with a rectal temperature probe and a pulse oximeter probe. The animal is then enveloped in a sterile adhesive drape to allow manipulation of the animal by the surgeons without contamination.

7.3.1.1.3 Anaesthesia. Isoflurane anaesthesia (5% for induction and approximately 2% for maintenance) is used throughout the procedure. Additionally, doxapram (2 mg/kg) is available for sublingual administration if needed to stimulate respiration.

7.3.1.2 Surgical procedure

The prepared animal is moved to the surgery table equipped with a circulating warm water blanket and placed in ventral recumbency on a sterile drape. Using a number 15 scalpel blade, an approximate 3-cm incision is made in the scapular area perpendicular to the spine, and a subcutaneous pocket is formed caudal to the incision using blunt dissection. The animal is then placed on its left side to allow access to the dorsal incision as well as the right jugular vein. The sterile adhesive drape allows the surgeon to fully manipulate the animal without contamination. A small incision is made in the skin over the right jugular vein, and a 3–6-Fr tapered polyurethane catheter with attached skin parallel vascular access port is routed subcutaneously over the right shoulder from the dorsal incision to the ventral incision. The external jugular vein is isolated at the bifurcation of the two mandibular branches using blunt dissection with mosquito haemostatic forceps, and two non-absorbable suture tags (4-O braided polyester) are placed around the vessel. The jugular vein is ligated with the cranial suture, and the vessel is retracted. The caudal suture is retracted as well to restrict blood flow. The adventitia of the vessel is grasped with micro-dissecting forceps, and a small venotomy is made using Castroviejo scissors. The catheter is advanced approximately 3 cm into the jugular vein. Suture beads are located at 3 and 3.5 cm from the tip of the catheter. The catheter is secured in place by tying the caudal suture around the catheter and vein as well as securing the cranial ligation suture to the catheter between the suture beads. The port is accessed with a Huber needle and patency verified. The port is placed into the dorsal subcutaneous pocket such that there is no tension from the port on the incision that may lead to pressure necrosis or incision dehiscence. Unlike most vascular access port placements, the port is not secured to the underlying musculature: the piglets grow very rapidly, and there is concern that a fixed port may lead to catheter dislodgement from the jugular vein. The vascular access port is then accessed using an appropriate external catheter. An introducer needle is placed through the skin over the septum of the skin parallel port and advanced into the port until the bottom of the port is felt. The stilette of the introducer needle is then removed. The nylon dosing catheter is placed into the port through the introducer needle and advanced into the lumen of the implanted polyurethane catheter approximately 3 cm. The introducer needle is then removed from the port, and the nylon dosing catheter trimmed to approximately 15 cm. An appropriate locking connector is attached to the nylon dosing catheter, and an injection cap (needleless access style) is used to close the system. The infusion system is then locked with approximately 0.3 mL of 100 IU/mL heparinised saline.

For most surgical skin closures, it is recommended to use a monofilament suture to prevent 'wicking' of pathogens into the incision. The skin of the piglet is very fragile, however, and a monofilament suture such as nylon tends to cut the skin; therefore a 4-O braided polyester suture is used in a horizontal mattress suture pattern to close both incisions. The horizontal mattress pattern aids in decreasing tension on the incision. The incisions and the sutures are then sealed with a high-viscosity cyanoacrylate tissue adhesive. This forms a barrier over the incision and aids in preventing the 'wicking' properties of the braided suture. Once the tissue adhesive is completely dry, anaesthesia is discontinued and the animal is fitted with an infusion jacket with a pocket to house the external dosing catheter apparatus.

The above-described procedure could also be followed to allow the implantation of pumps that could be attached to the catheter. The osmotic pump or refillable battery-operated pump could be placed subcutaneously in the scapular area and attached to the catheter as described. This would allow continuous infusion; however, there are limitations on volume and rates when using these pumps, as well as demands on the stability of the formulation when it is at subcutaneous body temperature rather than room temperature or body surface temperatures.

7.3.1.3 Maintenance

7.3.1.3.1 Recovery procedures. It is critical to return the piglets to the sow as soon as possible, as prolonged separation from the sow can negatively affect development and lead to heat loss. Lubricating eye ointment is applied, and the animal is placed into an incubator until fully recovered. It is imperative that the animal be completely recovered and mobile prior to placing it back with the sow. The piglet must not be immobile when in the cage with the sow, or otherwise it could be inadvertently injured by the sow.

The animal's incision sites are checked a minimum of twice daily, and jackets are changed frequently as the animal grows. Sutures should be removed 10–14 days post-operatively.

7.3.1.3.2 Housing. The temperature of the animal room is elevated depending upon the age of the youngest litter in the room. Piglets younger than one month require a room temperature of 28 ± 3°C, and ages 1–2 months require temperatures of 26 ± 3°C. The animals are housed with the sow in specially designed caging. The caging is designed to have a creep area where a heat lamp provides additional heat for animals up to 2 weeks of age. The creep area also allows the piglets to move away from the sow. The access holes on the creep boards have to be enlarged to allow the animals with jackets to easily gain access to the creep area. The jackets have pockets located on the dorsal surface that house the dosing apparatus, and these would inhibit some of the animals from gaining access to the creep area (Figure 7.3). Environmental enrichment is provided for the sow with commercially available toys, which can be hung from the cage sides or placed on the bottom of the cage.

Figure 7.3 Sow with piglets post-natal day 7. Jacketed piglets had catheters implanted on post-natal day 3.

7.3.1.3.3 Checks and troubleshooting. The animals and equipment are checked a minimum of twice daily. Commonly seen problems include jackets that need adjustment or changing, replacement of external dosing catheters that are destroyed by the sow or siblings, and occasional swollen incision sites (the vascular access port site is the most common location.) Replacement of the external dosing catheters requires manual restraint of the piglet and the skin over the port to be aseptically prepared. A new sterile dosing catheter can then be placed, as described in Section 7.3.1.2.

7.3.1.4 Record keeping

During the surgical procedure, an anaesthesia record is maintained that documents the animal's health status prior to surgery, during the procedure, and through the recovery period. Vital signs such as body temperature, heart rate, and pulse oximetry are monitored continuously and recorded approximately every five minutes. Anaesthesia type and rate, as well as oxygen flow rates, are also recorded. The anaesthesia record is also used to document personnel responsible for all aspects of the surgical procedure and any adverse events that may occur. This record is generally hand recorded.

Records during dosing must include all equipment and materials used to administer the dose; dose volume to be delivered, with expected and actual rates; time of administration; and any clinical observations noted during dosing. These records may be computer or hand generated.

7.3.2 Non-surgical models: Juvenile canine (beagle)

7.3.2.1 Methods of restraint

Restraint devices vary and are dependent upon the size of the animal and the vessel chosen for infusion. Animals are generally restrained in a small-sized veterinary examination bag and placed on a thoracic positioner from post-natal day 14 until a weight of approximately 2 kg (males post-natal day 38 and females post-natal day 42). Animals too small to be properly restrained in the veterinary examination bag are placed in a rodent litter box to limit mobility (see Figure 7. 4).

Animals approximately 2–4 kg are placed in a medium-sized examination bag on a thoracic positioner (males PND 63 and females PND 73).

Once a catheter is placed in a cephalic or saphenous vein, a splint is secured to immobilise the limb during infusion. Disposable splint material could be customised for the size and shape of individual animals. As the size and mobility of the animals increase, Elizabethan plastic collars can be used as necessary to keep the animal from reaching the catheter or extension lines. The combination of examination bag, thoracic positioner, splint, and collar is used until the animals attain a weight of approximately 4 kg, at which time they are placed in a canvas sling for dosing (Stiltzlein et al. 2011).

7.3.2.2 Methods of vascular access

The cephalic vein is the primary site for the placement of a 24-gauge, ¾" over-the-needle catheter, while the saphenous and jugular veins are practical secondary locations. The hair is clipped from the insertion site, and the skin is prepared using chlorhexadine scrub and isopropyl alcohol prior to catheter placement.

Figure 7.4 Restraint for animals too small for a veterinary examination bag.

7.3.2.3 Housing

The temperature of the animal room is elevated depending upon the age of the youngest litter in the room. Puppies younger than one week require a room temperature of 27° ± 3°C, younger than three weeks 26° ± 3°C, and younger than five weeks require temperatures of 24° ± 3°C. The animals are housed with the dam in specially designed caging with a whelping box, and a heat lamp is provided in the whelping box area for animals less than 2 weeks of age. Environmental enrichment is provided for the dam using commercially supplied chew toys and balls and is also provided for the puppies at the time of weaning. On post-natal day (PND) 49, the puppies are weaned from the dam and gang-housed together until PND 77. On PND 77, the litter is separated into two approximately equal-sized groups, which will be group-housed through PND 90. On PND 91, animals are housed individually. Once animals are individually housed, they begin group exercise sessions outside of their home cages at least twice weekly.

7.3.2.4 Checks and troubleshooting

The animals are monitored continuously during the dosing period. The infusion sites are checked for swelling during dosing, which would indicate extravascular dose administration and require placement of a catheter in an alternate location for the remainder of the dose administration. Additionally, the animals are observed for signs of distress or discomfort during infusion and adjustments made if necessary. Previous infusion sites are also checked daily for any signs of irritation.

7.4 Equipment

7.4.1 Surgical models: Juvenile Göttingen minipig

7.4.1.1 Surgical facilities

Catheter implantation as described in this chapter is considered a minor surgical procedure, as a body cavity is not penetrated. This does not, however, diminish the need for dedicated surgical facilities to conduct this procedure. Implantation of a foreign body increases the chances for surgical complications. In human cardiovascular implantation surgery, infection is the most common complication and ultimately leads to device removal (Wilkoff 2008).

Dedicated surgical facilities should be located in an area that will minimise traffic and thus decrease the potential for contamination. The facilities must have limited fixed equipment and be constructed of materials that are easily cleaned and sanitised. The design should include separate areas for instrument and pack preparation, pre-surgical preparation of the animals, surgeon preparation, surgery, and recovery. Each area must be equipped appropriately to maximise surgical support for the animals (NAS 2011).

7.4.1.2 Catheters/vascular access ports

Equipment for vascular infusion is highly variable (Healing and Smith 2000), and when developing a new model, all options should be explored. In developing the surgical juvenile minipig model, various types of vascular access ports and catheters have been evaluated, as described below.

Traditional-style ports (Figure 7.5) are accessed with a Huber needle. This allows easy access to the vascular access port; however, daily access with the Huber needle can cause excessive damage to the fragile skin of the piglet.

A tapered catheter (Figure 7.6) externalised and placed into a jacket pocket can be adapted with a blunted needle and injection cap. This externalised catheter can be prone to kinking and could easily be pulled out if the animal escaped from the jacket. To repair this, a second surgical procedure would have to be performed.

The port used in our laboratory is a low-profile, skin-parallel vascular access port that minimises the occurrence of pressure necrosis of the skin over the port with a flexible infusion catheter that minimises skin injury at the port access site. Additionally, if the animal does escape from the jacket or otherwise disrupts the external dosing equipment, it can be replaced without surgical intervention.

7.4.1.3 External equipment

The external equipment used at our laboratory for this surgical model begins with the external dosing catheter, which is the part of the system for accessing the subcutaneous vascular access port. An 18-g introducer

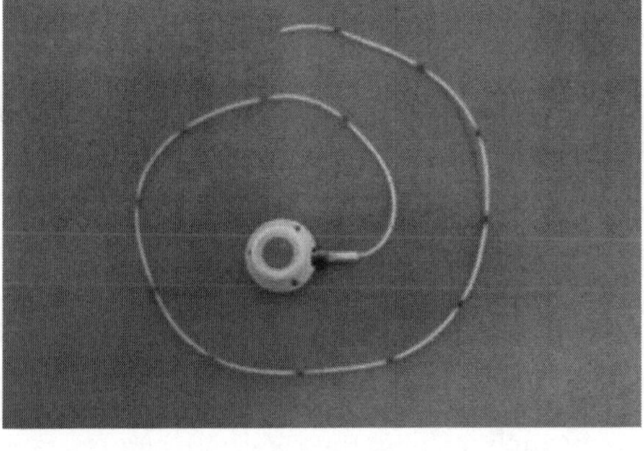

Figure 7.5 Traditional vascular access port.

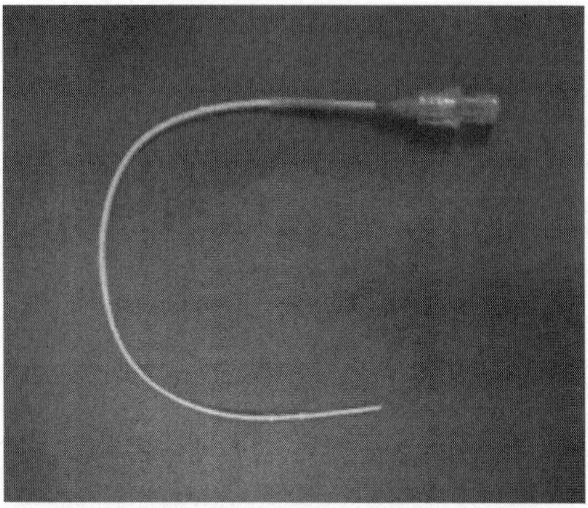

Figure 7.6 Tapered catheter with injection cap.

is used to place the 20-g nylon dosing catheter through the skin and into the implanted catheter via the skin parallel port. A locking luer adaptor is then attached to the catheter, and the system is closed using a needleless-style injection cap. During the bolus dosing, the only equipment necessary would be a B-D® blunt plastic cannula and appropriate-sized syringe. The external dosing catheter is removed from the jacket pocket, and the injection cap is swabbed with alcohol prior to all injections. The jackets are custom made by the manufacturer in various sizes, and each size is highly adjustable to accommodate a large weight range.

7.4.2 *Non-surgical models: Juvenile canine (beagle)*

Figure 7.7 shows some of the equipment used with juvenile dogs.

7.4.2.1 *Restraining devices*

Various types of restraining devices can be used with the juvenile dog and need to be modified continually as the animals grow. These include veterinary examination restraint bags, thoracic positioners, Elizabethan collars, and splints. Slings are generally used for the larger animals.

7.4.2.2 *Vascular access*

A 24-G x ¾" intravenous catheter is used for animals up to 4 kg, and 22-G × 1" catheters for animals heavier than 4 kg. The catheters are placed and then attached to an appropriate-sized prefilled syringe with attached 36" microbore extension set. The catheters and extension sets are secured to the limb or about the neck with medical tape.

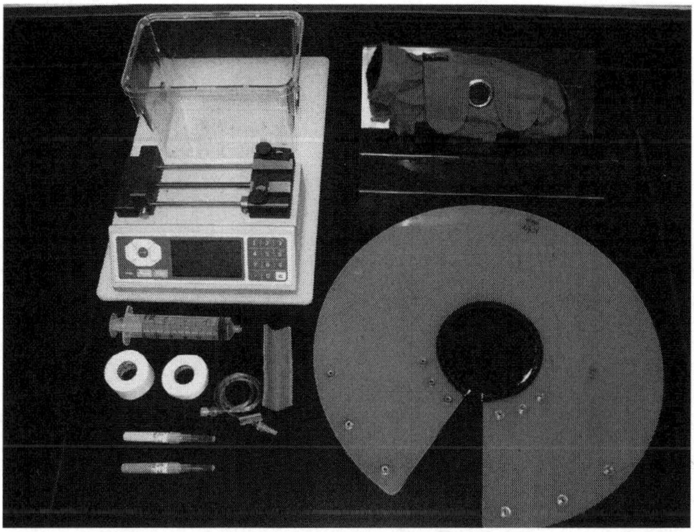

Figure 7.7 Dosing equipment for juvenile dogs.

7.4.2.3 Infusion pumps

Various syringe pumps could be used for this model depending upon the necessary delivery rate. Syringe pumps should be calibrated before use on each study at rates that represent the range of settings and syringe sizes that will be utilised. Yearly manufacturer maintenance routines are recommended also.

7.5 Background data

7.5.1 Surgical model: Juvenile Göttingen minipig

7.5.1.1 Complications associated with the surgical process

The surgical process is a stressful event for the juvenile minipig. They are removed from their dam and littermates, in most instances experience slight hypothermia, and are returned to their cages with external dosing equipment and jackets.

Clinical pathology (particularly white blood cell parameters) is affected by the healing process and the presence of a foreign body, and potentially by infection if bacterial contamination occurs during the surgical procedure or upon return to the home cage.

Histopathologic changes are an expected outcome of any surgical procedure and are discussed in detail in Section 7.5.3.

7.5.1.2 Issues associated with study design

The study design for juvenile animals typically involves control and test article animals represented in each litter. This minimises the number of dams needed to complete the study, and thus reduces the number of animals overall. However, with the 'within-litter' design, the control animals or siblings in low dose groups could potentially be exposed to test article metabolites shed in the urine or faeces. Depending upon litter size, there is also the potential issue of multiple siblings being represented in each study group.

Also, depending on the actual dosing schedule required for the study, the normal light/dark cycles may be disrupted for dosing and study-related events.

7.5.1.3 In-life changes associated with the model

Decreased body weight gains are expected in all jacketed animals, including juveniles. Table 7.2 illustrates body weight changes noted in surgically catheterised jacketed juvenile minipigs in a method development study conducted at our laboratory. The IV infusion group underwent surgery and was jacketed on PND 3. Dosing was initiated on PND 4 for all groups. The data shows a decrease in body weight gains immediately after surgery compared to the IV bolus and oral gavage animals, and then a gradual return to normal. Experience has shown that jacketed animals that have not undergone surgery experience similar decreases. Expected clinical signs include swelling and reddening of the surgical sites as part of the normal healing process. Some reddening of the skin at jacket

Table 7.2 Juvenile Minipig Body Weights in Route-of-Administration Comparison Study

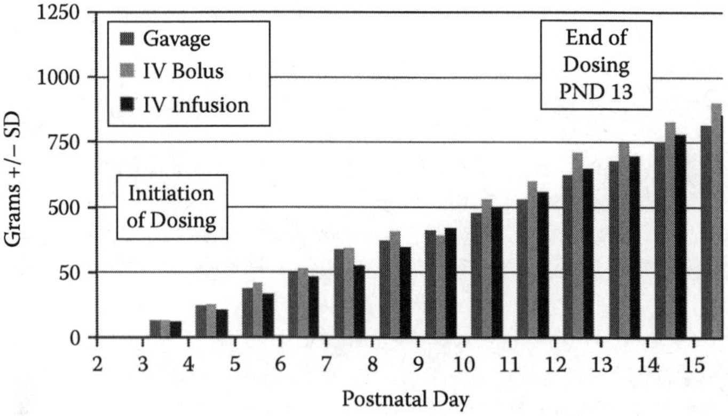

pressure points may be seen, and it is imperative to observe the growth of the animals and change jackets frequently to avoid damage to the skin.

7.5.2 Non-surgical model: Juvenile canine (beagle)

7.5.2.1 Issues associated with study design

As discussed in Section 7.5.1.2, the 'within-litter' study design for juvenile studies can be responsible for metabolite exposure of the control animals. Utilizing a one-group-per-litter design may offset this, but more dams are required for this design to eliminate the problem of too many siblings represented in any one group.

Extravascular administration of the test article through a misplaced or dislodged catheter could cause unexpected histopathological changes. Multiple catheter placement sites further complicate necropsy tissue collection and histology slide preparation.

7.5.2.2 In-life changes associated with the model

Brief daily restraint for intravascular dosing should not affect the growth of the juvenile dog. Initial signs of stress such as struggling during restraint will dissipate when the animal acclimates to the dosing procedure. As the animal grows and the restraint method changes, stress responses may be seen again until the animal acclimates.

Clinical signs may include irritation at the injection sites, particularly if the test article is irritating.

7.5.3 Pathology considerations

Despite the benign effects reported for prolonged catheter administrations, experience has revealed a variety of responses to infusions given in a standard toxicology laboratory environment. Given that establishment of the infusion portal represents an invasive procedure, some level of host response should be expected. Under optimal circumstances, those responses are transient and of minimal clinical significance. However, complicating factors can result in tissue reactions and alterations in homeostasis to the degree that experimental interpretations are influenced and the clinical well-being of the subject is adversely affected.

Pathological alterations associated with the process of establishing and maintaining an infusion portal may be categorised as local, distant, or systemic.

7.5.3.1 Local reactions

A local inflammatory and repair reaction should be expected even when infusion portals are established under the best of aseptic conditions. In the absence of complications, the local reaction may consist of slight inflammatory cell infiltration, fibroplasia, and capillary proliferation preceding formation of mature fibrous connective tissue around the entry portal. In the presence of complications such as bacterial contamination, the local reaction may be so pronounced as to threaten the continued existence of the entry portal and may result in systemic reactions as well as lesions in distant sites.

When performing histopathologic evaluation of the local reactions, it is beneficial to subdivide the tissue reaction into morphologic components, as opposed to the use of blanket terms such as 'chronic active inflammation'. Careful recording of the lesion components along with subjective severity grading of those components will aid in distinguishing nonspecific, technique-, or contaminant-related tissue reactions from those reactions due to administration of test articles. Experience suggests that recording and grading the following histomorphologic elements is helpful: neutrophilic infiltration, lymphocytic or mononuclear cell infiltration, histiocytic infiltration, necrosis, fibroplasia, capillary proliferation, and fibrosis. It should be stressed that *fibroplasia,* an active fibroproliferative process, should be distinguished from the mature fibrous connective tissue that comprises *fibrosis.* Additional histologic features, particularly direct evidence of bacterial invasion or colonization, should also be recorded and the severity graded.

Along with the benefits that result from recording the components of lesions, there is also a benefit from grading the severity of the overall reactive process in each site. The pathologist can attain this goal by making an additional histopathology entry that summarises the overall severity of the various components of the reaction, for example, 'catheter reaction: minimal, mild, moderate, or severe'. This combined approach to data recording gives an indication of the exact nature of the tissue reaction in each site, yet also allows group-to-group comparisons of the overall severity of the reactions.

The exact location of histomorphologic alterations is as important as the precise nature of those alterations. Protocols that specify histopathologic evaluation of infusion sites will result in disappointing and possibly confusing results unless the laboratory Standard Operating Procedures (SOPs) specify exactly how those infusion sites are to be processed and examined microscopically. Using a standard intravenous catheter placement into a major vein as an example, the following specific sites should be processed and examined histologically:

- Cutaneous penetration site
- Subcutaneous catheter tract
- Vascular penetration site
- Blood vessel from intravascular catheter course (if there is an extended intravascular course)

- Blood vessel at catheter tip
- Blood vessel downstream from catheter tip.

Two samples from the catheter tip are recommended (Weber et al. 2011):

- Rodents and rabbits: 0.3 and 0.5–1.0 cm distal (downstream) from catheter tip
- Non-rodents: 0.5 and 1.0 cm distal from catheter tip

Examination of these sites, with careful recording of the reaction components and the severity of those components at each site, will aid in distinguishing effects of test article administration from changes due to nonspecific catheter reactions or complications such as bacterial contamination.

Terminology for infusion-related histomorphologic changes has been recommended by the Infusion Technology Organisation (ITO) (Weber et al. 2011). The list quoted of suggested diagnostic terms is not exhaustive, and should not preclude recording additional changes that may help characterise the tissue reaction. Histomorphologic changes must be recorded in an organised, detailed fashion that will distinguish changes associated with the test article from those associated with catheter placement or complications such as bacterial infection. Catch-all terms such as 'chronic inflammation' will commonly prove to be inadequate.

A technical caveat is indicated with regard to identification of the intravascular location of the catheter tip. During the process of fixation, the tissue may contract substantially while the catheter material remains unchanged, with the result that the catheter tip in the fixed specimen that is received in the histology laboratory is located some distance from the in-life location of the catheter tip. This has an obvious deleterious effect on detection of tissue reactions at the catheter tip or downstream endovascular reactions. The location of the catheter tip should be identified in the fresh necropsy specimen before the tissue specimen is immersed in fixative. Marking is easily accomplished by application of surgical ink, colloidal carbon ('India ink'), or placement of a surgical suture. Multicoloured surgical inks should be a standard supply in laboratories that conduct infusion studies.

Reaction to long-term catheterization is influenced by the chemical constituents of the catheter material. In a study of jugular vein catheterization in horses for 14 or 30 days, silastic catheters provoked the least degree of vascular response (Spurlock et al. 1990). Polytetrafluoroethylene catheters provoked the most intense tissue reaction, and polyurethane catheters provoked an intermediate reaction.

Infusion phlebitis is a common local complication of long-term infusions, particularly peripheral parenteral nutrition (PPN) (Lewis 1985).

Acidity of the nutrient solutions is one of the greatest risk factors, to the point that solution acidity is often considered the cause of infusion phlebitis. Simple pH measurements may be an inadequate measure of acidity, as pH indicates only the dissociated hydrogen ions and not the reservoir of undissociated acid (Lebowitz et al. 1971) Titratable acidity is a more accurate indication of true acidity, thus is necessary background information for interpretation of histopathologic lesions that are suspected to be related to infusion solution acidity. In a study of the effects of titratable acidity on experimentally induced infusion phlebitis (Kuwahara et al. 1996), solutions with different titratable acidity (0.16 to 12 mEq/L) were infused into the ear veins of rabbits for 6 hours at an infusion rate of 10 mL/kg/h. Even at a pH of 4.0, a 10% glucose solution caused only minimal infusion phlebitis because the glucose solution had a titratable acidity of 0.16 mEq/L. When the titratable acidity of the 10% glucose solutions was adjusted to 3 mEq/L with citrate and NaOH, the degree of infusion phlebitis was increased. The results suggest that (1) the titratable acidity of infusion solutions plays an important role in the potential to induce phlebitis when the pH is low, (2) when pHs are similar, the phlebitic potential of infusion solutions depends on the titratable acidity, and (3) the phlebitic potential of infusion solutions cannot be estimated by pH or titratable acidity alone. In a follow-up study (Kuwahara et al. 1998) it was determined that osmolality of the infusion solution is an important factor in the development of infusion phlebitis, regardless of infusion volume or rate, and that dilution is effective in reducing the phlebitic potential of infusion solutions.

Experience in the authors' laboratory indicates that early vascular reactions to catheterization range from minimal intimal thickening (Figures 7.8 and 7.9) to more substantial lesions that include loss of intimal lining with associated proliferation of subintimal tissues (Figure 7.10).

The intravascular catheter tract may be surrounded by a coagulum of fibrin, erythrocytes, and leukocytes (Figure 7.11), which may embolise to the lungs. On occasion the vessel wall is significantly damaged by presence of the catheter (Figure 7.12), resulting in necrosis of the vessel wall and inflammation in the perivascular tissues.

Vascular lesions associated with prolonged catheterization tend to have a more quiescent microscopic appearance, typically consisting of villous intimal proliferation that may or may not be associated with thrombus formation (Figures 7.13 and 7.14).

More pronounced chronic, catheter-associated lesions consist of intimal erosion with villous or plaque-like proliferation of subintimal tissues (Figures 7.15 and 7.16).

As with short-term catheterization, the intravascular course of the long-term catheter may be surrounded by fibrin, leukocytes, and coagulated blood (Figure 7.17), with the attendant potential for pulmonary embolism.

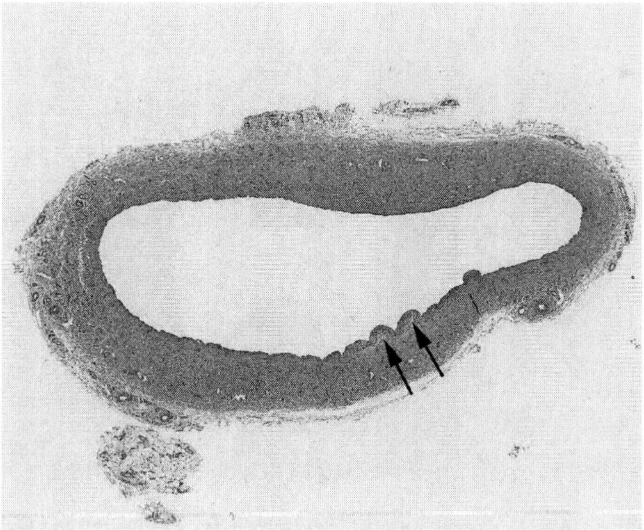

Figure 7.8 (See colour insert.) Jugular vein infusion site from beagle dog, Day 5. Note multifocal areas of intimal proliferation (arrows). (H&E stain, 25×.)

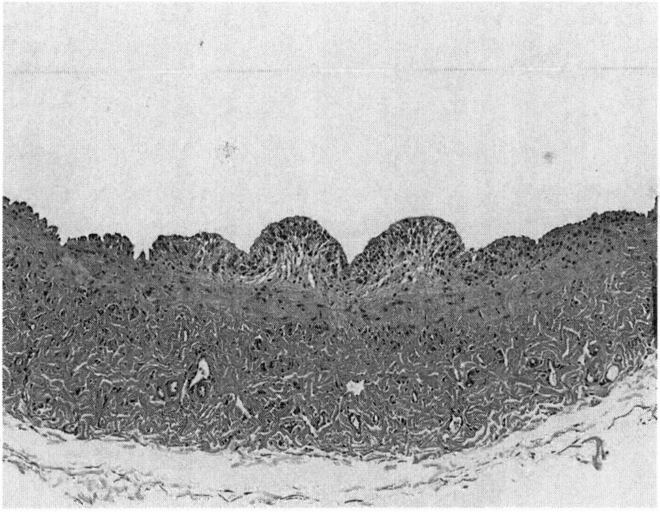

Figure 7.9 (See colour insert.) Jugular vein infusion site from beagle dog, Day 5. Note multifocal intimal proliferation. (H&E stain, 100×.)

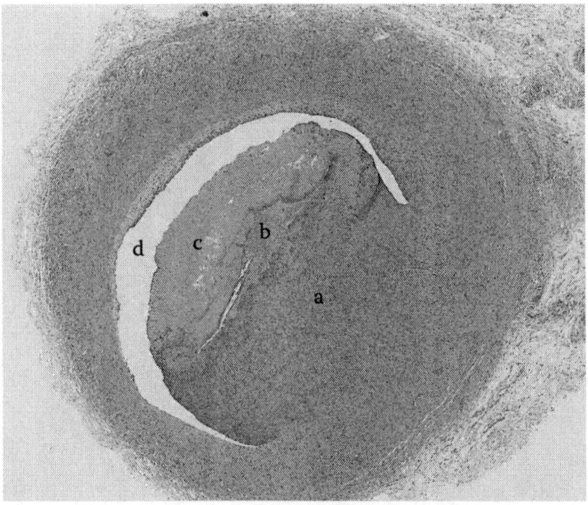

Figure 7.10 (See colour insert.) Jugular vein infusion site from beagle dog, Day 5. Note pronounced segmental intimal proliferation (a) with catheter tract (b), luminal accumulation of fibrinous material (c), and severely compromised lumen of jugular vein (d). (H&E stain, 25×.)

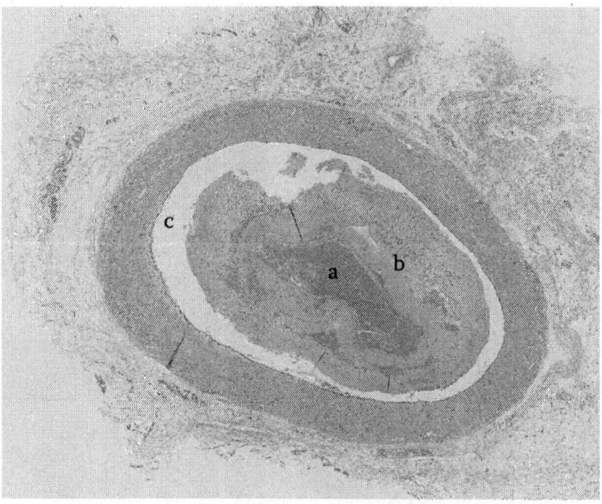

Figure 7.11 (See colour insert.) Jugular vein infusion site from beagle dog, Day 5. Note catheter tract (a) surrounded by zone of fibrin and leukocytes (b) and severely compromised femoral vein lumen (c). (H&E stain, 25×.)

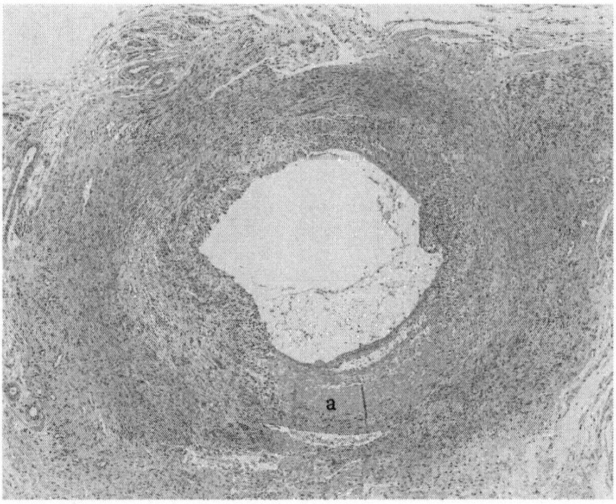

Figure 7.12 (See colour insert.) Jugular vein infusion site from beagle dog, Day 5. Note extensive disruption of vein wall with diffuse inflammatory cell infiltration and area of necrosis (a). H&E stain, 50×.

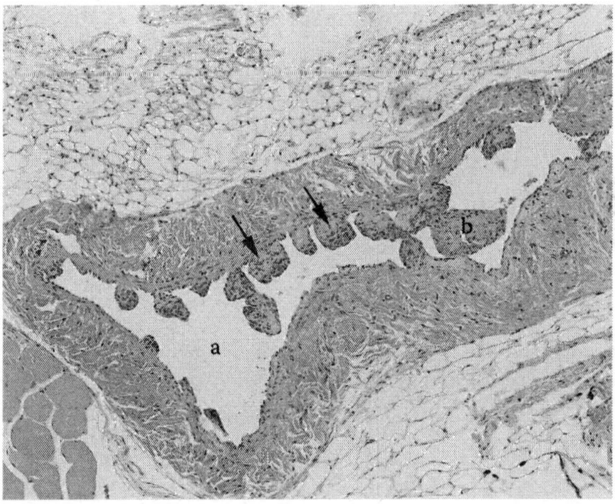

Figure 7.13 (See colour insert.) Femoral vein infusion site from Sprague-Dawley rat, Day 14. Note multifocal intimal proliferation forming villous projections (arrows) into femoral vein lumen (a) and a small thromboembolus (b) composed of fibrin and leukocytes. (H&E stain, 100×.)

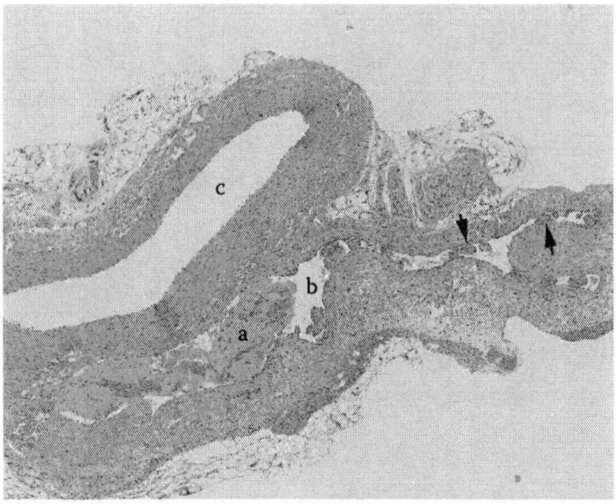

Figure 7.14 (See colour insert.) Femoral vein infusion site from Sprague Dawley rat, Day 14. Note intimal proliferation forming frond-like luminal projections (arrows) and aggregations of fibrin mixed with leukocytes (a) in femoral vein lumen (b). Adjacent femoral artery (c) is unaffected. (H&E stain, 50×.)

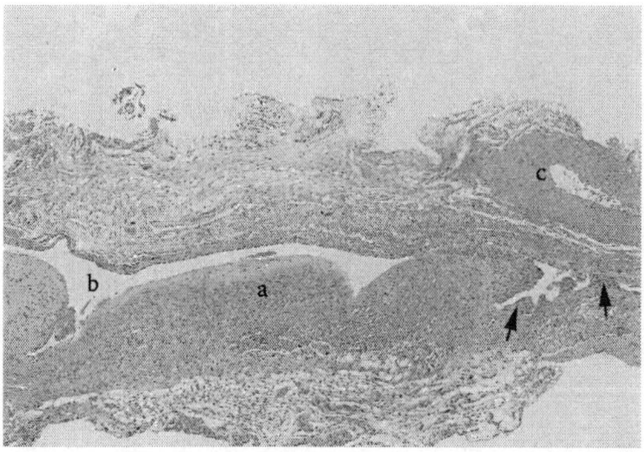

Figure 7.15 (See colour insert.) Femoral vein infusion site from Sprague-Dawley rat, Day 14. Note plaque-like area of intimal proliferation (a) with villous projections (arrows) into femoral vein lumen (b). The adjacent femoral artery (c) is unaffected. (H&E stain, 50×.)

There is a continual risk of bacterial contamination of catheterised vessels, which may result in severe alterations in vascular morphology as well as inflammation that extends into surrounding tissues (Figure 7.18).

7.5.3.2 *Distant reactions*

Distant reactions, defined as changes in sites removed from the infusion portal, may be focal, multifocal, or disseminated.

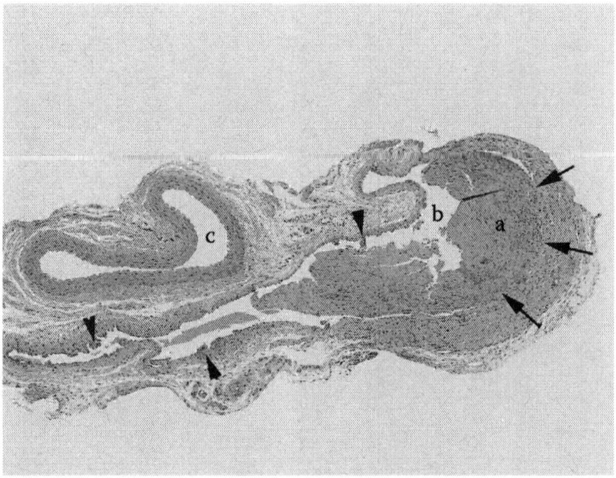

Figure 7.16 (See colour insert.) Femoral vein infusion site from Sprague Dawley rat, Day 14. Note segmental endothelial disruption (arrows) with attached mural thrombus (a). Villous intimal proliferation (arrowheads) is present. The lumen of the femoral vein (b) is moderately compromised. The femoral artery is unaffected (c). (H&E stain, 50×).

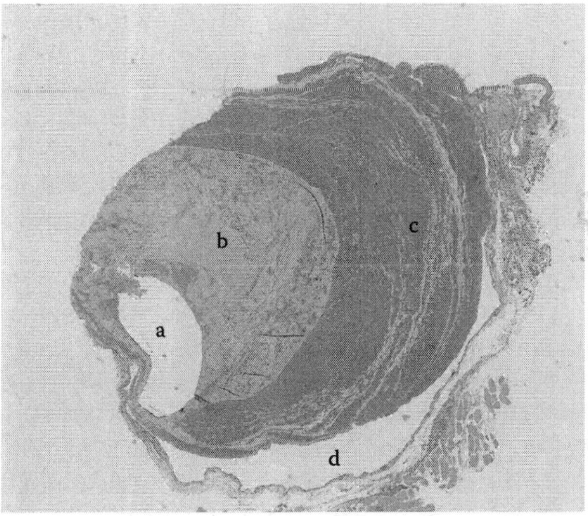

Figure 7.17 (See colour insert.) Specimen of femoral vein taken 2 cm anterior to infusion site of Sprague Dawley rat, Day 14. Note catheter tract (a) surrounded by zone of fibrin, necrotic debris, and leukocytes (b) with external coagulum (c) composed of erythrocytes and fibrin (c). Lumen of femoral vein (d) is severely compromised. (H&E stain, 25×).

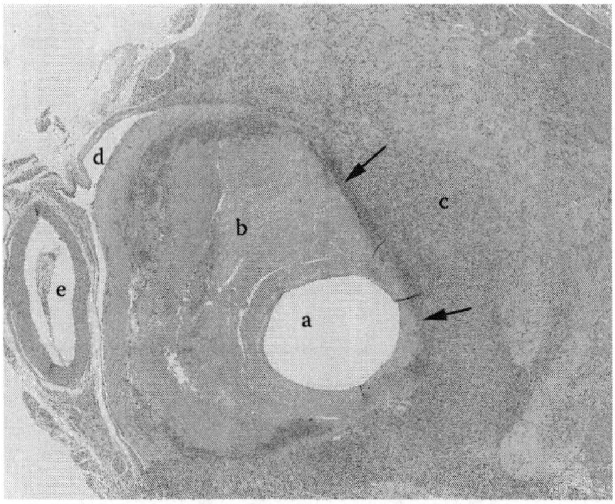

Figure 7.18 (See colour insert.) Femoral vein infusion site from Sprague Dawley rat, Day 14. Note catheter tract (a) surrounded by zone of necrotic debris and fibrin (b). Extensive discontinuity in vessel wall (arrows) is associated with pronounced inflammatory cell infiltration (c). Lumen of femoral vein (d) is severely compromised. Femoral artery (e) is unaffected. (H&E stain, 25×.)

Among other functions, the lungs serve as a filter for particulate matter in the venous system. Clumps of debris or inflammatory cells associated with venous infusion portals will pass through the right side of the heart to the pulmonary arteries, and often become trapped in the pulmonary vascular bed. The majority of these lesions tend to be innocuous, but in extreme circumstances pulmonary function may be compromised. A more common event is showering of inflammatory debris containing pathogenic bacteria into the lungs, with subsequent proliferation and extension of the bacterial lesion in the lungs.

Cynomolgus monkeys that served as saline controls in continuous infusion studies had minor changes such as endothelial hyperplasia and intimal thickening at the injection site and low-grade interstitial inflammation of the lungs of 40–50% of the animals (Lilbert and Burnett 2003). Thrombi were present at the injection sites of 30% of males and 40% of females. Severe procedure-related lesions such as necrosis with abscess formation at injection sites and entry points and thromboemboli and probable infarcts in the lungs were observed in a low incidence (<10%).

The most frequently encountered lesions in beagle dogs that served as saline controls in continuous infusion studies were endothelial hyperplasia (67.4%) and intimal thickening (80.2%) at the injection sites, and interstitial inflammation of the lungs (52.6%) (Lilbert and Mowat 2004). Thrombosis was noted at the injection site of 27.9% of the dogs, but there was a higher incidence of thromboemboli and secondary pulmonary infarction in dogs than in the parallel study in cynomolgus monkeys reported above (Lilbert and Burnett 2003). The higher incidence of pulmonary complications may represent a greater propensity to thrombus formation and fragmentation in dogs than in monkeys. However, it was also noted that the intravenous catheters were placed in the femoral vein of monkeys, while the catheters in dogs were placed in the jugular vein. It was speculated that the greater volume of blood flow in the jugular vein and greater potential for turbulence in the blood flow may have been the cause of the higher incidence of serious pulmonary complications in dogs.

It is considered that the prevalence of pulmonary complications of long-term catheterization is related to the duration of catheterization. This has been demonstrated in a study of histopathologic changes in the rat associated with long-term vena caval catheterization, in which lung emboli were noted in 2 of 12 rats sacrificed at day 32 following cannulation, 3 of 6 rats sacrificed at day 42, and 5 of 7 rats sacrificed at day 45 (de Jong et al. 2001).

The process of introducing a needle or catheter into a vein may introduce hair shaft fragments into the bloodstream, and those fragments subsequently are trapped in the capillary bed of the lungs (Hottendorf and Hirth 1974; Schneider and Pappritz 1976). The hair shaft emboli become surrounded by a granulomatous inflammatory reaction that often obscures the hair shaft fragment, which is thus very difficult to detect in standard histologic sections. Lesions of this type should be examined by polarization microscopy, which reveals the hair shaft as a bright white structure.

Microscopically similar pulmonary changes result from the introduction of fragments of catheter material into the bloodstream. As with hair shafts, identification of catheter material within small pulmonary granulomas is aided by polarization microscopy. Distinguishing catheter material from hair shaft fragments may be challenging. Hair shaft fragments commonly contain small melanin granules, but the pigment granules are not universally present in hair shaft fragments.

7.5.3.3 Systemic reactions

When they occur, systemic reaction to catheter placement can be generally divided into (1) inflammatory haemogram, (2) stress haemogram, (3) immunologic effects, (4) acute phase reaction, and (5) catastrophic effects such as haemorrhage and hypovolaemic shock. The acute phase reaction is largely an innate immune system reaction, but it will be considered separately from specific immune system reactions in this discussion.

Inflammation is a component of the healing process, thus some element of inflammation should be expected following an invasive procedure such as establishment of an intravascular catheter. Under ideal circumstances the resultant inflammatory reaction is so minor that no perturbations of standard clinicopathologic parameters are noted. However, in some cases the associated inflammatory reaction is sufficiently pronounced to cause alterations in the haemogram or, in more pronounced examples, clinical chemistry parameters. Typical alterations seen with a mild inflammatory haemogram would include elevated white cell count (WBC) and neutrophil count. Alterations in clinical chemistry parameters such as globulin levels may be associated with more pronounced inflammatory reactions. Lower serum albumin and calcium levels may be associated with inflammation, but those factors are associated with the acute phase reaction (discussed below) rather than a direct result of inflammation.

Establishment of an intravascular portal may result in a transient stress reaction, typically manifested in the haemogram as slightly elevated neutrophil count and lower lymphocyte and eosinophil counts.

7.5.3.4 Acute-phase reaction

The acute phase response is the most common, possibly ubiquitous, response to the establishment of an infusion portal and is worthy of more detailed discussion within the context of this chapter. Acute inflammation in mammals is associated with a transient increase in a group of circulating proteins that are collectively known as *acute-phase proteins*. This group includes serum amyloid A protein (SAA), fibrinogen, C-reactive protein (CRP), haptoglobin, complement factors C3 and C9, haemopexin, ceruloplasmin, α_2-macroglobulin, CD14, α_1-antichymotrypsin (ACT), α_1-cysteine proteinase inhibitor (α_1CPI), and α_1-antitrypsin (AAT) (Fey et al. 1994). Circulating acute-phase proteins are responsible for many of the innate immune system responses, which are largely aimed at limiting tissue damage associated with inflammation and associated processes such as haemorrhage. The acute phase proteinase inhibitors (e.g. AAT, ACT, α_1CPI, and α_2M) reduce tissue damage caused by proteinases released by dead or dying leukocytes and other cells. Haemopexin and haptoglobin bind to haem and globin, respectively, which may be released by erythrolysis in inflammatory lesions. CRP acts as an opsonizing agent, activates complement factors, binds to IgG receptors on mammalian cells and phosphocholine in bacterial membranes, and recognises nuclear constituents in damaged cells. CRP has been shown to be a reliable indicator of inflammation in humans (Dupuy et al. 2003) as well as in dogs and minipigs (Eckersall 2006). Serum haptoglobin level has been shown to be a better indicator of systemic inflammation in swine, and serum levels of α_2-macroglobulin, haptoglobin, or fibrinogen are more accurate than CRP levels as indicators of inflammation in laboratory rats (Dasu et al. 2004; Chen et al. 2003; Giffen et al. 2003).

The influences of acute phase reactions are well known in human surgery and medicine. The acute phase response is consistently associated with stroke in humans, and has been used to monitor the clinical progress of stroke patients (Dziedzic 2008; Smith et al. 2006). Parturition triggers an acute phase response in women, with C-reactive protein peaking during the first postpartum week and serum alpha 1-antichymotrypsin peaking approximately 9 days postpartum (Friis et al. 2009). The acute phase reaction associated with bronchoscopy in humans results in peripheral neutrophilia, dysregulation of iron homeostasis, and increased levels of fibrinogen and C-reactive protein (Huang et al. 2006). The acute phase response associated with cardiac surgery results in a marked remodelling of HDL (Jahangiri et al. 2009). An acute phase response is the most common side effect after the first intravenous dose of nitrogen-containing bisphosphonates (N-BPs), which are used in therapy of metabolic bone diseases (Makras et al. 2011; Reid et al. 2010; Olson and Van Poznak 2007). This interaction appears to involve vitamin D, and may impact the immune system as well (Bertoldo et al. 2010).

Formation of uterine implantation sites, regardless of the number of implantation sites, results in an acute phase reaction in dogs (Vannucchi et al. 2002), which could simulate or augment the acute phase reaction associated with surgical implantation of a catheter. Presence of acute phase reactants in pregnant dogs is so distinctive and invariable that it may serve as an early indicator of pregnancy (Vannucchi et al. 2002).

The majority of attention relative to acute phase responses has been focused on positive responder molecules, but interpretation of infusion studies should also consider negative acute phase reactants. In the dog, for example, serum albumin and transferrin are negative acute phase reactants (Ceron et al. 2005). Serum albumin levels are commonly measured as part of a standard clinical pathology screen in toxicology laboratories, thus there is a potential impact of nonspecific acute phase reaction on this routinely analyzed parameter. In many species albumin serves as a binding protein for serum calcium, thus a reduction in serum albumin level is associated with a reduction in serum calcium.

Monitoring of acute phase reactants, perhaps by protein microarray technology, may serve as a general indicator of health in animal populations (Gruys et al. 2006). Given the potential for nonspecific acute phase reaction to complicate interpretation of infusion studies, it would seem reasonable to incorporate an assay for one or more acute phase reactants in the clinical pathology analysis.

7.6 Conclusion

Recent regulatory guidelines have increased the demand for juvenile animal models for pre-clinical toxicity testing. The potential toxicity of any new drug should be evaluated in an animal model by delivering it via the intended clinical route. Vessel size, rapid growth, and housing concerns need to be considered when choosing the appropriate species for vascular infusion. Alternatives to continuous infusion may be necessary, such as multiple daily bolus doses or infusions, or single daily prolonged infusions, to ensure the human equivalent of neonate to toddler age is evaluated.

An understanding of the limitations of the model and study design is necessary. Additionally, an extensive knowledge of normally occurring background clinical and pathological findings common to all vascular infusion studies is required. Accurate assessment of adverse changes caused by a drug is dependent upon the interpretation of local, distant, and systemic pathological alterations.

References

AAP. 1995. Informed consent, parental permission, and assent in pediatric practice. Committee on Bioethics, American Academy of Pediatrics. *Pediatrics* 95: 314–317.

ASR. 2009. Guidelines for training in surgical research with animals. Academy of Surgical Research. *J Invest Surg* 22: 218–225.

Beck MJ, Padgett EL, Bowman CJ, Wilson DT, Kaufman LE, Varsho BJ, Stump K, Nemec MD and Holson JF. 2006. Nonclinical juvenile toxicity testing. In *Developmental and Reproductive Toxicology, A Practical Approach*, 2nd edition, Hood RD (ed). Boca Raton, FL: CRC Press/Taylor & Francis Group.

Beck MJ, Setser JJ, Gleason TR, Edwards TR, Davis AL, Miller WR, Allis AD, Raubenolt CA, Raby CK, Fenton TS and Padgett EL. 2009. A postpartum developmental study of Göttingen minipigs over three defined pediatric stages. 25th Annual Meeting of the Academy of Surgical Research, New Orleans.

Bertoldo F, Pancheri S, Zenari S, Boldini S, Giovanazzi B, Zanatta M, Valenti MT, Dalle Carbonare L and Lo Cascio V. 2010. Serum 25-hydroxyvitamin D levels modulate the acute-phase response associated with the first nitrogen-containing bisphosphonate infusion. *Journal of Bone and Mineral Research* 25: 447–454.

Blumer JL. 1999. Off-label uses of drugs in children. *Pediatrics* 104: 598–602.

Ceron JJ, Eckersall PD. and Martynez-Subiela S. 2005. Acute phase proteins in dogs and cats: Current knowledge and future perspectives. *Vet Clin Pathol* 34: 85–99.

Chen HH, Lin JH, Fung HP, Ho LL, Yang PC, Lee WC, Lee YP and Chu RM. 2003. Serum acute phase proteins and swine health status. *Can J Vet Res* 67: 283–290.

Dasu MR, Cobb JP, Laramie JM, Chung TP, Spies M and Barrow RE. 2004. Gene expression profiles of livers from thermally injured rats. *Gene* 327: 51–60.

De Jong WH, Timmerman A and Van Raaij MT. 2001. Long-term cannulation of the vena cava of rats for blood sampling: local and systemic effects observed by histopathology after six weeks of cannulation. *Lab Anim* 35: 243–248.

Dupuy AM, Terrier N, Senecal L, Morena M, Leray H, Canaud B and Cristol JP. 2003. Is C-reactive protein a marker of inflammation? [In French]. *Nephrologie* 24: 337–341.

Dziedzic T. 2008. Clinical significance of acute phase reaction in stroke patients. *Front Biosci* 13: 2922–2927.

Eckersall PD. 2006. Acute phase proteins as markers of disease in companion and laboratory animals. ACVP Annual Meeting, Tucson.

EMA 2008. *Guideline on the Need for Non-Clinical Testing in Juvenile Animals of Pharmaceuticals for Paediatric Indications.*

FDA 2006. *Guidance for Industry: Nonclinical Safety Evaluation of Pediatric Drug Products.*

Fey G, Hocke G, Wilson D, Ripperger J, Juan T, Cui M-Z and Darlington G. 1994. Cytokines and the acute phase response of the liver. In *The Liver: Biology and Pathobiology*, 3rd edition, Arias I, Boyer J, Fausto N, Jacoby W, Schachter D and Shafritz D (eds). New York: Raven Press Ltd.

Friis H, Gomo E, Mashange W, Nyazema N, Kostel P, Wieringa F and Krarup H. 2009. The acute phase response to parturition: A cross-sectional study in Zimbabwe. *Afr J Reprod Health* 13: 61–68.

Giffen PS, Turton J, Andrews CM, Barrett P, Clarke CJ, Fung KW, Munday MR, Roman IF, Smyth R, Walshe K and York MJ. 2003. Markers of experimental acute inflammation in the Wistar Han rat with particular reference to haptoglobin and C-reactive protein. *Arch Toxicol* 77: 392–402.

Gruys E, Toussaint MJ, Niewold TA, Koopmans SJ, Van Dijk E and Meloen RH. 2006. Monitoring health by values of acute phase proteins. *Acta Histochem* 108: 229–32.

Halpern SA. 1988. *American Pediatrics: The Social Dynamics of Professionalism, 1880–1980*. Berkeley, CA: University of California Press.

Healing G and Smith D. 2000. *Handbook of Pre-clinical Continuous Intravenous Infusion*. London and New York: Taylor & Francis.

Hood RD. 2006. *Developmental and Reproductive Toxicology: A Practical Approach*. Boca Raton, FL: CRC Press.

Hottendorf GH and Hirth RS. 1974. Lesions of spontaneous subclinical disease in beagle dogs. *Vet Pathol* 11: 240–258.

Huang YC, Bassett MA, Levin D, Montilla T and Ghio AJ. 2006. Acute phase reaction in healthy volunteers after bronchoscopy with lavage. *Chest* 129: 1565–1569.

Hull RM. 1995. Guideline limit volumes for dosing animals in the preclinical stage of safety evaluation. Toxicology Subcommittee of the Association of the British Pharmaceutical Industry. *Hum Exp Toxicol* 14: 305–307.

Jahangiri A, De Beer MC, Noffsinger V, Tannock LR, Ramaiah C, Webb NR, Van Der Westhuyzen DR and De Beer FC. 2009. HDL remodeling during the acute phase response. *Arterioscler Thromb Vasc Biol* 29: 261–267.

Kearns GL, Abdel-Rahman SM, Alander SW, Blowey DL, Leeder JS and Kauffman RE. 2003. Developmental pharmacology: Drug disposition, action, and therapy in infants and children. *N Engl J Med* 349: 1157–1167.

Kuwahara T, Asanami S, Tamura T and Kubo S. 1996. Experimental infusion phlebitis: Importance of titratable acidity on phlebitic potential of infusion solution. *Clinical Nutrition* 15: 129–132.

Kuwahara T, Asanami S, Tamura T and Kubo S. 1998. Dilution is effective in reducing infusion phlebitis in peripheral parenteral nutrition: an experimental study in rabbits. *Nutrition* 14: 186–190.

Lebowitz MH, Masuda JY and Beckerman JH. 1971. The pH and acidity of intravenous infusion solutions. *JAMA* 215: 1937–1940.

Lewis GBH, Hecker JF. 1985. Infusion thrombophlebitis. *Br J Anaesth* 57: 220–223.

Lilbert J and Burnett R. 2003. Main vascular changes seen in the saline controls of continuous infusion studies in the cynomolgus monkey over an eight-year period. *Toxicol Pathol* 31: 273–80.

Lilbert J and Mowat V. 2004. Common vascular changes in the jugular vein of saline controls in continuous infusion in the beagle dog. *Toxicol Pathol* 32: 694–700.

Makras P, Anastasilakis AD, Polyzos SA, Bisbinas I, Sakellariou GT and Papapoulos SE. 2011. No effect of rosuvastatin in the zoledronate-induced acute-phase response. *Calcif Tissue Int* 88: 402–408.

NAS. 2011. *Guide for the Care and Use of Laboratory Animals*. Washington, DC: The National Academies Press.

Olson K and Van Poznak C. 2007. Significance and impact of bisphosphonate-induced acute phase responses. *J Oncol Pharm Pract* 13: 223–229.

Reid IR, Gamble GD, Mesenbrink P, Lakatos P and Black DM. 2010. Characterization of and risk factors for the acute-phase response after zoledronic acid. *J Clin Endocrinol Metab* 95: 4380–4387.

Schneider P and Pappritz G. 1976. Hairs causing pulmonary emboli: A rare complication in long-term intravenous studies in dogs. *Vet Pathol* 13: 394–400.

Smith CJ, Emsley HC, Vail A, Georgiou RF, Rothwell NJ, Tyrrell PJ and Hopkins SJ. 2006. Variability of the systemic acute phase response after ischemic stroke. *J Neurol Sci* 251: 77–81.

Spurlock SL, Spurlock GH, Parker G and Ward MV. 1990. Long-term jugular vein catheterization in horses. *J Am Vet Med Assoc* 196: 425–430.

Stiltzlein EB, Enama TT, Fenton TS, Setser JJ, Miller W, Eapen AK, Padgett EL and Allis AD. 2011. Intravenous infusion and restraint in juvenile beagle dogs. 62nd Annual AALAS National Meeting, San Diego, CA.

Vannucchi CI, Mirandola RM and Oliveira CM. 2002. Acute-phase protein profile during gestation and diestrous: proposal for an early pregnancy test in bitches. *Anim Reprod Sci* 74: 87–99.

Weber K, Mowat V, Hartmann E, Razinger T, Chevalier H-J, Blumbach K, Green O, Kaiser S, Corney S, Jackson A and Casadesus A. 2011. Pathology in continuous infusion studies in rodents and non-rodents and ITO (Infusion Technology Organisation)–recommended protocol for tissue sampling and terminology for procedure-related lesions. *J Toxicol Pathol* 24: 113–124.

Wilkoff BL. 2008. Materials and design challenges for cardiovascular implantable electronic devices. Materials and Processes for Medical Devices Conference, Cleveland, OH.

chapter eight

Marmoset

Dr. Sven Korte
Covance Laboratories GmbH, Germany

Dr. Piotr Nowak
Covance Laboratories GmbH, Germany

Dr. Jörg Luft
Covance Laboratories GmbH, Germany

Dr. Birgit Niggemann
Covance Laboratories GmbH, Germany

Contents

8.1 Introduction

The common marmoset, *Callithrix jacchus*, is one of the smallest non-human primate species (adult body weight approximately 300–500 g) commonly used in biomedical research. The natural habitat is in the northeastern part of Brazil. Marmosets are bred mostly in self-sustaining breeding colonies today. Marmoset characteristics have enabled the species to be used in a wide range of research, including as models for human disease, physiology, drug metabolism, general toxicology, and reproductive biology (Zühlke and Weinbauer 2003; Austad and Fischer 2011). Owing to this extensive use, a sizeable amount of background information has become available over the last decade. In some pre-clinical studies, particularly when cynomolgus (*Macaca fascicularis*) and rhesus (*Macaca mulatta*) monkeys do not meet investigators' needs, marmosets may successfully be used as an alternative non-human primate species. As many test items need to be administered intravenously, in some cases prolonged or continuous infusion is required.

8.1.1 Choice and relevance of the species

The marmoset monkey is an alternative non-human primate model in drug development (Abbott et al. 2003; Mansfield 2003; Orsi et al. 2011; Korte et al. 2004) and has also been used as a model for induced Parkinson's and Huntington's diseases, multiple sclerosis, and rheumatoid arthritis

(Vierboom et al. 2010). For biopharmaceutical drugs, nonhuman primates are the relevant animal model in 80% of the cases (Chapman et al. 2007, 2009, 2010). When it comes to justifying using marmosets instead of macaques, tissue cross-reactivity evaluation may eliminate these other nonhuman primate (NHP) species, with the marmoset monkey remaining the only relevant model.

In comparison with continuous intravenous dosing of other NHP species (such as macaques), marmosets differ in having a much lower body weight and smaller body size. Ambulatory pump systems for larger primates, consisting of a pump, backpack system, batteries, and delivery fluid can weight as much as 300 to 400 grams. The maximum feasible weight for a free-ranging delivery system in marmosets needs to be much lighter in weight (no more than 10% of body weight, which is approximately 30 g).

Prolonged infusion studies in restrained marmosets have been conducted with a limited dosing time of 1.5 hours (or twice daily for 1 hour). This differs from macaques, where we allowed restrained infusion up to 4 hours.

If port surgeries are required, the device is commonly placed subcutaneously in macaques and extracutaneously in the marmoset.

8.1.2 *Ethical and regulatory guideline considerations*

Prior to conduct of surgery, the study protocol will need to be reviewed and approved by an Institutional Animal Care and Use Committee (IACUC), and all study tasks are performed in accordance with facility Standard Operating Procedures (SOPs).

The standard social housing for marmosets in our laboratory is according to ETS 123 (2007), as described in Table 8.1.

To minimise the risk of damaging the implanted port catheter system, marmosets should be single housed. Therefore, in accordance with

Table 8.1 Dimensions of Marmoset Cages as Specified by the New European Guideline Requirements for Housing of Nonhuman Primates (ETS 123, 2007)

	Minimum ETS 123 guideline requirements*
Floor space [m^2]	0.5
Height [m]	1.5**
Volume [m^3]	0.75***

*For 1 or 2 animals plus offspring up to 5 months old (plus 0.2 m^3 per additional animal over 5 months).

**Top of enclosure at least 1.8 m above ground.

***Calculated, not indicated in the guideline.

the ETS 123 Guideline, which allows going below space requirements for scientific reasons, the cage space is limited to the top part of the standard cage (Figure 8.1) to prevent excessive three-dimensional movements by this highly agile primate species.

8.1.3 *Criteria for choosing the surgical or the non-surgical technique*

Factors to be taken into consideration regarding invasive or non-invasive infusion are

- Number of intravenous administrations
- Duration of each administration
- Complete dosing regime
- Duration of follow-up studies (ideally with the same technique being used in all studies)

Basically three approaches need to be considered:

- Bolus (and slow bolus)
- Restrained dosing
- Ambulatory infusion with freely moving animals

Figure 8.1 Marmoset room, each cage with an attached balcony.

Since the route of administration and dosing regime throughout pre-clinical studies should be as consistent as possible with the intended dosing in humans, continuous infusion is required in certain substance classes (such as antibiotics) to maintain constant dose delivery over hours or days (24 hours, 7 days/week). Dosing volumes of up to 10 mL/kg can be given intravenously within 5 minutes (daily) using a needle or butterfly system. Other therapeutic areas require lower volumes to be administered over a longer period. Three options should be considered:

- A surgically or temporarily inserted indwelling catheter
- An extracutaneous port catheter system for delivery in a restraint bed (twice daily for a maximum of 1 hour)
- A combination of the port catheter system with a lightweight backpack system, with the animal carrying an ambulatory pump system

Schnell (2000) provides a very profound description of catheter access models and surgical procedures for infusion with detailed surgical descriptions. He was able to use the catheters for only one day because of dexterity of the animals in manipulating the catheters. The same author also describes the usage of subcutaneously or intraperitoneally implantable osmotic pumps (Alzet®, DURECT Corporation, USA) for local capillary adsorption with delivery rates of 0.25–10 μL/h and delivery duration of 1 day to up to 4 weeks. Implantable minipumps such as the Alzet osmotic device are not appropriate for long-term infusions in toxicity studies in the marmoset. The pump for this species has a limited reservoir capacity, and delivery rates produce variable concentrations of the test item in the blood and consequently variable pharmacological effects (Schnell 2000). Furthermore, this system requires surgery for replacement. Schnell (2000) describes technical limitations of this pump system and suggests the development of lightweight technologies for refining continuous intravenous infusion methods.

Most recently, technology has enabled researchers to equip the marmoset with an independent, non-tethered infusion system. This approach was driven mainly by the implementation of European housing conditions, animal welfare aspects, and stress reduction initiatives. In addition, the marmoset exhibits a very complex behavioural repertoire that would not be compatible with tethered infusion, where the animal is connected with a tube and swivel to a cage-mounted infusion pump.

8.1.4 *Limitations of available models*

One of the predominant factors limiting the use of the marmoset for specialised administration systems is body size (adult body length without tail 12 to 18 cm) and body weight (adult body weight generally

300–500 g). The small size and weight requires modification of techniques and diagnostics adopted from cynomolgus monkeys.

The marmoset has a circulating mean blood volume between 19 and 31 mL (mean of 65 mL/kg), whereas the cynomolgus monkey (Diehl et al. 2001), with an adult body weight of 3 to 6 kg, has a mean blood volume of 210 to 420 mL (mean of 70 mL/kg). Therefore, blood sampling in marmosets is limited owing to the small blood volume and small vein diameter.

Because of the small size of the marmoset, the area available for repeated puncture with a needle or an indwelling catheter—especially after tissue damage has occurred—is limited.

Continuous intravenous dosing (including daily administration for hours over many weeks) requires an implantable port catheter system. The maintenance of a port catheter system always has the inherent risk of infection. Therefore, highly skilled surgery, proper animal handling, and sophisticated maintenance are mandatory requirements. These aspects need to be taken into consideration.

8.2 Best practice

In Europe three administration procedures (bolus, prolonged-restrained, and continuous-free-ranging) are available. Large-volume or long-term administration should be conducted with port and backpack-carrying animals for welfare reasons. In the United States tethering is an additional option.

When low volumes or single dosing is required, port surgery should be avoided and is not considered a refinement.

8.2.1 Surgical models

8.2.1.1 Management of a port catheter system

For dose durations over 30 minutes/day the feasible infusion rates shown in Table 8.2, delivered through a port catheter system, are recommended at our laboratory.

The assessment of an accurate body weight on the day prior to first dose administration is vital to allow precise dose calculations and infusion pump setup.

8.2.1.2 Restrained marmosets implanted with a port catheter system

For longer infusion durations up to a maximum of 1.5 hours, the animal (implanted with a port catheter system) is inserted into a small woollen sock with one open end for easy breathing. Placing the marmoset in the sock can easily be achieved with the sock placed over a tube and the monkey being slid into the tube. Then the animal, fixed in the sock, is placed in a restraint bed (Figure 8.2). At least three days of acclimation to the restraint method should be conducted before the first test item administration, and the animals should be monitored for the entire duration of the dosing. This fixation method provides gentle but efficient restraint of this agile animal. Breathing pattern and body temperature can be monitored to ensure the well-being of the marmoset.

The significant differences between continuous intravenous infusion in marmoset monkeys and that in the more frequently used cynomolgus monkeys are shown in Table 8.3.

Table 8.2 Recommended Daily Dose Volumes for Marmoset Monkeys with Implanted Port Catheter System under Restraint[1] or with a Backpack[2] System

Duration of infusion in marmoset	0.5 hours[1]	0.5–1 hours[1]	1.5 hours[1]	2–5 hours[2]	7 days/24 hours (max. for 5 weeks)[2]
mL/kg/body weight	10	15	15 (up to 20)	-	-
mL/kg/minute	0.3	< 0.3	< 0.3	-	-
mL/day	-	-	-	-	0.72–4.80*
Maximum technical feasible flow rate (5 mL reservoir)	-	-	-	0.03 to 1.0 mL/hr	

*Refit of backpack pumps (after delivery of 4.8 of 5.0 mL).

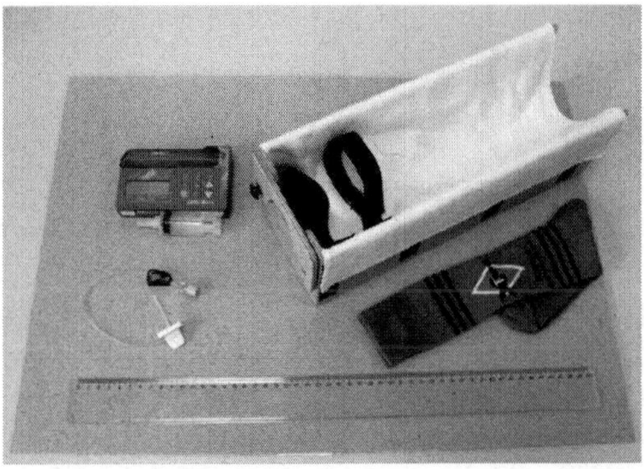

Figure 8.2 Marmoset restraint system consisting of a small-sized sock and a restraint bed.

Table 8.3 Technical Aspects of Continuous Intravenous Infusion in Marmoset Compared to the Cynomolgus Monkey

	Cynomolgus monkey	Marmoset monkey
Minimum body weight before undergoing implantation	3.0 kg*	350 g*
Weight of the empty pump system used for backpack infusion	Orchesta™ Model 500 V (including batteries) 200 g	Infu-Disk™ 12 g
Weight of the backpack (jacket)	S size: 200 g XL size: 300 g	13 g
Possible reservoir volume	10–250 mL	5 mL
Ratio of complete infusion system: body weight	1:8 (for short-term infusion 1:7)	1:8 (for short-term infusion 1:7)
Implantation of port	Subcutaneous	Extracutaneous
Sedation for port surgery	Inhalation anaesthesia with isoflurane and oxygen	Inhalation anaesthesia with isoflurane and oxygen
Recommended maximum infusion volume over 24 hours (once weekly)	60 mL/kg	60 mL/kg
Recommended maximum infusion volume over 24 hours (daily)	40 mL/kg	40 mL/kg
Maximum study duration (assuming 24 hours/day 7 days/week dosing)	39 weeks	5 weeks
Daily restraint time (experience)	4 hours (restraint chair)	1.5 hours or 2 times daily 1 hour (restraint bed with sock)
Minimum training period for restraint device	3 days	3 days

*also for pregnant animals.

8.2.1.3 *The backpack model implanted with port catheter system using a gas-driven minipump*

Animals should be trained for backpack infusion (Figure 8.3) for at least 3 days prior to the start of dosing, and the duration of training should be as late as required for dosing. An ideal training should consist of acclimatization to the empty backpack prior to surgery, followed by a second training set in the week before the first test item administration. This procedure will result in the animals being less stressed when additional samplings and diagnostic procedures are conducted on day 1 of the

Figure 8.3 Marmoset with an implanted port catheter system and a back-mounted infusion pump.

study. The weight of the backpack (17 g, see Section 8.4), together with the gas-driven infusion pump (12 g plus 5 mL reservoir, see Section 8.4) should not exceed 1/7 of the nominal body weight of the animal. It is therefore recommended to implant only animals with a body weight of at least 350 g. The animals need to be monitored at least twice daily, with a complete check of the port catheter once weekly. A net wound dressing and medical tape featuring a zinc oxide adhesive to cover the body is used as an additional fixation of the port system following the implantation. The tape should not be attached to the hair or skin of marmosets, and the wound dressing needs to be replaced twice weekly. Additionally, the area of port should remain free of hair, allowing port accessibility for cleaning and needle fixation. A very sharp clipper is used to minimise the likelihood of skin irritation. The area around the port should not be cleaned daily with antiseptic spray, because this would quickly lead to skin irritation. A 5-IU/mL heparin solution (250 µL) is used to rinse the port and protect the infusion system from occlusion. To avoid contamination of the chamber of the port with blood, this heparin solution is always injected into the port but not withdrawn. To verify the position of the catheter following surgery as well as identify possible blockage of the catheter during study conduct, radiographic techniques can be used if deemed necessary.

The position of the catheter tip is confirmed at time of necropsy. Following these procedures, a 3-week recovery period from surgery until start of dosing and an infusion period for up to 5 weeks have been accomplished (Zühlke et al. 2003).

8.2.2 Non-surgical model

Non-surgical infusion with marmosets is limited to a short dosing period and/or single doses (Schnell 2000). Where single animals might be dosed intravenously for up to 15 minutes in a restraint device (Figure 8.2) using an indwelling catheter in the femoral or caudal vein, repeated daily dosing has limitations because of the associated tissue damage.

8.3 Practical techniques

8.3.1 Surgical models

8.3.1.1 Preparation

Robust animals need to be selected to facilitate post-surgical recovery and the carrying of additional weight. Selected animals will undergo a veterinarian-conducted clinical examination. In addition, inclusion criteria require that there be no adverse findings in haematology and clinical pathology.

8.3.1.1.1 Presurgical preparation. All surgical procedures are carried out under sterile conditions in a surgical suite. All relevant body parts (right thigh with pubis region, right abdominal region, and middle dorsum for port positioning) are shaved and disinfected with 10% povidone-iodine solution.

8.3.1.1.2 Premedication. The marmoset receives an intramuscular pre-medication with the dissociative hypnotic drug ketamine (15 mg/kg). However, consideration is given to the muscle necrosis that occurs at the site of injection when ketamine is used (local myotoxic indication, Davy et al. 1987). Additionally, atropine sulfate (0.035 mg/kg) is given as a parasympatholytic (anticholinergic) subcutaneously to reduce secretion, dilate airways, and reduce the vomiting reflex. With this treatment, a possible obturation of the respiratory tract is avoided.

8.3.1.1.3 Anaesthesia. Inhalation anaesthesia is maintained by administration of isoflurane and oxygen with varied concentrations around 1.5% by volume using a small facemask. If intubation can be achieved, the assisted ventilation allows for better monitoring. When repositioning the animals during surgery to ventral recumbency, attention needs to be paid to avoid dislocating the tubus. During surgery, respiration is assisted using a respirator suitable for very small tidal volumes. All commonly available inhalation agents can be used to achieve anaesthesia in marmosets, although properly calibrated vaporisers suitable for the small tidal volume of the marmoset should always be used. Isoflurane is used at our laboratory for primates. Alternatively, injection anaesthesia can be used.

Injection anaesthesia is induced by medetomidine (0.05 mg/kg), a sedative with analgesic effects, in combination with ketamine (already given as a premedication at an intramuscular dose of 3 mg/kg) enhancing synergic effects. In cases where an injection anaesthesia is required, this balanced injection anaesthesia for marmosets is used in our laboratory. Medetomidine can be antagonised with atipamezole.

Ketamine (10 mg/kg) combined with xylazine (0.5 mg/kg) provides a surgical anaesthesia plane lasting 30 to 40 minutes (Flecknell 1987). Prolonging anaesthesia with xylazine causes delayed recovery and is therefore not recommended in marmosets. Xylazine can be antagonised with yohimbine.

8.3.1.2 Surgical procedure

The animal is fixed in a dorsal position on a warming pad (regulated temperature 37°C). The right thigh, especially its proximal part, is fixed to slightly raise the right inguinal region for easier access to the femoral vein.

8.3.1.2.1 Surgical approach to the vena femoralis. The animal is positioned in dorsal recumbency. After the preparation of the surgical site, a 10-mm-long incision is made in the skin from the proximal part of the femur in the inguinal region towards the sternum to expose the femoral vein. The vein is dissected bluntly from the adjacent tissue to obtain full exposure of the vein at a length of at least 5 mm. Traumatizing the adjacent nerve or the artery needs to be avoided. The tissues are kept moist using warm sterile saline solution. Two ligatures of the femoral vein are placed proximal and distal at least 5 mm apart (Ethibond Excel® 4-0 X 935, or Vicryl® 5-0). The distal part of the vessel is ligated. The proximal ligature is placed but not closed to control bleeding. This ligature is closed later after the catheter has been placed inside the vein. Using straight-bladed microscissors, a small transverse cut is made into the vein between the two ligatures in the distal third of the distance between both ligatures. The lumen of the vessel is dilated using the catheter introducer to insert the catheter, which is selected based on compatibility tests with the test item. During the insertion, bleeding is controlled by slightly pulling on the ligature. The catheter is inserted into the vein to a length of approximately 25 mm. The catheter is placed in the femoral vein and is advanced via the external and common iliac vein so that the tip of the catheter is situated in the *vena cava caudalis*. Once the catheter is completely inserted, the proximal ligation—thus far serving to manipulate vessel and control bleeding—is fastened. To prevent the catheter from dislocation, both ligatures are tied together. Approximately 8 mm latero-distal of the insertion, a subcutaneous pocket is bluntly prepared to place the catheter sling.

Subsequently, the subcutaneous tissue is adapted following the flexion of the catheter using a continuous suture technique with absorbable material (Vicryl® Rapide 5.0, 45 cm). For the cutaneous suture an intracutaneous technique is used. As a general note it must be emphasised that marmosets often remove superficial sutures, and therefore subcutaneous sutures should be used for wound closure in addition to intracutaneous closure of the skin.

For the positioning of the vascular access port, the animal is positioned in ventral recumbence and the surgical field (on the back of the animal) is prepared in the same way as for the inguinal region. Approximately 25 mm cranial from the base of the tail a minimal skin incision (less than 3 mm) is made by using the tip of a number 11 blade. A blunt tissue guide needle is inserted and advanced cautiously subcutaneously to the inguinal pocket. The catheter is then advanced from the inguinal guide needle and, avoiding damage to adjacent structures, it is finally diverted through the dorsal exit. The proximal part of the catheter is shortened and connected to the port. For the protection of the catheter, a sleeve is placed over the port and the conus, preventing kinking and rupture of the catheter at the conus. The port is secured with three sutures (Vicryl Rapid Plus 4.0). A sterile compress and a highly elastic tubular net bandage protect the extracutaneous portion of the catheter and the port.

The functionality of the system is tested by flushing it with a small volume of 0.9% NaCl solution. Finally, the system is filled with 5 IU/mL heparinised saline solution to prevent blockage, depending on the volume of reservoir of the port used (in our case 0.2 mL).

Inhalation anaesthesia is ended, and the animal is ventilated artificially with pure oxygen until it awakes.

8.3.1.3 Post-surgical procedures

Animals are treated with flunixin meglumine, a non-steroidal antiphlogistic with a strong analgesic effect, at doses of 2.5–5 mg/kg. Antibiotics are administered according to laboratory standards. Animals are placed into an incubator to reduce thermal losses and enhance a smooth recovery. To ensure that no post-surgery accident occurs from the prolonged reaction time after sedation and excessive movement, the available space is minimised for the first 60 minutes. Then marmosets are returned to their previous social environment with a limited cage size as soon as they have regained consciousness and normal mobility. Social contact within the group is pivotal for the well-being of the marmoset and enhances fast recovery.

When complications occur, the correct final position as well as the leak tightness of the port catheter system can be checked via radiography. A non-ionic contrast medium (0.2 mL Optiray®) can be used. Our laboratory uses a digital X-ray machine.

Long-term catheter-associated infections develop in most cases by luminal access. An antibiotic lock technique can be applied. A long-lasting broad-spectrum antibiotic belonging to the group of the 3rd-generation cephalosporins can be administered into the port catheter system.

Because of possible interference between test item and antibiotic, the usage of antibiotics has to be carefully considered in regulatory toxicology studies.

8.3.2 Non-surgical model

The recommended volume for daily intravenous bolus (within a minute) administration is 2.5 mL/kg, with a maximum of 10 mL/kg.

Several tube-like restraint systems have been described (Schnell 2000), all of which allow free breathing at the front part, as well as fixation. At our laboratory, plastic tubes 20 cm in length and 8 cm in diameter are used for hygienic reasons (Figure 8.4). In the middle and end of the tube, a sliding shield allows fixation of the animals. For stress reduction during dosing, one recommendation is to reduce light intensity to the upper part of the device where the animal's head is. A feasible approach for tail vein access can be achieved by closing the lower end of the tube with only

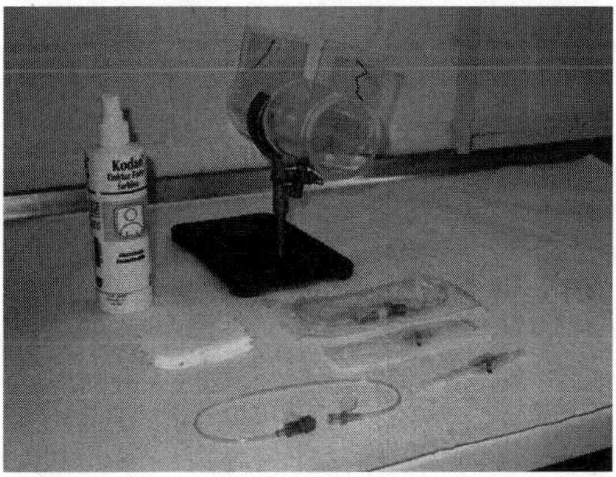

Figure 8.4 Setup for intravenous bolus injection using a restraint tube. Possible dosing sites are brachial, femoral, and caudal veins.

the shaven tail protruding outside. We recommend a three-day training session just prior to the start of the dosing regime and for the duration of dosing (e.g. 5 or 15 minutes).

8.4 Equipment

8.4.1 Surgical models

Surgical models are required for continuous intravenous administration (e.g. 24 hours) or repetitive daily intravenous dosing over a long period of time, as well as for backpack infusions.

8.4.1.1 Surgical facilities

Surgeries are carried out under very strict sterile conditions in a surgical suite built to standards and under conditions applied in human surgery (Figure 8.5). The surgical team consists of the surgeon assisted by another

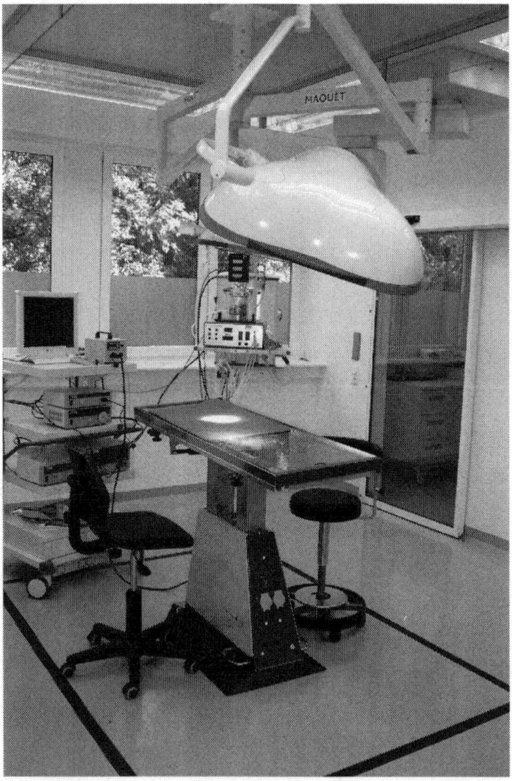

Figure 8.5 Surgery suite.

person, both working under complete sterility. A second assistant handles non-sterile work and monitors anaesthesia and body functions. Sterilised instruments are used for each animal.

8.4.1.2 Catheters/vascular access ports

Anatomy, physiology, and social aspects of marmosets need to be taken into consideration for choosing the most appropriate port. Port catheter systems used in marmosets are adapted mostly from ports used in small rodents. Compared to the rat, the marmoset has little subcutaneous adipose tissue. Therefore, the port is more likely to cause pressure on nerve tissue and blood vessels. This can cause ischemic irritation of the skin and necrosis at the implantation site. Therefore, the area where cutis and subcutis are separated by the port needs to be as small as possible. Similarly, all surgical procedures should be designed to minimise trauma as far as possible.

Besides size and shape, the volume of the port reservoir (inner chamber) plays a crucial role. Port systems for this species should have volumes not exceeding 0.15 mL, because only small volumes of formulated test item are infused, and its time in the port catheter system should be limited for reasons of test item stability—at 38°C—and hygiene.

Ports can be ordered separately from catheters, which are then connected during surgery, or ports with pre-connected catheters can be used.

We suggest not using pre-connected catheters, for two reasons:

1. Surgery with pre-connected ports necessitates implanting the port first, and then tunnelling the catheter from the port implantation site towards the inguinal region. Thus the catheter tip is pushed through septic areas, after which it will be inserted into the vein during preparation of the surgical site.
2. The length of the catheter cannot be adapted to the length needed. Cutting the catheter at the tip prior to implantation should be avoided since the catheter tip is rounded to minimise turbulence, which would be increased by the sharp edge of a cut catheter tip. Blood turbulence at the sharp-edged catheter tip would increase the possibility of thrombosis.

We therefore recommend using a port system and catheters that are assembled during the surgical procedure. This allows starting surgery at the *vena femoralis* with only very little possible exposure of the catheter tip to non-sterile conditions before implantation. In addition, it is possible to cut the catheter at the end that will be connected to the port, allowing the correction of the length without having a cut edge in the blood vessel. This requires accurate manipulation at the port site, but has huge advantages compared to the pre-assembled port catheter system.

Catheters used in marmoset studies should not exceed 3.0 French (outer diameter 0.9 mm, inner diameter 0.5 mm). The catheter is inserted into the *vena* at a length of 1–2 cm. Ligatures are placed proximal and distal to the insertion site in a standard. The preferred material is polyurethane, but silicone can also be used.

8.4.1.3 External equipment

The first citation of the development of a specialised marmoset jacket can be found at Ruiz De Elvira and Abbott (1986) and was made out of Velcro®. The marmoset jacket (13 g, 11 cm length, 8 cm width, Figure 8.3) used for continuous infusion at our laboratory has been developed together with Lomir Biomedical Inc. (New York), and adaptations have been made to meet requirements for minimal weight and optimal comfort (softer material than Velcro). Instead of a metal zipper, Velcro is used for the jacket and pouch closure. The openings for the neck and arms are rounded to avoid any sharp edges. Using the activity monitoring system Actiwatch Mini® (CamNtech Ltd., Cambridge, UK) it can be demonstrated that the animals' activity is reduced only by approximately 10% when wearing the 30-g infusion backpack system over 48 hours.

8.4.1.4 Infusion pumps

Continuous infusion in the marmoset requires a backpack system with a minimal weight pump (pump empty: 12 g + 5 g volume). The Infu-Disk™ technology (Med-e-Cell, USA)) fulfils these requirements. The sterile, pre-set pump with delivery rates from 0.03 to 4.0 mL/hour and an accuracy above 98% has a delivery volume of 5 mL. Infu-Disk is battery powered, and the flow rate is pre-set by the supplier. The core of the disposable pump is the E-Cell™, which produces a continuous flow of gas. The gas, separated from the fluid by the diaphragm, displaces the fluid volume. The external pump carried by the marmoset can be connected to the port system and can be replaced regularly. Pumps can be obtained with various delivery rates to match the animal's weight. In regulatory toxicology studies, body weight groups need to be specified to allocate specific infusion rates to each group. To deliver the correct volume over a certain period of time, delivery rates have to be determined based on the weight of the marmoset. The pumps are supplied with extensive flowcharts for viscosity, temperature and calculations for blood pressure that need to be compensated for to ensure correct delivery. Care should be taken to limit temperature changes during the operation of the pump to not more than 5°C, and fluid delivery pressures should be kept constant because sudden changes in temperature and pressure will result in large changes of delivery rate. Delivering fluids with vapour pressures greater than 80 mm Hg will significantly affect the delivery rate of the Infu-Disk pump. The expected flow rate for pumps delivering fluids with high vapour pressure is difficult to predict. Fluid contact surfaces of the Infu-Disk are medical grade plastics and elastomers compatible with a variety of fluids.

To test the feasibility and the practicability of the system under real-life conditions (e.g. pumping against blood pressure), tests were performed at Covance using the maximum technical flow rate/hour (4.80 mL/day for 14 days) with regular replacement of the pump, and minimum flow rate/hour (5.04 mL/week for 21 days) with replacement once weekly.

The Infu-Disk pump fulfils the requirements for jacketed continuous infusion in freely moving marmoset monkeys. Alternative small portable programmable infusion devices are becoming available for small animal ambulatory models.

8.4.2 Non-surgical models

8.4.2.1 Restraining devices

The restraint bed (23 cm length, 12 cm width, 12 cm height) (Figure 8.2) used for infusion was individually designed for this type of study. The box has a circular opening on one side with a vertical sliding door for head fixation. Woven fabric is fixed on both outside ends of the box with Velcro, creating a soft hanging bed comparable to a hammock. The marmoset, being restrained in the small woollen sock, is vertically placed in the restraint bed with the port on the back facing upwards. The sock with the marmoset is then restrained with two Velcro strips. With proper fixation, the snout of the marmoset is now close to the opening cut at the end of the sock, leaving the eyes covered. The infusion pump can now be connected to the port.

8.4.2.2 Vascular access

The vascular access at surgery is limited to the *vena femoralis*. In relation to catheter size other veins are too small in this small animal model.

8.4.2.3 Infusion pumps

For infusion in restrained marmosets, generally all standard pump systems can be used. However, owing to the low volume required for this species, the Orchesta™ 500-V pump (Solomon Scientific: www.solsci.com) has shown to be of high practicability. Settings are easy to enter, and various bag sizes (from 5 mL onwards) help minimise test item usage. The delivery volume is ranged from 0.1 to 100.0 mL/hr, delivered by 7-µL peristaltic micropulses. The Model 500 may be purchased for stand-alone use and later upgraded for an Orchesta network use by purchasing the Orchesta automation software and wireless hardware.

8.5 Background data

8.5.1 Surgical models

8.5.1.1 Complications associated with the surgical process

Complications associated with implantations of catheters and port systems seen in marmosets are generally the same as seen in other species. Besides sterile surgery, handling the port catheter system, such as exchanging the pumps or making injections into a port, must be done under highly sterile/aseptic conditions to prevent infection. It must be anticipated that the small size and somewhat fragile nature of the marmoset makes this species more susceptible to complications after infections have occurred compared to larger lab animals or rodents.

8.5.1.2 In-life changes

Common clinical background findings seen in implanted marmosets are slight to moderate injuries, necrotic or swollen skin, missing upper dermal layer, oedema, and lesions at extremities. On rare occasions the port catheter system was found blocked, and therefore was flushed with 0.9% NaCL/heparine solution (5-IU/mL).

Individual variations in body weights were noted throughout the study period in animals of both sexes (Table 8.4). This is likely related to the post-surgical stress and start of daily handling for dosing, as well as to repeated diagnostic procedures.

The predominant clinical pathology findings attributed to port implantation were increases in aspartate aminotransferase and alkaline phosphatase values. These serum enzyme changes are considered to be related to the surgery. Additional electrolyte parameters were increased in some animals by occasional fluid supplements (such as PRANG™ Electrolyte Replenisher: LBS Biotechnology).

Increased reticulocytes were attributed to the blood sampling regime, and not to the dose administration technique. There were no apparent

Table 8.4 Marmoset Body Weight Gain (g) of Control Animals (0.9% NaCl) over the Course of a 14-Day (Twice Daily for 1 Hour) Infusion Study with Port Catheterized Animals

Sex	Pre-surgery (prior to port implantation)	10 Days after surgery (day 1 of study)	Day 7 of study	Day 14 of study
Male Mean	376	361	353	340
S.D.	27	39	21	33
Female Mean	368	360	337	356
S.D.	26	22	19	30

Table 8.5 Selected Haematology Data (Control Animals 0.9% NaCL, n = 4 males, n = 4 females) over the Course of a 14-day (twice daily for 1 hour) Infusion Study with Port-Catheterized Animals

Parameter	Pre-surgery (prior to port implantation)	8 Days after surgery (2 Days prior to start of dosing)	Day 15 of study
Mean reticulocyte count [‰]	♂ 78 ♀ 65	♂ 65 ♀ 63	♂ 98 ♀ 90
Mean red blood cell count [10 E12/L]	♂ 6.3 ♀ 6.7	♂ 5.9 ♀ 6.3	♂ 5.6 ♀ 6.0
Mean haemoglobin [mmol/L]	♂ 8.8 ♀ 9.2	♂ 8.4 ♀ 9.1	♂ 7.9 ♀ 8.5
Mean total white blood cell count [10 E9/L]	♂7.2 ♀ 7.1	♂ 9.4 ♀ 9.7	♂ 6.8 ♀ 9.4
Mean white blood cell count – lymphocytes [%]	♂ 56 ♀ 45	♂ 43 ♀ 53	♂ 51 ♀ 39

trends observed indicating an infection over the course of the infusion studies (Table 8.5).

8.5.1.3 Pathology

The skin region where the catheter enters the skin (catheter site), the administration site, and the endothelium adjacent to the tip of the catheter should be examined in addition to routinely examined tissues (Figures 8.6a–c).

Background findings in marmoset monkeys that underwent continuous intravenous infusion using a port catheter system were as follows. Epidermal scabs at the catheter site, subcutaneous oedema, haemorrhage, acute and subacute inflammation, fibrosis, and ulcer are predominantly seen. Application site lesions include subcutaneous oedema, haemorrhage, and vascular acute, subacute, and chronic inflammation as the primary lesions. These changes tend to be slightly increased in severity with longer presence of the catheter system in the marmosets (e.g. six weeks' implantation versus one week). In summary, all histopathological lesions seen in the skin around the implanted catheter and the vein are generally consistent with the histopathological findings encountered in dermal implanted material and after repeated venipuncture. Other histopathological findings (Kaspareit et al. 2006) are regarded as background lesions in marmosets.

8.6 Conclusion

Using an implanted port for infusion instead of daily intravenous punctures reduces venous tissue damage and has positive animal welfare aspects.

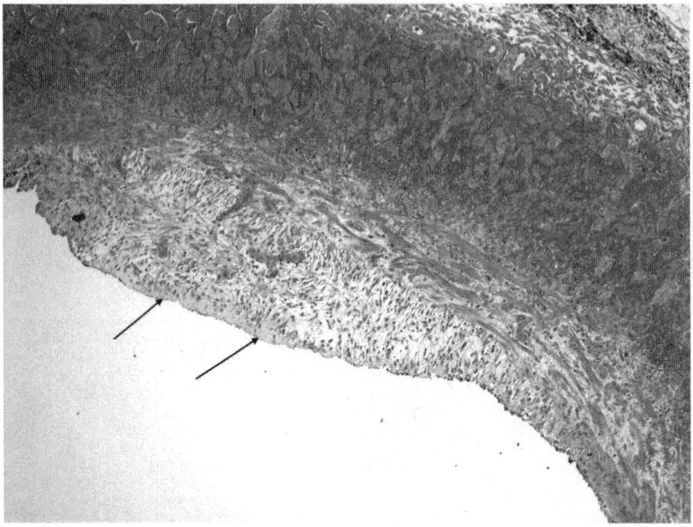

Figure 8.6a Vein, intimal proliferation.

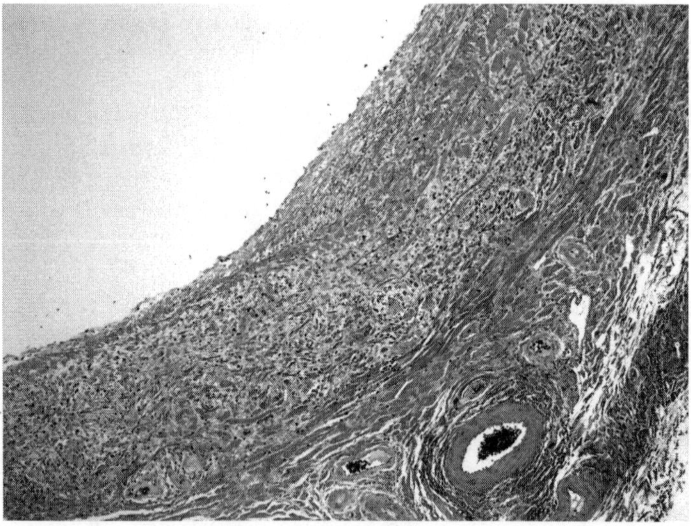

Figure 8.6b Vein, inflammation, vascular wall.

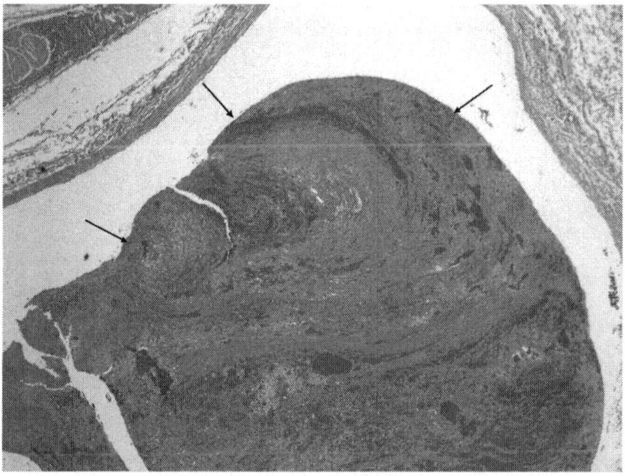

Figurc 8.6c Vcin, thrombus.

With an average body weight of only about 300–500 g, the marmoset requires lightweight dosing equipment. If the individual body weight allows wearing the backpack infusion, this is considered the preferred approach in case intermittent or long-term administration is required.

If the individual body weight prohibits wearing the backpack infusion system, intermittent daily dosing for up to 1.5 hours can be conducted with port-catheterised animals in a restraint bed.

It must be emphasised that the studies described are much more sophisticated than comparable administration regimes in macaques. Therefore, the marmoset serves only as an alternative model. In the future, research needs to focus on the practicability of subcutaneously implanted miniaturised port systems. Miniaturised implanted port systems might allow social housing since the animals would not be able interfere with the port catheter system.

The new technology using the mini-pump system in a backpack enables researchers now to conduct daily or even weekly continuous intravenous infusion in the marmoset. The recommendations provided here outline a feasible approach for implementation of this administration regime, and the requirements needed to ensure the well-being of the highly agile and sensitive marmoset.

References

Abbott DH, Barnett DK et al. 2003. Aspects of common marmoset basic biology and life history important for biomedical research. *Comp. Med.* 53(4): 339–550.

Austad SN, Fischer KE. 2011. The development of small primate models for aging research. *ILARJ* 52(1): 78–88.

Chapman KL, Pullen N, Graham M, Ragan I. 2007. Preclinical safety testing of monoclonal antibodies: The significance of species relevance. *Nat. Rev. Drug Discov.* 6: 120–126.

Chapman KL, Pullen N et al. 2009. Preclinical development of monoclonal antibodies: Considerations for the use of non-human primates. *mAbs* 1(5): 505–516.

Chapman KL, Pullen N et al. 2010. The future of non-human primate use in mAb development. *Drug Discov. Today* 15(S/6): 235–242.

Davy CW, Trennery PN, Edmunds JG, Altman JFB, Eichler DA. 1987. Local myotoxicity of ketamine hydrochloride in the marmoset. *Laboratory Animals* 17: 60–67.

Diehl KH, Hull R, Morton D, Pfister R, Rabemampianina Y, Smith D et al. 2001. A good practice guide to the administration of substances and removal of blood, including routes and volumes. *J. Appl. Toxicol.* 21: 15–23.

ETS. 2007. European convention for the protection of vertebrate animals used for experimental and other scientific purposes, Appendix A, (ETS No. 123). *Official Journal of the European Communities* K, 2525.

Flecknell PA. 1987. *Laboratory Animal Anesthesia*. p. 156. London: Academic Press.

Kaspareit J, Friderichs-Gromoll S, Buse E, Habermann G. 2006. Background pathology of the common marmoset (*Callitrix jacchus*) in toxicological studies. *Exp. Toxic. Pathol.* 57: 405–410.

Korte S, Zühlke U, Kaspareit J, Friderichs-Gromoll S, Müller W, Vogel F, Weinbauer GF. 2004. Reference control data for the common marmoset (*Callithrix jacchus):* A comparison with macaques. *The Toxicologist* 78: 205.

Mansfield K. 2003. Marmoset models commonly used in biomedical research. *Comp. Med.* 53(4): 383–392.

Schnell CR. 2000. Surgical preparation and multidose infusion toxicity studies in the marmoset. In *Handbook of Pre-clinical Continuous Intravenous Infusion.* Healing G and Smith D (eds.), pp. 210–222. London and New York: Taylor & Francis.

Orsi A, Rees D, Andreini I, Venturella S, Cinelli S, Oberto G. 2011. Overview of the marmoset as a model in nonclinical development of pharmaceutical products. *Regul. Toxicol. Pharmacol.* 59(1): 19–27.

Ruiz De Elvira MC, Abbott H. 1986. A backpack system for long-term osmotic minipump infusions into unrestrained marmoset monkeys. *Laboratory Animals* 20: 329–334.

Vierboom MP, Breedveld E, Kondova I, 't Hart BA. 2010. Collagen-induced arthritis in common marmosets: a new nonhuman primate model for chronic arthritis. *Arthritis Res. Ther.* 12(5): R200.

Zühlke U, Weinbauer G. 2003. The common marmoset (*Callithrix jacchus*) as a model in toxicology. *Toxicol. Pathol, Suppl.* (31): 123–127.

Zühlke U, Korte S, Niggemann B, Fuchs A, Müller W. 2003. Feasibility of long-term continuous intravenous infusion in unrestrained marmoset monkeys. *The Toxicologist* 72: S–1.

Index

Note: Page numbers ending in 'f' refer to figures. Page numbers ending in 't' refer to tables.

E

N

O

S

T

V